Study Guide to Accompany

Atkins, Jones and Laverman's

CHEMICAL PRINCIPLES
The Quest for Insight

Seventh Edition

John Krenos
Rutgers, The State University of New Jersey

w.h.freeman
Macmillan Learning

ISBN-13: 978-1-319-01755-2
ISBN-10: 1-319-01755-X

Printed in the United States of America

First Printing

Macmillan Learning
One New York Plaza
New York, NY 10004-1562

www.macmillanhighered.com

CONTENTS

PREFACE

The *Study Guide* accompanies the textbook *Chemical Principles: The Quest for Insight,* Seventh Edition, by Peter Atkins, Loretta Jones, and Leroy Laverman—an authoritative and thorough introduction to chemistry for students anticipating careers in health, science, engineering, or technology-related disciplines. We follow the order of topics in the textbook. The parallels between the symbols, concepts, and style of this supplement and the textbook enable the reader to move easily back and forth between the two.

Much of the *Study Guide* is presented in outline style with material highlighted by bullets, arrows, tables, and figures. In general, **boldface bullets** offset major items of importance for each **Topic,** and **arrows** provide explanatory and descriptive material. **Worked-out examples** are designed to reinforce concepts. The telegraphic style may help a student obtain a broad perspective of large blocks of material in a relatively short period of time, and should prove particularly useful in preparation for quizzes and examinations. We believe that the *Study Guide* will be most useful in conjunction with a careful reading of the text. While we have covered much of the material in the text, we have not attempted to be encyclopedic. To help obtain a broad overview of the material, important equations are highlighted in boxes. Students may find this aspect of the guide particularly useful. Where appropriate, we have introduced supplementary tables either to amplify material in the text, to clarify it further, or to summarize a body of material. Sprinkled throughout the guide are **Notes**; these are designed primarily to point out common pitfalls. We maintain the format changes introduced in the previous edition, namely, a more readable open format. The content is also updated to closely follow textbook changes in content and pedagogical approaches.

ACKNOWLEDGMENTS

The *Study Guide* author is indebted to the many individuals who have contributed to it significantly. Chiefly among them is Joseph Potenza, the former co-author, and developer of the telegraphic style.

The thorough and incisive reading of many early drafts by Beth Van Assen is gratefully appreciated. Her critical suggestions led to improved readability and scientific accuracy. Many examples and important sections of descriptive text were clarified and expanded with her help. Her encouragement, patience, and persistence were essential.

Two of our colleagues at Rutgers critically reviewed several topics, and both deserve special praise. Harvey Schugar, an expert on the chemistry of d-block elements, broadened our knowledge and appreciation of inorganic chemistry. Spencer Knapp, an expert on synthetic organic chemistry, carefully critiqued the organic chemistry topics, directing us to a modern approach that complements the textbook.

The help and encouragement of the staff at Macmillan Learning is gratefully acknowledged. Special thanks to Michelle Russel Julet for choosing us to author the first edition of this guide. Heidi Bamatter (Development Editor) guided the successful scheduling process for this edition. Jodi Isman (Project Editor) supervised the day-to-day editing process and was always available at a moment's notice for essential guidance. Jodi wisely chose Kate Daly to painstakingly edit the manuscript. Kate made significant improvements to readability and suggested new valuable content as well.

Finally, the efforts of the authors of the textbook, Peter Atkins, Loretta Jones, and Leroy Laverman in creating a new approach to teaching general chemistry are acknowledged and appreciated. Beginning an introductory chemistry textbook with quantum theory is logical, but challenging. The authors succeed by treating the fundamentals of chemistry (an "as needed" review) in a separate section and by minimizing the coverage of much material (mostly historical) presented in other books. This approach leads to the treatment of the major topics in chemistry enlightened and enlivened in the first instance by the molecular viewpoint. The grouping of material by **Focus** sections rather than by chapters (as traditionally done) is an innovative and significant approach in this edition.

Focus 1   ATOMS
Topic 1A: INVESTIGATING ATOMS

1A.1 The Nuclear Model of the Atom

- **Background**

  → J. J. Thomson found that charged cathode-ray particles, which are now called **electrons**, were the same regardless of the metal used for the cathode. He concluded that they are part of the foundation of all atoms.

  → Thomson measured a value of e/m_e, the ratio of the magnitude of the electron's charge e to its mass m_e . Values for e and m_e were not known until the physicist Robert Millikan carried out experiments that enabled the calculation of the value of e.

  → Fundamental unit of charge: $e = 1.602 \times 10^{-19}$ C. Mass of the electron: 9.109×10^{-31} kg.

  → The charge of $-e$ is "one unit" of negative charge, and the charge of e is "one unit" of positive charge.

  → Based on the scattering of alpha particles on platinum foil, Ernest Rutherford proposed a nuclear model of the atom. Later work showed that the nucleus of an atom contains particles called **protons,** each of which has a charge of $+e$ (responsible for the positive charge), and **neutrons** (uncharged particles).

  → The number of protons in the nucleus is different for each element and is called the **atomic number,** Z, of the element (*Fundamentals* B). The total charge on an atomic nucleus of atomic number Z is $+Ze$ and, for the atoms to be electrically neutral, there must be Z electrons around it.

- **Subatomic particles**

  → Electron: mass ($m = 9.109\,383 \times 10^{-31}$ kg)   charge ($-e = -1.602\,177 \times 10^{-19}$ C)

  → Proton: mass ($m = 1.672\,622 \times 10^{-27}$ kg)   charge ($e = 1.602\,177 \times 10^{-19}$ C)

  → Neutron: mass ($m = 1.674\,927 \times 10^{-27}$ kg)   charge $= 0$

- **Nucleus**

  → *Nucleons* (protons and neutrons) occupy a small volume at the center of the atom. The binding energy of the nucleus is attributed to a strong force (nuclear) acting over a very short distance.

  → The radius of the nucleus (assumed to be spherical) is given roughly by $r_{nuc} = r_0\,A^{1/3}$, where $r_0 \approx 1.3 \times 10^{-15}$ m $= 1.3$ fm and A is defined below.

- **Atom**

  → Atomic number: $Z = N_p =$ number of protons in the nucleus

  → Atomic mass number: $A = N_p + N_n =$ number of protons and neutrons in the nucleus

  → Uncharged atom: $N_p = N_e$ (number of protons equals the number of electrons)

  → *Electrons* occupy a much larger volume than the nucleus and define the "size" of the atom. The binding energy of the electrons is attributed to a weak force (coulomb) acting over a much longer distance than the strong force.

1A.2 Electromagnetic Radiation

- **Oscillating amplitude of electric and magnetic field**
  - → Waves are characterized by *wavelength* λ (lambda) and *frequency* ν (nu).
  - → Waves always travel at the speed of light (constant for a given medium)
  - → No known upper or lower limit of frequency or wavelength

Distance Behavior (fixed time *t*)   **Time Behavior (fixed position *x*)**

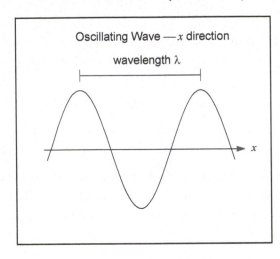   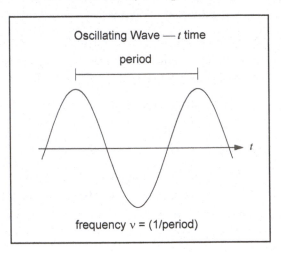

Speed of light (distance / time) = wavelength / period = wavelength × frequency

$$\boxed{c = \lambda \nu}$$ speed of light = wavelength × frequency

SI units: $(m \cdot s^{-1})$ (m) (s^{-1}) $[1 \text{ Hz (hertz)} = 1 \text{ s}^{-1}]$

- → The speed of light c (c_0 in vacuum $\approx 3.00 \times 10^8$ m·s⁻¹) depends on the medium it travels in. Medium effects on wavelength in the visible region are small (beyond three significant figures).
- → Visible radiation or visible light: 700 nm to 400 nm.
- → Ultraviolet radiation: < 400 nm (< 200 nm vacuum ultraviolet)
- → Infrared radiation: > 700 nm
- → Infrared radiation: > 700 nm
- → Visible spectrum colors (pneumonic): ROY G BIV (red, orange, yellow, green, blue, indigo, violet)
- → See **Table 1A.1** in the text: Color, Frequency, and Wavelength of Electromagnetic Radiation

1A.3 Atomic Spectra

- **Spectral lines**
  - → Discharge lamp of hydrogen

    H_2 + electrical energy → H + H*   (* ≡ asterisk denotes excited atom)

    $H^* \rightarrow H^{(*)} + h\nu$ ($^{(*)}$ ≡ denotes a less excited atom)

- **Lines form a discrete pattern**
  - → Discrete energy levels

- **Hydrogen atom spectral lines**
  - → Spectral lines imply discrete energy level for electrons in atoms
  - → Johann Rydberg's general equation
  - → $n_2 = n_{upper}$ and $n_1 = n_{lower}$

$$v = \mathcal{R} \left(\frac{1}{n_1^2} - \frac{1}{n_2^2} \right) \quad n_1 = 1,\ 2,\ \ldots \quad n_2 = n_1 + 1,\ n_1 + 2,\ \ldots$$

$\mathcal{R} = 3.29 \times 10^{15}$ Hz
Rydberg constant

  - → Rydberg expression reproduces pattern of lines in H atom emission spectrum. The value of $\mathcal{R}$ is obtained empirically. (Font: Monotype Corsiva)

Note: Lines with a common n_1 can be grouped into a *series* and some have special names:

$n_1 = 1$ (Lyman),	2 (Balmer),	3 (Paschen),	4 (Brackett),	5 (Pfund)
121.6 nm	656.3 nm (red)	1.875 μm	4.05 μm	7.46 μm
102.6 nm	486.1 nm (blue)	1.282 μm	2.62 μm	4.65 μm
97.3 nm	434.0 nm (indigo)	1.094 μm	2.16 μm	3.74 μm
95.0 nm	410.2 nm (violet)	1.005 μm	1.94 μm	3.30 μm
....	....	....	....	....

series limit: 91.2 nm   364.7 nm   0.820 μm   1.46 μm   2.28 μm
$n_2 = \infty$

spectrum: vacuum UV   vis → UV   IR   IR   IR

Infrared: *near*   (0.8 μm to 20 μm)   and   *far*   (50 μm to 1000 μm)
(0.7 μm to 50 μm)   (50 μm to 100 μm)

Note: Different limits for the infrared are determined mainly by instrumentation and/or the light source used.

Topic 1B: QUANTUM THEORY

1B.1 Radiation, Quanta, and Photons

- **Black body**
  - → Perfect absorber and emitter of radiation
  - → Intensity of radiation for a series of temperatures leads to two laws (Stefan–Boltzmann and Wien's).

→ Stefan–Boltzmann law:

$$\frac{\text{Power emitted (watts)}}{\text{Surface area (meter}^2)} = \text{constant} \times T^4$$

→ Wien's law: $\boxed{T\lambda_{\text{max}} = \text{constant}}$ where constant $= 2.88$ K·mm

Wavelength corresponding to maximum intensity $= \lambda_{\text{max}}$

At higher temperature, maximum intensity of radiation shifts to lower wavelength.

- **Energy of a quantum (packet) of light (generally called a photon)**

  → Postulated by Max Planck to explain black body radiation

  → Resolved the "ultraviolet catastrophe" of classical physics, which predicted intense ultraviolet radiation for all heated objects ($T > 0$)

  → Quantization of electromagnetic radiation

$$\boxed{E = h\nu}$$ Photon energy = Planck's constant × photon frequency

SI units: (J)   ($h = 6.6261 \times 10^{-34}$ J·s)   (s^{-1})

- **Photoelectric effect**

  → Ejection of electrons from a metal surface exposed to photons of sufficient energy

  → Indicates that light behaves as a particle

$$\boxed{E_{\text{K}} = h\nu - \Phi}$$ E_{K} = kinetic energy of the ejected electron, Φ = threshold energy (work function) required for electron ejection from the metal surface, and $h\nu$ = photon energy

  → $h\nu \geq \Phi$ required for electron ejection

- **Bohr frequency condition**

$$\boxed{h\nu = E_{\text{upper}} - E_{\text{lower}}}$$ Relates photon energy to energy difference between two energy levels in an atom

1B.2 The Wave–Particle Duality of Matter

- **Wave behavior of light**

  → Diffraction and interference effects of superimposed waves (*constructive* and *destructive*)

- **Wave behavior of matter**

  → Proposed by Louis de Broglie, who considered the properties of matter of mass m traveling with velocity v.

  → Such matter behaves as a wave with a characteristic wavelength.

  → de Broglie wavelength λ for a particle with linear momentum $p = mv$ $\boxed{\lambda = \dfrac{h}{mv}}$

  → Wave character of electrons is verified by electron diffraction.

1B.3 The Uncertainty Principle

- **Complementarity of location (*x*) and momentum (*p*)**

  → Uncertainty in x is Δx; uncertainty in p is Δp.

  → Limitation of knowledge (p and x cannot be determined simultaneously)

  $$\boxed{\Delta p \Delta x \geq \hbar/2}$$ Heisenberg uncertainty principle, where $\hbar = h/2\pi$

  → $\boxed{\hbar \text{ is called "h bar"}}$ $\hbar = 1.0546 \times 10^{-34}$ J·s

  → Refutes classical physics on the atomic scale

Topic 1C: WAVEFUNCTIONS AND ENERGY LEVELS

1C.1 The Wavefunction and Its Interpretation

- **Classical trajectories**

  → Precisely defined paths

- **Wavefunction ψ**

  → Gives *probable* position of particle with mass m

- **Born interpretation**

  → Probability of finding particle in a region is proportional to ψ^2.

- **Schrödinger equation in one-dimension**

  → Allows calculation of ψ by solving a differential equation

  → $H\psi = E\psi$; H is called the Hamiltonian

  → H represents the sum of kinetic energy and potential energy in a system

  $$\boxed{-\frac{\hbar^2}{2m}\nabla^2\psi + V(x)\psi = E\psi}\qquad \boxed{\nabla^2 = \frac{d^2}{dx^2}}$$

1C.2 The Quantization of Energy

- **Particle in a box**

  → Mass m confined between two rigid walls a distance L apart

  → $\psi = 0$ outside the box and at the walls (boundary condition)

  → The *operator* for a one-dimensional system is

  $$\boxed{\nabla^2 = \frac{d^2}{dx^2}}$$

→ In the box, $V(x) = 0$. The resulting differential equation is

$$-\frac{\hbar^2}{2m}\frac{d^2\psi}{dx^2} = E\psi$$

→ Solving with the boundary condition and normalizing the function yields

$$\psi_n(x) = \left(\frac{2}{L}\right)^{1/2} \sin\left(\frac{n\pi x}{L}\right) \qquad n = 1,\ 2,\ \dots$$

→ $\psi_n(x) =$ wavefunction that satisfies the Schrödinger equation between the box limits.

→ n is a *quantum number* ($n = 1, 2, 3, 4, 5, \dots$)

Note: A node is a point in the box where $\psi = 0$ and ψ *changes* sign. $\boxed{\psi^2 \geq 0,\ \text{always}}$

→ The resulting energy levels follow

$$E_n = \frac{n^2 h^2}{8mL^2}$$

where $E_n =$ allowed energy values of a particle in a one-dimensional box

Note: $n = 1$ gives the zero-point energy E_1. $E_1 \neq 0$ implies residual motion
(consistent with the *uncertainty principle*).

→ The energy difference between two neighboring levels is $\boxed{\Delta E = E_{n+1} - E_n = \frac{(2n+1)h^2}{8mL^2}}$

- **Probability as a function of position in the box**

  → Plot is shown for the first three levels, ψ^2 as a function of the dimensionless variable x/L

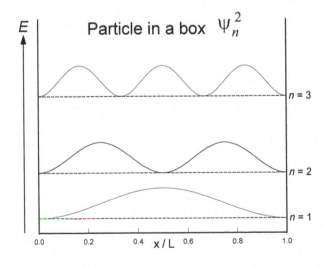

Topic 1D: THE HYDROGEN ATOM

1D.1 Energy Levels

- **Charge on an electron**

  $\rightarrow$ $q = -e = -1.602\,177\,33 \times 10^{-19}$ C

- **Vacuum permittivity**

  $\rightarrow$ $\varepsilon_0 = 8.854\,187\,817 \times 10^{-12}$ $C^2 \cdot N^{-1} \cdot m^{-2}$

- **Coulomb potential energy** $\boxed{V(r) = \dfrac{q_1 q_2}{4\pi\varepsilon_0 r}}$

$$V(r) = \frac{\text{product of charges on particles}}{4\pi \times \text{permittivity of free space} \times \text{distance between charges}} \qquad \text{(SI system)}$$

Units of $V(r)$: $J = \dfrac{C^2}{(C^2 \cdot N^{-1} \cdot m^{-2})\,m} = N \cdot m = (kg \cdot m \cdot s^{-2})\,m = kg \cdot m^2 \cdot s^{-2}$

$\rightarrow$ For the H atom, $V(r) = \dfrac{(-e)(+e)}{4\pi\varepsilon_0 r} = -\dfrac{e^2}{4\pi\varepsilon_0 r}$

$\rightarrow$ Solution of the Schrödinger equation using $V(r)$ above leads to energy levels E_n given below.

- **H atom and one-electron ion energy levels (He^+, Li^{2+}, Be^{3+}, etc.)**

  $\rightarrow$ Solutions to the Schrödinger equation

$$\boxed{\begin{array}{l} E_n = -h\mathcal{R}\left(\dfrac{Z^2}{n^2}\right) \qquad Z = 1 \ \& \ n = 1, 2, 3, \ldots \qquad (n \text{ is dimensionless}) \\[2em] \mathcal{R} = \dfrac{m_e e^4}{8h^3 \varepsilon_0^2} = 3.289\,842 \times 10^{15} \text{ Hz} \qquad \text{Units: } \dfrac{kg \cdot C^4}{(J \cdot s)^3 (C^2 \cdot N^{-1} \cdot m^{-2})^2} = s^{-1} \equiv Hz \end{array}}$$

$\rightarrow$ All quantum numbers are dimensionless. With $\mathcal{R}$ in units of frequency, the H atom energy-level equation has the same form as the Planck equation, $E = h\nu$.

$\rightarrow$ E_n = energy levels (states) of the H atom. Note the *negative* sign.

$\rightarrow$ $\mathcal{R}$ = Rydberg constant, calculated exactly using Bohr theory *or* the Schrödinger equation

$\rightarrow$ Z = atomic number, equal to 1 for hydrogen

$\rightarrow$ n = principal quantum number

$\rightarrow$ As n increases, energy increases, the atom becomes less stable, and energy states become more closely spaced (more dense).

$\rightarrow$ Integer n varies from 1 (ground state) to higher integers (excited states) to ∞ (ionization).

$\rightarrow$ Energies of H atom states vary from $-h\mathcal{R}$ ($n = 1$) to 0 ($n = \infty$). States with $E > 0$ are possible and correspond to an ionized atom in which the energy > 0 equals the kinetic energy of the electron.

1D.2 Atomic Orbitals (AOs)

- **Definition of AO**

  → Wavefunction (Ψ, psi) describes an electron in an atom.

  → Orbital (Ψ^2) holds 0, 1, or 2 electrons.

  → Orbital can be viewed as a *cloud* within which the point density represents the *probability* of finding the electron at that point.

  → Orbital is specified by *three* quantum numbers (n, ℓ, m_ℓ).

- **Wavefunction**

  → Fills all space

  → Depends on the *three* spherical coordinates: r, θ, ϕ

  → Written as a product of a radial [$R(r)$] and an angular [$Y(\theta,\phi)$] wavefunction; mathematically, $\psi(r,\theta,\phi) = R(r)\,Y(\theta,\phi)$

- **Wavefunction for the H atom 2s orbital** → $n = 2$, $\ell = 0$, and $m_\ell = 0$

$$R(r) = \frac{\left(2 - \dfrac{r}{a_0}\right)e^{-r/2a_0}}{(2a_0)^{3/2}} \qquad Y(\theta,\phi) = (4\pi)^{-1/2} \qquad a_0 = \frac{4\pi\varepsilon_0\hbar^2}{m_e e^2} = \left\{ \begin{array}{c} 5.291\,77\times10^{-11}\ \text{m} \\ \text{(Bohr radius)} \end{array} \right\}$$

$$\text{Units:}\quad R(r) = \text{m}^{-3/2} \qquad Y(\theta,\phi) = \text{none} \qquad a_0 = \frac{(\text{C}^2\cdot\text{N}^{-1}\cdot\text{m}^{-2})\,(\text{J}\cdot\text{s})^2}{\text{kg}\cdot\text{C}^2} = \text{m}$$

1D.3 Quantum Numbers, Shells, and Subshells

Three Quantum Numbers [n, ℓ, m_ℓ] Specify an Atomic Orbital

Symbol	Name	Allowed Values	Constraints
n	Principal quantum number	$= 1, 2, 3, \ldots$	Positive integer
ℓ	Orbital angular momentum quantum number	$= 0, 1, 2, \ldots, n-1$	Each value of n corresponds to n allowed values of ℓ.
m_ℓ	Magnetic quantum number	$= \ell, \ell-1, \ell-2, \ldots, -\ell$ $= 0, \pm1, \pm2, \ldots, \pm\ell$	Each value of ℓ corresponds to $(2\ell+1)$ allowed values of m_ℓ.

→ Quantum number n is related to the energy and "size" of the orbital.

→ Quantum number ℓ is related to the shape of the orbital.

→ Quantum number m_ℓ is related to the orientation of the orbital in space.

- **Terminology (nomenclature)**

  shell: AOs with the same n value

  subshell: AOs with the same n and ℓ values;

    $\ell = 0, 1, 2, 3$ equivalent to   s-, p-, d-, f-subshells, respectively, or

    s-orbital $\Rightarrow$ $\ell = 0$ $m_\ell = 0$

    p-orbital $\Rightarrow$ $\ell = 1$ $m_\ell = -1, 0,$ or $+1$

    d-orbital $\Rightarrow$ $\ell = 2$ $m_\ell = -2, -1, 0, +1,$ or $+2$

    f-orbital $\Rightarrow$ $\ell = 3$ $m_\ell = -3, -2, -1, 0, +1, +2,$ or $+3$

- **Physical significance of the wavefunction $\psi(r, \theta, \phi)$**

  → For all atoms and molecules, the square of the wavefunction, $\psi^2(r, \theta, \phi)$, is proportional to the probability of finding the electron at a point r, θ, ϕ. We can also regard $\psi^2(r, \theta, \phi)$ as the electron density at point r, θ, ϕ.

- **Plot of electron density for the 2s-orbital of hydrogen**

Computer-generated electron density dot diagram for the hydrogen atom 2s-orbital. The nucleus is at the center of the square and the density of dots is proportional to the probability of finding the electron. Notice the location of the spherical node.

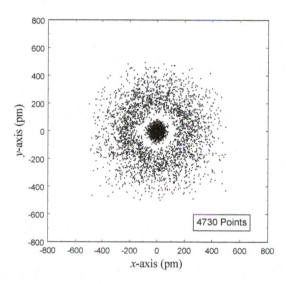

- **Concept of AOs** → Two interpretations are useful:

  1 Visualize an AO as a *cloud of points*, with the density of points in a given volume proportional to the probability of finding the electron in that volume.

  2 Visualize an AO as a *surface* (boundary surface) within which there is a given probability of finding the electron. For example, the 95% boundary surface, within which the probability of finding an electron is 95%.

1D.4 The Shapes of Orbitals

- **Boundary surfaces** → *Shapes* of atomic orbitals

    s-orbital   *spherical*

    p-orbital   *dumbbell* or *peanut*

    d-orbital   *four-leaf clover* or *dumbbell* with equatorial *torus* (doughnut)

- **Number and type of orbitals** → Follow from allowed values of quantum numbers

    s-orbitals: *s* means that $\ell = 0$; if $\ell = 0$, then $m_\ell = 0$. So, for each value of *n*, there is *one* s-orbital.

    > *one* for each *n* : 1s, 2s, 3s, 4s, …

    p-orbitals: *p* means that $\ell = 1$; if $\ell = 1$, then $m_\ell = -1, 0, +1$, and $n > 1$. So, for each value of $n > 1$, there are *three* p-orbitals, p_x, p_y, p_z.

    > *three* for each $n > 1$: $2p_x, 2p_y, 2p_z$; $3p_x, 3p_y, 3p_z$; …

    The *n*p-orbitals are referred to collectively as the 2p-orbitals, 3p-orbitals, and so on.

    d-orbitals: d means that $\ell = 2$; if $\ell = 2$, then $m_\ell = -2, -1, 0, +1, +2$, and $n > 2$. So, for each value of $n > 2$, there are *five* d-orbitals, d_{xy}, d_{xz}, d_{yz}, $d_{x^2-y^2}$, and d_{z^2}.

    > *five* for each $n > 2$: $3d_{xy}$, $3d_{xz}$, $3d_{yz}$, $3d_{x^2-y^2}$, and $3d_{z^2}$;
    >
    > $4d_{xy}$, $4d_{xz}$, $4d_{yz}$, $4d_{x^2-y^2}$, and $4d_{z^2}$; …

    The *n*d-orbitals are referred to collectively as the 3d-orbitals, 4d-orbitals, and so on.

- **Wavefunction Properties**

    → Positive and negative regions are designated as + and − on orbital diagrams

    ($n > 1$ (radial node), $\ell > 0$ (angular node))

    → Possess nodal surfaces defined by $\psi = 0$ (wavefunction changes sign).
    Example: All 2p orbitals have positive and negative lobes separated by a nodal plane.

1D.5 Electron Spin

- **Spin states**

    → An electron has *two* spin states in an atom, represented as ↑ and ↓ **or** α and β.

    → In the atom, spin states are described by the spin magnetic quantum number, m_s

    → For any electron in an atom, only two values of m_s are allowed: $+\frac{1}{2}, -\frac{1}{2}$.

    → Spin is an intrinsic property of an electron. The spin quantum number for a free electron is given by $s = \frac{1}{2}$.

1D.6 The Electronic Structure of Hydrogen

- **Degeneracy of orbitals**

    In the H atom, orbitals in a given shell are *degenerate* (have the same energy).

    For many-electron atoms, this is not true and the energy of a given orbital depends on *n* and ℓ.

- **Ground and excited states of H**

    In the ground state, $n = 1$, and the electron is in the 1s-orbital. The first excited state corresponds to $n = 2$, and the electron occupies one of the *four* possible orbitals with $n = 2$ (2s, $2p_x$, $2p_y$, or $2p_z$). Similar considerations hold for higher excited states.

- **Ionization of H**

    Absorption of a photon with energy $\geq h\mathcal{R}$ ionizes the atom, creating a free electron and a free proton. Energy in excess of $h\mathcal{R}$ appears as kinetic energy of the system.

- **Summary**

    The state of an electron in a H atom is defined by four quantum numbers $\{n, \ell, m_\ell, m_s\}$. As the value of n increases, the "size" of the H atom increases.

Topic 1E: MANY-ELECTRON ATOMS

1E.1 Orbital Energies

- **Many-electron atoms**

    → Atoms with more than one electron

    → Coulomb potential energy equals the sum of *nucleus-electron* **attractions** and *electron-electron* **repulsions**.

    → Schrödinger equation cannot be solved exactly.

    → Accurate wavefunctions are obtained numerically by using computers.

- **Variation of energy of orbitals**

    → For orbitals in the same shell but in different subshells, a combination of *nucleus-electron* attraction and *electron-electron* repulsion influences the orbital energies.

- **Shielding and penetration**

    → Qualitative understanding of orbital energies in atoms

- **Shielding**

    → Each electron in an atom is *attracted* by the nucleus and *repelled* by all the other electrons. In effect, each electron feels a *reduced* nuclear charge ($Z_{eff}\, e$ = effective nuclear charge). The electron in an occupied orbital is *shielded* to some extent from the nuclear charge and its energy is raised accordingly.

- **Penetration**

    → The **s-, p-, d-, … orbitals** have different shapes and different electron density distributions. For a given *shell* (same value of the principal quantum number n), **s-electrons** tend to be closer to the nucleus than **p-electrons**, which are closer than **d-electrons**. We say that, other things equal, **s-electrons** are more *penetrating* than **p-electrons**, and **p-electrons** are more *penetrating* than **d-electrons**.

    Note: Shielding and penetration can be understood qualitatively on the basis of the nucleus-electron potential energy term: $-(Ze)e/4\pi\varepsilon_0 r$, where Ze is the nuclear charge and r the nucleus-electron distance. *Shielded* electrons have the equivalent of a reduced Z value ($Z_{eff} < Z$) and therefore *higher* energy; *penetrating* electrons have the equivalent of a reduced value of *r* and therefore *lower* energy.

- **Review**

  → *Z* has three equivalent meanings:

    Nuclear charge (actually, *Ze*)

    Atomic number

    Number of protons in the nucleus

- **Consequences of *shielding* and *penetration* in many-electron atoms**

  → Orbitals with the same *n* and different ℓ values have *different* energies.

  → For a given *shell (n)*, subshell energies *increase* in the order: *n*s < *n*p < *n*d < *n*f.

    Example: A 3s-electron is lower in energy than a 3p-electron, which is lower than a 3d-electron.

  → Orbitals within a given *subshell* have the *same* energy.

    Example: The five 3d-orbitals are degenerate for a given atom.

  → *Penetrating* orbitals of higher shells may be lower in energy than less *penetrating* orbitals of lower shells.

    Example: A penetrating 4s-electron may be lower in energy than a less penetrating 3d-electron.

1E.2 The Building-Up Principle

- **Pauli exclusion principle**

  → No more than *two* electrons per orbital

  → *Two* electrons occupying a single orbital must have paired spins: ↑↓

  → *Two* electrons in an atom may *not* have the same *four* quantum numbers.

- **Terminology**

closed shell:	Shell with maximum number of electrons allowed by the exclusion principle
valence electrons:	Electrons in the outermost occupied shell of an atom; they occupy the shell with the largest value of *n* and are used to form chemical bonds.
electron configuration:	List of all occupied *subshells* or *orbitals*, with the number of electrons in each indicated as a numerical superscript. **Example:** Li: $1s^2 2s^1$.

- **Building-Up (*Aufbau*) Principle**

  → Order in which electrons are added to *subshells* and *orbitals* to yield the *electron configuration* of atoms

  → **The (*n* + ℓ) rule:** Order of filling subshells in *neutral atoms* is determined by filling those with the *lowest* values of (*n* + ℓ) first. Subshells in a group with the same value of (*n* + ℓ) are filled in the order of increasing *n* (*topic not covered in the text*).

- **Usual filling order of subshells (*n* + ℓ rule):** 1s < 2s < 2p < 3s < 3p < 4s < 3d < 4p < 5s < 4d < …

Note: Ionization or subtle differences in shielding and penetration can change the order of the energy of subshells. *Thus, energy ordering of subshells is not fixed absolutely, but may vary from atom to atom or from an atom to its ion.*

- **Orbitals within a subshell**
  - → Hund's rule
  - → Electrons add to *different* orbitals of a subshell with spins *parallel* until the subshell is half full.

- **Applicability**
  - → The building-up principle in combination with the *Pauli exclusion principle* and *Hund's rule* accounts for the *ground-state* electron configurations of atoms.
  - → The principle is generally valid, **but there are exceptions** (see the periodic table on SG page 14).

Topic 1F: Periodicity

1F.1 The General Structure of the Periodic Table

- **Order of filling subshells**
  - → Understanding the organization of the periodic table
  - → Straightforward determination of (most) electron configurations

- **Terminology** → See the following two tables

Groups:	Columns in the periodic table, labeled 1–18 horizontally
s-block elements:	Groups 1, 2; s-subshell fills
p-block elements:	Groups 13–18; p-subshell fills
d-block elements:	Groups 3–12; d-subshell fills
Transition elements:	Groups 3–11; d-subshell fills
Main-group elements:	Groups 1, 2 and 13–18
Lanthanides:	4f-subshell fills
Actinides:	5f-subshell fills

Note: A transition element has a partially filled d-subshell either as the element or in any commonly occurring oxidation state. Thus, Zn, Cd, and Hg with completely filled d-subshells are not transition elements. The $(n-1)$d electrons of the d-block elements are considered to be valence electrons. Groups 1, 2, and 13–18 are alternatively labeled with Roman numerals I–VIII, which correspond to the number of valence electrons in the element.

Valence shell:	Outermost occupied shell (highest n)
Period:	Row of the periodic table
Period number:	Principal quantum number of valence shell
Period 1:	H, He; 1s-subshell fills
Period 2:	Li through Ne; 2s-, 2p-subshells fill
Period 3:	Na through Ar; 3s-, 3p-subshells fill
Period 4:	K through Kr; 4s-, 3d-, 4p-subshells fill

- **Periodic table** → Two forms are displayed (one below, the other on the next page):

    The one below shows the elements, the other shows the *final* subshell filled.

    Look at the tables to understand the terminology.

Periodicity of Elements in the Periodic Table

Period ↓   **Group (1–18)** →

Period	1	2	3	4	5	6	7	8	9	10	11	12	13	14	15	16	17	18
1	H	2											13	14	15	16	17	He
2	Li	Be											B	C	N	O	F	Ne
3	Na	Mg	3	4	5	6	7	8	9	10	11	12	Al	Si	P	S	Cl	Ar
4	K	Ca	Sc	Ti	V	*Cr*	Mn	Fe	Co	Ni	*Cu*	Zn	Ga	Ge	As	Se	Br	Kr
5	Rb	Sr	Y	Zr	*Nb*	*Mo*	Tc	*Ru*	*Rh*	***Pd***	*Ag*	Cd	In	Sn	Sb	Te	I	Xe
6	Cs	Ba	Lu	Hf	Ta	W	Re	Os	Ir	*Pt*	*Au*	Hg	Tl	Pb	Bi	Po	At	Rn
7	Fr	Ra	Lr	Rf	Db	Sg	Bh	Hs	Mt	Ds	Rg	Cn	Uut	Fl	Uup	Lv	Uus	Uuo

Lanthanides	*La*	*Ce*	Pr	Nd	Pm	Sm	Eu	*Gd*	Tb	Dy	Ho	Er	Tm	Yb
Actinides	*Ac*	***Th***	Pa	U	*Np*	Pu	Am	*Cm*	Bk	Cf	Es	Fm	Md	No

- **Electron configurations of the elements** → See the Appendix section on *Ground-State Electron Configurations* in the text.

- **17 italicized elements**

    → Exceptions to the $(n + \ell)$ rule

    → Differ by the placement of *one* electron

    Example: Cr: $1s^2 2s^2 2p^6 3s^2 3p^6 4s^1 3d^5$ = [Ar] $4s^1 3d^5$ (not [Ar] $4s^2 3d^4$ as might be expected)

- **2 bold, italicized elements**
  - → Differ by the placement of *two* electrons (Pd, Th)
  - → **Pd**   We expect $[Kr]4d^8 5s^2$, but *actually* find $[Kr]4d^{10}$ with *no* 5s-subshell electrons.
  - → **Th**   We expect $[Rn]5f^2 7s^2$, but *actually* find $[Rn]6d^2 7s^2$ with *no* 5f-subshell electrons.

- **Recently named elements**
  - → **Rg** = Roentgenium $(Z = 111)$ [in honor of Wilhelm Röntgen (Roentgen)]
  - → **Cn** = Copernicium $(Z = 112)$ [in honor of Nicolaus Copernicus]
  - → **Fl** = Flerovium $(Z = 114)$ [in honor of Georgiy Flerov (1913-1990), a Russian physicist]
  - → **Lv** = Livermorium $(Z = 116)$ [in honor of the Lawrence Livermore Lab]

Note: **Cn** was originally given the symbol **Cp**, but that symbol was previously associated with the name *cassiopeium*, now know as lutetium (Lu). The symbol Cp is also used in organometallic chemistry to denote the cyclopentadienyl ligand. For these reasons, IUPAC disallowed the use of Cp as a future element symbol.

Note: There is evidence for the existence of certain isotopes of Elements 113, 115, 117, and 118. The claims require ratification.

Order of Filling Subshells in the Periodic Table

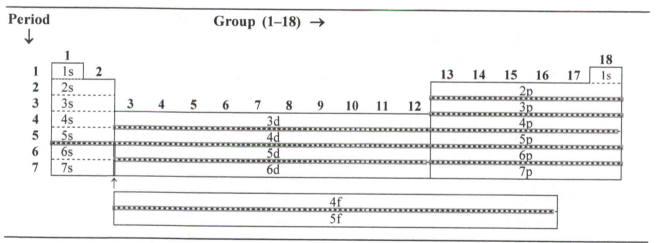

1F.2 Atomic Radius (*r*)

- **Definition of atomic radius**
  - → Half the distance between the centers of neighboring atoms (nuclei).
  - → For metallic elements, *r* is determined for the solid.
    Example: For solid Zn, 274 pm between nuclei, so *r* (atomic radius) = 137 pm
  - → For nometallic elements, *r* is determined for diatomic molecules (covalent bond).
    Example: For I_2 molecules, 266 pm between nuclei, so *r* (covalent radius) = 133 pm

- **General trends in radius with atomic number**
  - → *r* decreases from *left* to *right* across a period   (*effective* nuclear charge increases)
  - → *r* increases from *top* to *bottom* down a group   (change in valence electron principal quantum number and size of valence shell)

  Note: The atomic radius of noble-gas atoms can be obtained from low-temperature solids (high pressure for helium). These atoms are held together by weak, van der Waals forces (Focus 6), so their size defined in this way is actually *larger* than the covalent radius of the neighboring halogen atom. There are other definitions of size. If *all* atomic sizes are defined from *atomic wavefunctions*, then each noble-gas atom is *smaller* than its neighboring halogen atom.

- **Spatial plot of atomic wavefunctions for several noble gas atoms ($r^2 \psi^2$)**

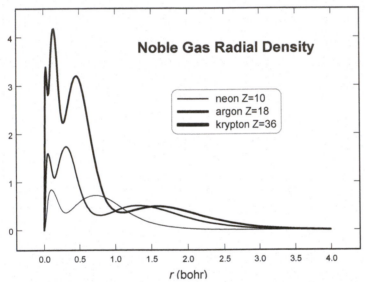

Notes:

Radial density $\propto r^2 \psi^2$

1 bohr $= a_0 \approx 52.9$ pm

Observe the appearance of shell structure.

  Note: Kr has the highest curve due to its greater noble gas radial density, followed by Ar and then Ne. The "size" of the atom is given by the most probable value of *r* for the valence shell.

1F.3 Ionic Radius

- **An ion's share of the distance between neighbors in an ionic solid (cation to anion)**
  - → Distance between the nuclei of a neighboring cation and anion is the sum of two ionic radii
  - → Radius of the oxide anion (O^{2-}) is 140 pm.
    Example: Distance between Zn and O nuclei in zinc oxide is 223 pm; therefore, the ionic radius of Zn^{2+} is 83 pm [$r(Zn^{2+}) = 223$ pm $- r(O^{2-})$].
  - → Cations are **smaller** than parent atoms, for example, Zn (133 pm) and Zn^{2+}(83 pm).
  - → Anions are **larger** than parent atoms, for example, O (66 pm) and O^{2-}(140 pm).

- **Isoelectronic atoms and ions**
  - → Atoms and ions with the same number of electrons
    Example: Cl^-, Ar, K^+, and Ca^{2+}

1F.4 Ionization Energy (I)

- I

  → Energy needed to remove an electron from a gas-phase atom in its lowest energy state.

- **Symbol**

  → Number subscripts (*e.g.*, I_1, I_2) denote removal of successive electrons. For a given species, $I_2 > I_1$ always.

- **General periodic table trends in I_1 for the main-group elements**

  → Increases from *left* to *right* across a period   (Z_{eff} increases)

  → Decreases from *top* to *bottom* down a group   (increase in principal quantum number n of valence electron)

- **Exceptions**

  → For Groups 2 and 15 in Periods 2–4, I_1 is *larger* than the neighboring *main group element* in Groups 13 and 16, respectively. Repulsions between electrons in the same orbital and/or extra stability of completed and half completed subshells are responsible for these exceptions to the general trends.

- **Metals toward the lower left of the periodic table (e.g., Cs, Ba)**

  → Have low ionization energies.

  → Readily lose electrons to form cations.

- **Nonmetals toward the upper right of the periodic table (e.g., F, O)**

  → Have high ionization energies.

  → Do not readily lose electrons.

1F.5 Electron Affinity (E_{ea})

- **Energy *released* when an electron is *added* to a gas-phase atom**

- **Periodic table trends in E_{ea} for the main-group elements**

  → Increases from *left* to *right* across a period.   (Z_{eff} increases)

  → Decreases from *top* to *bottom* down a group.   (increase in principal quantum number n of valence electron)

  → Generally, the same as for I_1 with the major exceptions displayed in **Figure 1F.12**

- **Summary of trends in r, I_1, and E_{ea}**

  → All depend on Z_{eff} and n of outer subshell electrons.

  → Recall that there are *many* exceptions to the general trends.
     Example: For electron affinity, the value for N is actually less than that for O.

  → The diagram on the next page indicates the general behavior of the three properties (r, I_1, and E_{ea}) with respect to both Z_{eff} and n (exceptions ignored, many in the case of electron affinity). The wavefunction definition for atomic size is used if noble gas atoms are included.

Main-Group Elements in the Periodic Table

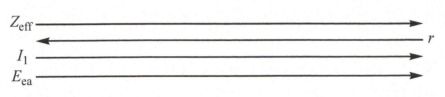

H							He
Li	Be	B	C	N	O	F	Ne
Na	Mg	Al	Si	P	S	Cl	Ar
K	Ca	Ga	Ge	As	Se	Br	Kr
Rb	Sr	In	Sn	Sb	Te	I	Xe
Cs	Ba	Tl	Pb	Bi	Po	At	Rn

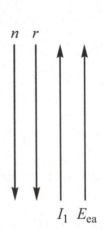

1F.6 The Inert-Pair Effect

- **Tendency to form ions two units lower in charge than expected from the group number relation**

  → Most pronounced in p-block elements near the bottom of the groups

  → Due in part to the different energies of the valence p- and s-electrons

  → Important for the *heavier* two members of Groups 13, 14, and 15; for example, Pb(IV) & Pb (II) or Sb(V) & Sb(III)

  → Valence s-electrons are called a "lazy pair."

1F.7 Diagonal Relationships

- **Diagonally related pairs of main-group elements often show similar chemical properties.**

  → Diagonal band of metalloids dividing metals from nonmetals

  → Similarity of Li and Mg (react directly with N_2 to form nitrides, N^{3-})

  → Similarity of Be and Al (both react with acids and bases)

1F.8 The General Properties of the Elements

- **s-block elements (Groups 1 and 2)**

  → All are reactive metals (except H); they form *basic* oxides (O^{2-}), peroxides (O_2^{2-}), or superoxides (O_2^-).

- **Compounds of s-block elements are ionic, except for beryllium.**

  Notes: *Hydrogen* (a nonmetal) is placed by itself, or *more usually* in Group 1, because its electronic configuration ($1s^1$) is similar to those of the alkali metals: [noble gas] ns^1.

  Helium, with electronic configuration $1s^2$, is placed in Group 18 because its properties are similar to those of neon, argon, krypton, and xenon, all with filled subshells.

- **p-block elements (Groups 13–18)**

  → Members are metals, metalloids, and nonmetals.

  Metals: Group 13 (Al, Ga, In, Tl); Group 14 (Sn, Pb); Group 15 (Bi)

  Note: These metals have relatively low I_1, but *larger* than those of the *s* block and *d* block.

  Metalloids: Group 13 (B); Group 14 (Si, Ge); Group 15 (As, Sb); Group 16 (Te, Po)

  Note: Metalloids have the *physical* appearance and properties of *metals* but behave *chemically* as *nonmetals*.

  Nonmetals: Group 14 (C); Group 15 (N, P); Group 16 (O, S, Se); Groups 17 & 18 (All)

  Notes: High E_{ea} in Groups 13–17; these atoms tend to *gain* electrons to complete their subshells.

  Group 18 noble-gas atoms are generally *nonreactive*, except for Kr and Xe, which form a few compounds.

- **d-block elements (Groups 3–12)**

  → All are metals (*most* are transition metals), with properties intermediate to those of s-block and p-block metals.

  Note: *All* Group 12 cations retain the filled d-subshell. For this reason, these elements (Zn, Cd, Hg) are *not* classified as transition metals. Recall that in this text d-orbital electrons are considered to be valence electrons for *all* the d-block elements.

- **Transition metals**

  → Form compounds with a variety of oxidation states (oxidation numbers)
  → Form alloys
  → Facilitate subtle changes in organisms

- **f-block elements (lanthanides and actinides)**

  → Rare on Earth
  → Lanthanides are incorporated in *superconducting* materials.
  → Lanthanides are used in electronic devices such as plasma TVs, disk drives, and mobile phones.
  → Actinides are all *radioactive* elements, most do not occur naturally on Earth.

Focus 2   MOLECULES

Topic 2A: IONIC BONDING

2A.1   The Ions That Elements Form

- **Cations**

  → Remove outermost electrons in the order $n\text{p}, n\text{s}, (n-1)\text{d}$

- **Metallic s-block elements and metallic p-block elements in Periods 2 and 3**

  → Form cations by losing electrons down to the noble-gas core

  Examples: Mg, $[\text{Ne}]\,3\text{s}^2 \rightarrow$ magnesium(II), Mg^{2+}, [Ne];

  Al, $[\text{Ne}]\,3\text{s}^2\,3\text{p}^1 \rightarrow$ aluminum(III), Al^{3+}, [Ne]

- **Metallic p-block elements in Periods 4 and higher**

  → Form cations with complete, typically unreactive d-subshells

  Example: Ga, $[\text{Ar}]\,4\text{s}^2\,3\text{d}^{10}\,4\text{p}^1 \rightarrow$ gallium(III), Ga^{3+}, $[\text{Ar}]\,3\text{d}^{10}$

- **Metallic d-block elements**

  → Lose s-electrons and often a variable number of d-electrons

  Examples: iron(II), Fe^{2+}, $[\text{Ar}]\,3\text{d}^6$ and iron(III), Fe^{3+}, $[\text{Ar}]\,3\text{d}^5$

- **Many metallic p-block elements**

  → May lose either their *p*-electrons or all their *s*- and *p*-electrons in the valence shell

  Example: Sn, $[\text{Kr}]\,5\text{s}^2\,4\text{d}^{10}\,5\text{p}^2 \rightarrow$ tin(II), Sn^{2+}, $[\text{Kr}]\,5\text{s}^2\,4\text{d}^{10}$ *or* tin(IV), Sn^{4+}, $[\text{Kr}]\,4\text{d}^{10}$

- **Anions**

  → Add electrons until the next noble-gas configuration is reached.

  Examples: Carbide (methanide), C^{4-}, $[\text{He}]\,2\text{s}^2\,2\text{p}^6$ or [Ne]   {octet}

  Hydride, H^-, 1s^2 or [He]   {duplet}

2A.2   Lewis Symbols (Atoms and Ions)

- **Valence electrons**

  → Depicted as dots; a pair of dots represents two paired electrons; single dots represent unpaired electrons

  → Lewis symbols for neutral atoms in the first two periods of the Periodic Table:

  H·   He: Li·   Be: ·B·   ·C·   :N·   :O·   :F·   :Ne:

  Note: Ground-state structures for B and C are :B·   and :C·. B and C have one and two unpaired electrons, respectively.

- **Variable valence**

  → Ability of an element to form two or more ions with different oxidation numbers

→ Displayed by many d-block and p-block elements

Examples: Lead in lead(II) oxide, PbO, and in lead(IV) oxide, PbO_2
Iron in iron(II) oxide, FeO, and in iron(III) oxide, Fe_2O_3

→ Lewis symbols for ions: use brackets where necessary to show charge is associated with an ion as a whole.

Example: Ca^{2+}, $2 \left[:\ddot{\underset{..}{C}}l: \right]^-$

2A.3   The Energetics of Ionic Bond Formation

• **Ionic bond** → Electrostatic attraction (coulombic) of *oppositely* charged ions

• **Ionic model** → Energy for the *formation* of ionic bonds is supplied *mainly* by coulombic attraction of *oppositely* charged ions. This model gives a good description of bonding between the ions of metals (particularly *s*-block) and those of nonmetals.

• **Ionic solids and ionic crystals**

→ Are assemblies of cations and anions arranged in regular arrays.

→ The ionic bond is *nondirectional*; each ion is "bound" to *all* its neighbors.

→ Typically have *high* melting and boiling points and are brittle, for example NaCl.

→ Form electrolyte solutions if they dissolve in water.

• **Formulas of compounds composed of monatomic ions**

→ Formulas are predicted by assuming that atoms forming cations lose all valence electrons and those forming anions gain electrons in the valence subshell(s) until each ion has an octet of electrons or a duplet in the case of H, He, and Be.

→ For cations with variable valence, the oxidation number is used.
Example: Iron(III) chloride, $FeCl_3$

→ Relative numbers of cations and anions are chosen to achieve electrical neutrality, using the smallest possible integers as subscripts.
Example: Iron(III) oxide, Fe_2O_3

2A.4   Interactions between Ions

• **In an ionic solid**

→ All the cations repel each other, and all the anions repel each other.

→ Each cation is attracted to all the anions to a greater or lesser extent.

→ In this view, an ionic bond is a "global" characteristic of the entire crystal.

- **Coulomb potential energy between two ions, E_P**

$$E_{P,12} = \frac{(z_1 e) \times (z_2 e)}{4\pi\varepsilon_0 r_{12}} = \frac{z_1 z_2\, e^2}{4\pi\varepsilon_0 r_{12}}$$

  e is the fundamental charge; z_1 and z_2 are the charge numbers of the two ions; r_{12} is the distance between the centers of the ions; ε_0 is the vacuum permittivity.

- **Madelung constant (A) for a three-dimensional crystal lattice**

$$E_P = -A \times \frac{|z_1 z_2| N_A e^2}{4\pi\varepsilon_0 d}$$

  A is a positive number; d is the distance between the centers of nearest neighbors in the crystal; N_A is the Avogadro constant (number of species per mol).

  → A depends on the *arrangement* of ions in the solid (Madelung constant).

  → E_P is always negative and A is positive.

  → Therefore, ionic crystals with **small ions** (short interionic distances), **large values of A**, and **highly charged ions** tend to have large attractive energies.

- **Values of A**
  → 1.747 56 (NaCl structure) [$d = 279.8$ pm for NaCl (actual value near 0 K)]
  → 1.762 67 (CsCl structure) [$d = 348$ pm for CsCl (sum of ionic radii)]

- **Lattice energy**
  → Energy required to totally convert the solid-phase ions of the crystal into gas-phase ions infinitely far apart

- **Potential energy of an ionic solid (opposite sign = lattice energy)**
  → Takes attractive and repulsive interionic interactions into account
  → Is estimated by the **Born-Mayer** equation

- **Born-Mayer equation (Not Covered in Text)**

$$E_{P,min} = -\frac{N_A\, |z_A z_B| e^2}{4\pi\varepsilon_0 d}\left(1 - \frac{d^*}{d}\right) A$$

  The constant d^* is commonly taken to be 34.5 pm. It results from the repulsive effects of overlapping electron charge clouds.

- **More accurate repulsive energy correction**

  $E_P{}^* \propto e^{-d/d^*}$ where d^* is a constant (34.5 pm).

Topic 2B: COVALENT BONDING

2B.1   Lewis Structures

- **Covalent bond**
    - → Pairs of electrons *shared* between two atoms.
    - → Located between two neighboring atoms and *binds* them together.
      Examples: Nonmetallic elements such as H_2, N_2, O_2, F_2, Cl_2, Br_2, I_2, P_4, and S_8

- **Rules**
    - → Atoms attempt to complete duplets or octets by sharing pairs of valence electrons.
    - → Valence of an atom is the number of bonds it can form.
    - → A line (–) represents a shared pair of electrons.
    - → A lone pair of nonbonding electrons is represented by two dots (:).
      Example: H–H, (single bond) duplet on each atom (valence of hydrogen = 1) and :N≡N:,
      (triple bond and two lone pairs) octet on each atom (valence of nitrogen = 3)

- **Lone pairs of electrons**
    - → Electron pairs *not* involved in bonding.
      Example: The electrons indicated as dots in the Lewis structure of N_2 shown earlier.

- **Polyatomic species**
    - → Show which atoms are bonded (atom connectivity) and which contain lone pairs of electrons.
    - → Do not portray the *shape* of a molecule or ion.

- **Rules**
    - → Count total number of valence electrons in the species.
    - → Arrange atoms next to bonded neighbors.
    - → Use minimum number of electrons to make all single bonds.
    - → Count the number of nonbonding electrons required to satisfy octets.
    - → Compare to the actual number of electrons left.
    - → If lacking a sufficient number of electrons to satisfy octets, make *one extra bond* for each deficit pair of electrons.
    - → Sharing pairs of electrons with a neighbor completes the octet or duplet.
    - → Each shared pair of electrons counts as one *covalent bond* (line).

      | One shared pair | **single bond** | (–) | Bond order = 1 |
      | Two shared pairs | **double bond** | (=) | Bond order = 2 (multiple bond) |
      | Three shared pairs | **triple bond** | (≡) | Bond order = 3 (multiple bond) |

- **Bond order**
    - → Number of bonds that link a specific pair of atoms

- **Terminal atom**
    - → Bonded to only one other atom

- **Central atom**
  - → Bonded to at least two other atoms
- **Molecular ions**
  - → Contain *covalently* bonded atoms: NH_4^+, Hg_2^{2+}, SO_4^{2-}
- **Rules of thumb**
  - → Usually, the element with the lowest I_1 is a central atom, but *electronegativity* (introduced in **Topic 2D.1**) is a better indicator. For example, in HCN, carbon has the lowest I_1 and is the central atom. It is also less electronegative than nitrogen.
  - → Usually, there is a symmetrical arrangement about the central atom. For example, in SO_2, OSO is symmetrical, with S as the central atom and the two O atoms terminal.
  - → Oxoacids have H atoms bonded to O atoms; H_2SO_4 is actually $(HO)_2SO_2$, with two O atoms and two OH groups bonded to S.

    Examples: Ethyne (acetylene), C_2H_2 (10 valence electrons): H–C≡C–H

    Hydrogen cyanide, HCN (10 valence electrons): H–C≡N:

    Ammonium ion, NH_4^+ (8 valence electrons): $\left[\begin{array}{c} H \\ | \\ H{-}N{-}H \\ | \\ H \end{array}\right]^+$

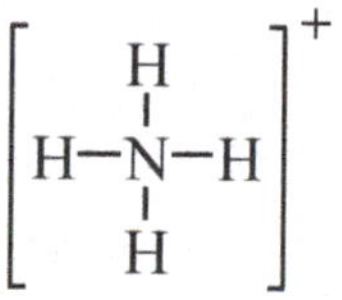

2B.2 Resonance

- **Multiple Lewis structures**
  - → Some molecules can be represented by different Lewis structures (*contributing structures*) in which the *locations* of the electrons, *but not the nuclei*, vary.
  - → Multiple Lewis structures used to represent a given species are called *resonance structures*.
- **Electron delocalization**
  - → Species requiring multiple Lewis structures often involve *electron delocalization* with some electron pairs distributed over *more* than two atoms.
- **Blending of structures**
  - → Double-headed arrows (⟷) are used to relate contributing Lewis structures, indicating that a blend of the contributing structures is a better representation of the bonding than any one structure alone.
- **Resonance hybrid**
  - → Blended structure of individual contributing Lewis structures

- **Resonance structures**
  - → Several examples follow:

    Example: N_2O, nitrous oxide, has $2(5) + 6 = 16$ valence electrons or eight pairs.

    $$:\ddot{N}=N=\ddot{O}: \quad \longleftrightarrow \quad :N\equiv N-\ddot{O}:$$

    Blending of two structures, both of which follow the octet rule.

    Note: The central N atom is the least electronegative atom (**Topic 2D.1**).

    Note: A triple bond to an O atom is found only in species such as BO^-, CO, NO^+, and O_2^{2+}.

    Example: C_6H_6, benzene, has $6(4) + 6(1) = 30$ valence electrons, or 15 pairs.

    Kekulé structures   Resonance hybrid

    Note: Carbon atoms lie at the vertices of the hexagon (the six C–H bonds radiating from each corner are not shown in the stick structures). The last structure depicts six valence electrons *delocalized* around the ring. The resonance hybrid between the two Kekulé structures suggests that all bonds between neighboring carbon atoms in benzene are equivalent, which, in contrast to an individual Kekulé structure, agrees with experiment.

2B.3 Formal Charge

- **Formal charge**
  - → An atom's number of valence electrons (V) minus the number of electrons assigned to it in a Lewis structure

- **Electron assignment in a Lewis structure**
  - → Atom possesses all of its lone pair electrons (L) and half of its bonding electrons (B)

    [B means shared bonding electrons]

  - → Formal charge $= V - (L + \frac{1}{2}B)$

- **Contribution of individual Lewis structures to a resonance hybrid**
  - → Structures with individual formal charges closest to **zero** usually have the lowest energy and are the major contributors.

    Example: N_2O: $:\ddot{N}=N=\ddot{O}: \quad \longleftrightarrow \quad :N\equiv N-\ddot{O}: \quad \longleftrightarrow \quad :\ddot{N}-N\equiv O:$

    $\quad\quad\quad\quad\quad\quad\; -1 \;\; +1 \;\;\; 0 \quad\quad\quad\quad 0 \;\; +1 \;\; -1 \quad\quad\quad\quad -2 \;\; +1 \;\; +1$

    (high formal charges)

- **Plausibility of isomers**
  - → The isomer with the lowest formal charges is *usually* preferred.

    Examples: N_2O: $:\overset{..}{N}=N=\overset{..}{O}:$ and $:\overset{..}{N}=O=\overset{..}{N}:$
                                   $\quad\; -1\;\; +1\;\;\; 0$ $\quad -1\;\; +2\;\; -1$
                         (low formal charges)   (high formal charges)

    HCN: $H-C\equiv N:$ and $H-N\equiv C:$
                  $\;\; 0\;\; 0\;\; 0$ $\;\; 0\; +1\; -1$

    Note: In each example, the first structure with lower formal charges is preferred.

Summary

- **Formal charge**
  - → Indicates the extent to which atoms have gained or lost electrons in a Lewis structure (covalent bonding)
  - → Exaggerates the *covalent* character of bonds by assuming that electrons are shared equally
  - → Structures with the lowest formal charges usually have the lowest energy (major contributors to the resonance hybrid).
  - → Isomers with lowest formal charges are usually favored.

- **Oxidation number**
  - → Exaggerates the *ionic* character of bonds

    Example: Carbon disulfide, $:\overset{..}{S}=C=\overset{..}{S}:$ Here, C has an oxidation number of +4 [all the electrons in the double bonds are assigned to the more electronegative S atoms (**Topic 2D.1**] and a formal charge of zero (the electrons in the double bonds are shared equally between the C and S atoms).

Topic 2C: BEYOND THE OCTET RULE

2C.1 Radicals and Biradicals

- **Radicals**
  - → Species with an unpaired electron
  - → All species with an odd number of electrons are radicals.
  - → Are highly reactive, important for reactions in the upper atmosphere, cause rancidity in foods, degradation of plastics in sunlight, and perhaps contribute to human aging

Examples: CH₃ (methyl radical), OH (hydroxyl), OOH (hydrogenperoxyl),
NO (nitric oxide), NO₂ (nitrogen dioxide), O₃⁻ (ozonide ion)

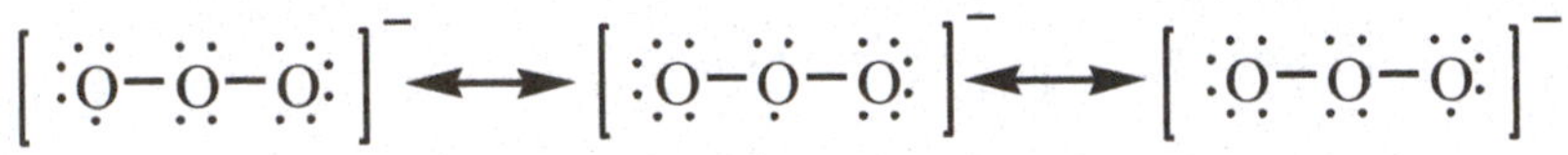

$$\left[\ddot{\text{:O}} - \ddot{\text{O}} - \ddot{\text{O}}\text{:} \right]^{-} \longleftrightarrow \left[\ddot{\text{:O}} - \ddot{\text{O}} - \ddot{\text{O}}\text{:} \right]^{-} \longleftrightarrow \left[\ddot{\text{:O}} - \ddot{\text{O}} - \ddot{\text{O}}\text{:} \right]^{-}$$

Note: Ozonide ion does not follow the octet rule.

- **Biradicals**
  - → Species containing *two* unpaired electrons

    Examples: O (oxygen atom) and CH₂ (methylene) [unpaired electrons on a single atom]
    O₂ (oxygen molecule) and larger organic molecules [on different atoms]

- **Antioxidant**
  - → Species that reacts rapidly with radicals before they have a chance to do damage
    Examples: Vitamins A, C, and E; coenzyme Q; substances in coffee, orange juice, chocolate

2C.2 Expanded Valence Shells

- **Expanded valence shells**
  - → More than eight electrons associated with an atom in a Lewis structure (expanded octet)
    A *hypervalent compound* contains an atom with more atoms attached to it than is permitted by the octet rule. Empty *d*-orbitals are utilized to permit octet expansion.
  - → Electrons may be present as bonding pairs or lone pairs.
  - → Are characteristic of nonmetal atoms in *Period 3 or higher*.
  - → *First- and second-period elements do not utilize expanded valence shells.*

- **Variable covalence**
  - → Ability to form different numbers of covalent bonds
  - → Elements showing variable covalence include those in the table:

Valence Shell Occupancy

Elements	8 electrons	10 electrons	12 electrons
P	PCl₃ / PCl₄⁺	PCl₅	PCl₆⁻
S	SF₂	SF₄	SF₆
I	IF	IF₃	IF₅

Example: Major Lewis structures for sulfuric acid, H_2SO_4 (32 valence electrons)

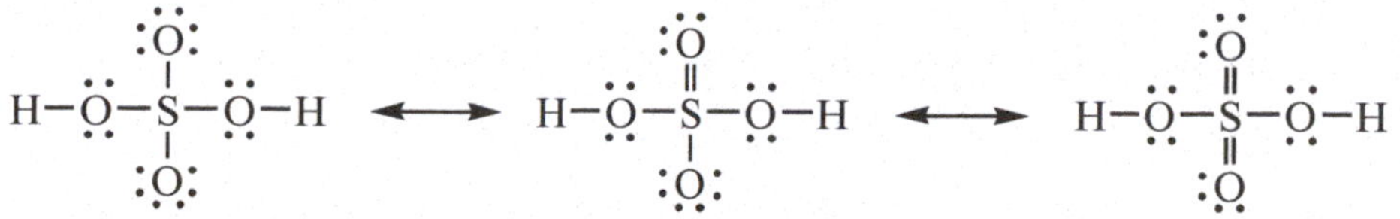

In the structure on the right, all atoms have zero formal charge and the S atom is surrounded by 12 electrons (expanded octet). *Because it has the lowest formal charges, this structure is expected to be the most favored one energetically and the one to make the greatest contribution to the resonance hybrid.* An additional Lewis structure has one double bond (not shown). Similar examples include SO_4^{2-}, SO_2, and S_3O (S is the central atom).

2C.3 Incomplete Octets

- **Incomplete octet**
  - → Fewer than eight valence electrons on an atom in a Lewis structure

  Example: BF_3,

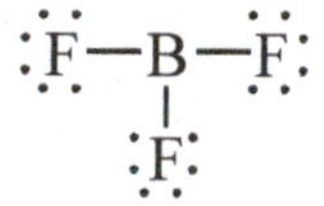

  The single-bonded structure with an incomplete octet makes the major contribution.

  - → Other compounds with incomplete octets are BCl_3 and $AlCl_3$ both vapor at high temperature.

  - → At room temperature, aluminum chloride exists as the dimer, Al_2Cl_6, which follows the octet rule (see **Margin Figure 17**, page 93, in the text).

- **Coordinate covalent bond**
  - → One in which both electrons come from one atom

    Example: Donation of a lone pair by one atom to form a bond that completes the otherwise incomplete octet of another atom. In the Lewis structure for the reaction of BF_3 with NH_3, both electrons in the B–N bond are derived from the N lone pair.

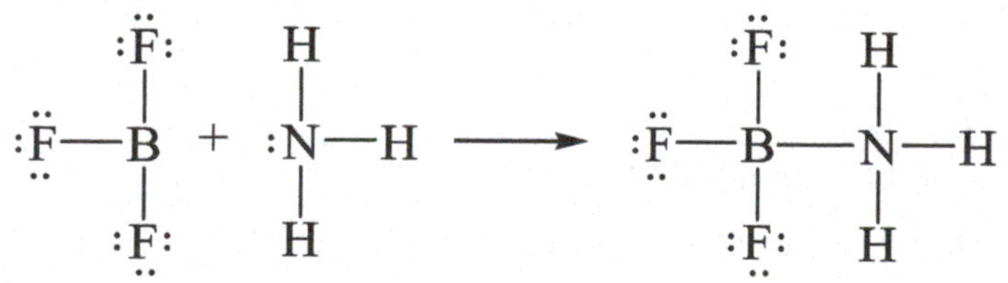

Topic 2D: THE PROPERTIES OF BONDS

2D.1 Correcting the Covalent Model: Electronegativity

- **Bonds**
  - → *All molecules* may be viewed as resonance hybrids of pure covalent and ionic structures.

    Example: H_2 The structures are $H{-}H \longleftrightarrow H^+[H{:}]^- \longleftrightarrow [{:}H]^- \, H^+$

    Here, the two ionic structures make equal contributions to the resonance hybrid, but the single covalent structure is of major importance.

- **Partial charges**
  - → When the bonded atoms are different, the ionic structures are not energetically equivalent.
  - → Unequal sharing of electrons results in a polar covalent bond.

    Example: HF The structures are

$$H{-}F \longleftrightarrow H^+\left[{:}\ddot{\underset{..}{F}}{:}\right]^- \longleftrightarrow [{:}H]^-\left[\ddot{\underset{..}{F}}{:}\right]^+$$

  Note: The electron affinity of F is greater than that of H, resulting in a small negative charge on F and a corresponding small positive charge on H:

$$E_{ea}(F) > E_{ea}(H) \implies {}^{\delta+}H{-}F^{\delta-} \text{ (partial charges, } |\delta+| = |\delta-|)$$
$$E(H^+F^-) \ll E(H^-F^+) \implies H^-F^+ \text{ is a very } minor \text{ contributor}$$

- **Electric dipole**
  - → A partial *positive* charge separated from an equal but *negative* partial charge

- **Electric dipole moment (μ)**
  - → Magnitude of an electric dipole: partial charge times distance between charges
  - → Units: debye (D)
  - → 4.80 D ≡ an electron (−) separated by 100 pm from a proton (+)
  - → $\boxed{\mu = (4.80 \text{ D}) \times \delta \times (\text{distance in pm}/100 \text{ pm})}$

- **Electronegativity (χ)**
  - → Electron-attracting power of an atom when it is bonded to another atom
  - → Mulliken scale: $\chi = \frac{1}{2}(I_1 + E_{ea})$
  - → Follows same periodic table trends as I_1 and E_{ea}
  - → *Increases from left to right and from bottom to top*
  - → Pauling numerical scale based on bond energies [$\chi(F) = 3.98$] is used in text (qualitatively similar to Mulliken's scale).
  - → See **Figure 2D.2** in the text for a tabulation of values for selected elements.

- **Rough rules of thumb for A-B bonds:**

$$(\chi_A - \chi_B) \geq 2 \quad \text{Bond is } \textit{essentially } \text{ionic}$$
$$0.5 \leq (\chi_A - \chi_B) \leq 1.5 \quad \text{Bond is } \textit{polar } \text{covalent}$$
$$(\chi_A - \chi_B) \leq 0.5 \quad \text{Bond is } \textit{essentially } \text{covalent}$$

- **Electric dipole moment (μ) conventions**

  → In the original convention, the vector points towards the negative charge. Many chemists continue to use this convention.

  → In the modern convention, the vector points towards the positive charge. Physicists employ this convention.

2D.2 Correcting the Ionic Model: Polarizability

- **Ionic bonds**

  → All have *some* covalent character.
  A cation's positive charge attracts the electrons of an anion or atom in the direction of the cation (*distortion* of spherical electron cloud).

- **Highly *polarizable* atoms and ions**

  → Readily undergo a *large* distortion of their electron cloud
  Examples of polarizable species: Large anions and atoms such as I^-, Br^-, Cl^-, I, Br, and Cl

- **Polarizing power**

  → Property of ions (and atoms) that cause distortions of electron clouds

  → Increases as size decreases and increases as charge of a cation increases
  Examples of species with significant polarizing power:
  The small and/or highly charged cations Li^+, Be^{2+}, Mg^{2+}, and Al^{3+}

- **Significant covalent bonding character**

  → Bonds between highly polarizing cations and highly polarizable anions have significant covalent character.

  → The Be^{2+} cation is highly polarizing and the Be–Cl bond has significant covalent character (even though there is an electronegativity difference of 1.59).

  → In the series AgCl to AgI, the bonds become more covalent as the polarizability (also size) of the anion increases ($Cl^- < Br^- < I^-$).

2D.3 Bond Strengths

- **Dissociation energy (D)**

  → Energy is required to separate bonded atoms in neutral molecules.

  → Bond breaking is *homolytic*, which means that each atom retains half of the bonding electrons.

  → Determines the *strength* of a chemical bond.

  → The *greater* the dissociation energy, the *stronger* the bond.

  → D is defined exactly for diatomic (two-atom) molecules.

→ For polyatomic (greater than two-atom) molecules, D also depends on the other bonds in the molecule. However, for many molecules, this dependence is slight.

- **Values of D (Table 2D.1)**

  → Vary from about 139 kj·mol^{-1} (I–I single bond) to 1062 kj·mol^{-1} (C≡O triple bond)

- **D for several diatomic molecules**

$$H–H(g) \rightarrow H(g) + H(g) \qquad D = 424 \text{ kJ·mol}^{-1} \text{ (strong)}$$
$$H–F(g) \rightarrow H(g) + F(g) \qquad D = 543 \text{ kJ·mol}^{-1} \text{ (strong)}$$
$$F–F(g) \rightarrow F(g) + F(g) \qquad D = 146 \text{ kJ·mol}^{-1} \text{ (weak)}$$
$$I–I(g) \rightarrow I(g) + I(g) \qquad D = 139 \text{ kJ·mol}^{-1} \text{ (weak)}$$
$$O=O(g) \rightarrow O(g) + O(g) \qquad D = 484 \text{ kJ·mol}^{-1} \text{ (strong)}$$
$$N≡N(g) \rightarrow N(g) + N(g) \qquad D = 932 \text{ kJ·mol}^{-1} \text{ (very strong)}$$

- **Bond strength**

  → For polyatomic molecules, bond strength is defined as the *average* dissociation energy for one type of bond found in different molecules.

  For example, the tabulated C–H single bond value is the *average* strength of such bonds in a selection of organic molecules, such as methane (CH_4), ethane (C_2H_6), and possibly ethene (C_2H_4).

  Note: Values of average dissociation energies in text **Table 2D.2** are actually those for the property of *bond enthalpy* (see **Topic 4E**), measured at 298.15 K. The bond *dissociation energies* of diatomic molecules given in **Table 2D.1** apply at 0 K (absolute zero).

- **Factors influencing bond strength**

  → Bond multiplicity   (C≡C > C=C > C–C)

  → Resonance   (C=C > C⋯C (benzene) > C–C)

  → Lone pairs on neighboring atoms (F–F < H–H)

  → Atomic radii   (HF > HCl > HBr > HI)

  Note: The smaller the radius, the stronger the bond.

2D.4 Bond Lengths

- **Bond length**

  → Internuclear distance, at the potential energy minimum, of two atoms linked by a covalent bond

  → Helps determine the overall size and shape of a molecule

  → Evaluated by using spectroscopic or x-ray diffraction (for solids) methods

  → For bonds between the same elements, length is *inversely* proportional to strength.

- **Factors influencing bond length**

  → Bond multiplicity   (C≡C < C=C < C–C)

  → Resonance   (C=C < C⋯C (benzene) < C–C)

  → Lone pairs on neighboring atoms (F–F > H–H)

  → Atomic radii   (HF < HCl < HBr < HI)

Note: The smaller the radius, the shorter the bond.

Note: These trends are the *opposite* of the ones for bond strength.

- **Covalent radius**

  → Contribution an atom makes to the length of a covalent bond

  → *Half* the distance between the centers (nuclei) of neighboring atoms joined by a covalent bond (*for like atoms*)

  → Covalent radii may be added to estimate bond lengths in molecules.

  → Tabulated values are *averages* of radii in polyatomic molecules.

  → *Decreases* from left to right in the periodic table

  → *Increases* going down a group in the periodic table

  → Decreases for a given atom with increasing multiple-bond character

  Example: Use the covalent radii given in **Figure 2D.11** in the text to *estimate* the several bond lengths in the acetic acid molecule, CH_3COOH.

Lewis Structure:	Bond type	Bond length estimate (pm)	Actual (pm)
	C–H	77 + 37 = 114	≈109
	C–C	77 + 77 = 154	≈150
	C–O	77 + 66 = 143	134
	C=O	67 + 60 = 127	120
	O–H	66 + 37 = 103	97

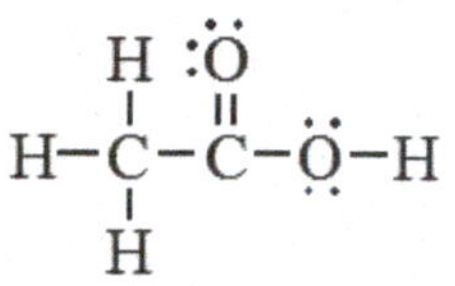

Lewis Structure:

```
      H  :O:
      |   ||
  H - C - C - O: - H
      |       ··
      H
```

Topic 2E: THE VSEPR MODEL

2E.1 The Basic VSEPR (Valence-Shell Electron-Pair Repulsion) Model

- **Valence electrons about central atom(s)**

  → Control the shape of a molecule

- **Lewis structure**

  → Shows distribution of *valence* electrons in bonding pairs (bonds) and as lone pairs or unpaired electrons

- **Bonds, lone pairs**

  → Regions of high electron density that repel each other (Coulomb's law) by rotating about a central atom, thereby maximizing their separation

- **Bond angle(s)**

  → Angle(s) between bonds joining atom centers

- **Multiple bonds**
  - → Treated as a *single* region of high electron concentration in VSEPR

- **Electron arrangement**
  - → Ideal locations of bond pairs and lone pairs about a central atom (angles between electron pairs define the geometry)

Typical Electron Arrangements in Molecules with *One* Central Atom

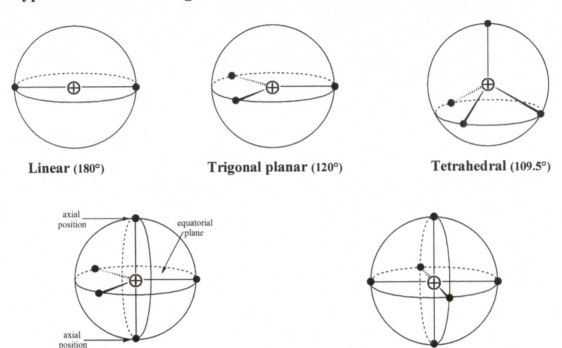

Linear (180°)   **Trigonal planar** (120°)   **Tetrahedral** (109.5°)

Trigonal bipyramidal (120° equatorial, 90° axial-equatorial)   **Octahedral** (90° and 180°)

- **VSEPR formula, AX_nE_m**   (See **Topic 2E.2** in the text)
  - → A = central atom
  - → X_n = n atoms (*same* or *different*) bonded to central atom
  - → E_m = m lone pairs on central atom
  - → Formula is useful for generalizing types of structures.

  Note: Molecular shape is defined by the location of the *atoms alone*.
  The bond angles are ∠ X–A–X.

  Some Examples: Linear, BeF_2 (AX_2); trigonal planar, BF_3 (AX_3); tetrahedral, CH_4 (AX_4); trigonal bipyramidal, PCl_5 (AX_5); octahedral, SF_6 (AX_6)

Typical Shapes of Molecules with *One* Central Atom*

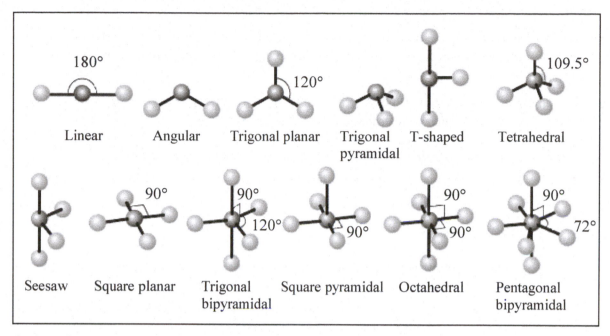

* These figures depict the locations of the atoms only, *not* lone pairs.

- **VSEPR (sometimes pronounced "vesper") method** (See **Toolbox 2E.1** in the text)

  → Write the Lewis structure(s). If there are resonance structures, pick *any* one.

  → Count the number of electron pairs (bonding and nonbonding) around the central atom(s). Treat a multiple bond as a *single* unit of high electron density.

  → Identify the *electron arrangement*. Place electron pairs as far apart as possible.

  → Locate the *atoms* and classify the *shape* of the molecule.

  → Optimize *bond angles* for molecules with *lone pairs* on the central atom(s) with the concept in mind that repulsions are in this order:

    lone pair-lone pair > lone pair-bonding pair > bonding pair-bonding pair.

  Example: Use the VSEPR model to predict the *shape* of tetrafluoroethene, C_2F_4.

  Lewis structure:

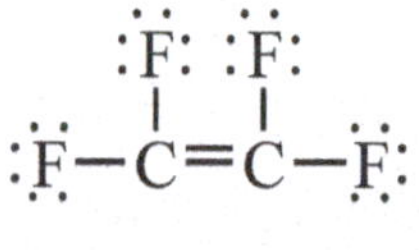

  There are three concentrations of electron density around each C atom, so the shape is *trigonal planar* about each C atom. C_2F_4 is an AX_3AX_3 species.

2E.2 Molecules with Lone Pairs on the Central Atom

- **VSEPR method for molecules with lone pairs ($m \neq 0$)**

  → Lone-pair electrons have preferred positions in certain electron arrangements.

→ In the *trigonal bipyramidal* arrangement, lone pairs prefer the *equatorial* positions, in which electron repulsions are minimized.

→ In the *octahedral* arrangement, all positions are *equivalent by symmetry*. The preferred electron arrangement for *two lone pairs* is the occupation of *opposite* corners of the octahedron.

→ Lone pairs contribute to the electron arrangement about a central atom and influence the shape of molecules, which is determined by the location of the nuclei.

Example: Use the VSEPR model to predict the shape of trichloroamine, NCl_3.

Lewis structure:

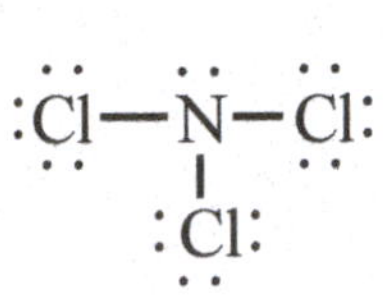

$$:\ddot{C}l-\ddot{N}-\ddot{C}l: \\ | \\ :\ddot{C}l:$$

The four electron pairs around N yield a *tetrahedral* electron arrangement, while the shape is *trigonal pyramidal* (like a badminton birdie). Because lone pair-bonding pair repulsion is greater than bonding pair-bonding pair repulsion, the bond angles [$\angle$ Cl–N–Cl] will be less than the tetrahedral value, 109.5°.

2E.3 Polar Molecules

● **Polar molecule**

→ Molecule with a *nonzero* dipole moment

● **Polar bond**

→ Bond with a *nonzero* dipole moment

● **Molecular dipole moment**

→ *Vector sum* of bond dipole moments

● **Representation of dipole**

→ Single arrow with head pointed toward the *positive* end of the dipole, the H atom in the following example.

Example: Lewis structure: $H-\ddot{C}l:$ and polar (dipole) representation: $H \leftarrow Cl$

● **Predicting molecular polarity**

→ Determine molecular *shape* using VSEPR theory.

→ Estimate electric dipole (bond) moments (text Section 2E.3) or use text **Figure 2E.7** to decide whether the molecular *symmetry* leads to a *cancellation* of bond moments (*nonpolar molecule*, no dipole moment) or not (*polar molecule*, nonzero dipole moment).

Example: Is XeF_2 a polar or nonpolar molecule?

Lewis structure: :F—Xe—F: The *electron* arrangement is *trigonal bipyramidal*; the *shape* is *linear* (lone pairs prefer the equatorial positions). The bond angle is 180°, the bond dipoles cancel, and the molecule is *nonpolar*: F→Xe←F.

Topic 2F: VALENCE-BOND (VB) THEORY

2F.1 Sigma (σ) and Pi (π) Bonds

- **Two major types of *bonding* orbitals**
  - → σ and π

- **σ-orbital**
  - → Has no *nodal surface* containing the interatomic (bond) axis
  - → Is cylindrical or "sausage" shaped
  - → Formed by overlapping in several ways: two s-orbitals, an s-orbital and a p-orbital end to end, two p-orbitals end to end, a certain hybrid orbital and an s-orbital, a p-orbital end to end with a certain hybrid orbital, or two certain hybrid orbitals end to end

- **π-orbital**
  - → *Nodal plane* containing the interatomic (bond) axis.
  - → Two quasi-cylindrical shapes (lobes), one above and the other below the nodal plane.
  - → Formed from side-by-side overlap of two *p*-orbitals.

- **Schematic of σ-orbital formation**
  - → Two 1s-orbitals of H atoms combine to form a σ-orbital of H_2.

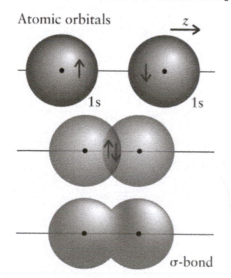

- **Schematic of π-orbital formation**
  - → Two $2p_x$-orbitals overlap side by side.

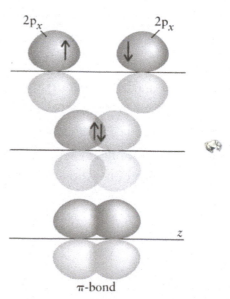

- **Electron occupancy**
  - → σ- and π-orbitals can hold 0, 1, or 2 electrons, corresponding to no bond, a half-strength bond, or a full covalent bond, respectively.
  - → In any orbital, the spins of two electrons must be *paired* (opposite direction of arrows).

- **Types of bonds according to valence-bond theory**
  - → **Single bond** is a σ-bond.   [H_2, for example]
  - → **Double bond** is a σ-bond plus one π-bond.   [O_2, for example]
  - → **Triple bond** is a σ-bond plus two π-bonds.   [N_2, for example]

2F.2 Electron Promotion and the Hybridization of Orbitals *and*

2F.3 Other Common Types of Hybridization

- **Hybrid orbitals**
  - → Produced by mixing (hybridizing) orbitals of a central atom
  - → A construct consistent with observed shapes and bonding in molecules

Hybridization Schemes

Electron arrangement around the central atom	Hybrid orbitals (number)	Angle(s) between σ-bonds	Example(s) [central atoms]
Linear	sp (two)	180°	$\underline{Be}Cl_2$, $\underline{C}O_2$
Trigonal planar	sp^2 (three)	120°	$\underline{B}F_3$, $\underline{C}H_3^+$
Tetrahedral	sp^3 (four)	109.5°	$\underline{C}Cl_4$, $\underline{N}H_4^+$
Trigonal bipyramidal	sp^3d (five)	120°, 90°, and 180°	$\underline{P}Cl_5$
Octahedral	sp^3d^2 (six)	90° and 180°	$\underline{S}F_6$

- **Determination of hybridization schemes**
  - → Write the Lewis structure(s). If there are resonance structures, pick *any* one.
  - → Determine the number of lone pairs and σ-bonds on the central atom.
  - → Determine the number of orbitals required for hybridization on the central atom to accommodate the lone pairs and σ-bonds.
  - → Determine the hybridization scheme from the preceding table.
  - → Use VSEPR rules to determine the molecular *shape* and *bond angle(s)*. (π-bonds are formed from the overlap of unhybridized atomic p-orbitals in a side-by-side arrangement.)

Example: Describe the bonding in BF_3 using VB theory and hybridization. The shape is trigonal planar (VSEPR). The hybridization at B is sp^2 (see table). The three σ-bonds are formed from overlap of $B(sp^2)$ orbitals with $F(2p)$ orbitals ($B(sp^2)$–$F(2p)$ bonds).

Note: Detailed studies indicate that both s- and p-orbitals take part in the bond formation of terminal atoms, such as F.

- **Central atoms**

  → Each central atom can be represented using a hybridization scheme, as in the example that follows.

  Example: Use VB theory to describe the bonding in and shape of acetaldehyde, a molecule with two central atoms.

  Lewis Structure   **3-D Representation**

  Bonds: C–C σ-bond from overlap of a methyl C sp^3 with an aldehyde C sp^2 hybrid orbital
  C=O σ-bond from overlap of an aldehyde C sp^2 hybrid orbital with an O 2p orbital
  C=O π-bond from side-by-side overlap of C 2p and O 2p atomic orbitals
  C–H single bond (three) from overlap of an H $1s$ atomic orbital and a methyl C sp^3 hybrid orbital
  C–H single bond from the overlap of a H 1s atomic orbital and an aldehyde C sp^2 hybrid orbital

  Angles: Three ∠ H–C–H ~ 109.5°; three ∠ H–C–C ~ 109.5°; one ∠ C–C=O ~ 120°; one ∠ O=C–H ~ 120°; and one ∠ C–C–H ~ 120°

  Molecular shape: Tetrahedral at methyl carbon atom, trigonal planar at aldehyde C atom

2F.4 Characteristics of Multiple Bonds

- **Alkanes (C_nH_{2n+2}, n = 1, 2, 3, ...)**

  → Characteristics: tetrahedral geometry and sp^3 hybridization at C atoms; all C–C and C–H single (σ) bonds; rotation allowed about C–C single bonds
  Examples: Methane, CH_4; and propane, $CH_3CH_2CH_3$

- **Alkenes (C_nH_{2n}, n = 2, 3, 4, ...)**

  → Characteristics: one C=C double bond (σ plus π); other C–C bonds and all C–H bonds single (σ); trigonal-planar geometry and sp^2 hybridization at double-bonded C atoms; tetrahedral geometry at other C (sp^3) atoms; rotation *not* allowed about double C=C bond; rotation allowed about C–C single bonds
  Examples: Ethene (ethylene), $H_2C=CH_2$; and propylene, $CH_3CH=CH_2$

- **Alkynes (C_nH_{2n-2}, n= 2, 3, 4, ...)**

  → Characteristics: one C≡C triple bond (σ plus two π); other C–C bonds and all C–H bonds (σ); linear geometry and sp hybridization at triple-bonded C atoms; tetrahedral geometry at other C (sp^3) atoms

    Examples: Ethyne (acetylene), HC≡CH; 2-butyne (dimethylacetylene), $H_3CC≡CCH_3$

- **Benzene (C_6H_6)**

  → Characteristics: *Planar* molecule with hexagonal C framework; trigonal-planar geometry and sp^2 hybridization at C atoms; σ framework has C–C and C–H single bonds; 2p-orbitals (one from each C atom) that overlap side-by-side to form *three π-bonds (six π-electrons) delocalized over the entire six C atom ring*

  → See Lewis resonance structures in **Topic 2B.2** of this study guide.

- **Double bonds**

  → Consist of *one σ-* and *one π-bond* and are always *stronger* than a single σ-bond:
    C=C is *weaker* than two single C–C σ-bonds
    N=N is *stronger* than two single N–N σ-bonds
    O=O is *stronger* than two single O–O σ-bonds

  → Formed readily by Period 2 elements (C, N, O)

  → Rarely found in Period 3 and higher period elements

  → Impart rigidity to molecules and influence molecular shape

    Example: Describe the structure of the carbonate anion in terms of hybrid orbitals, bond angles, and σ- and π-bonds.

    Lewis resonance structures:

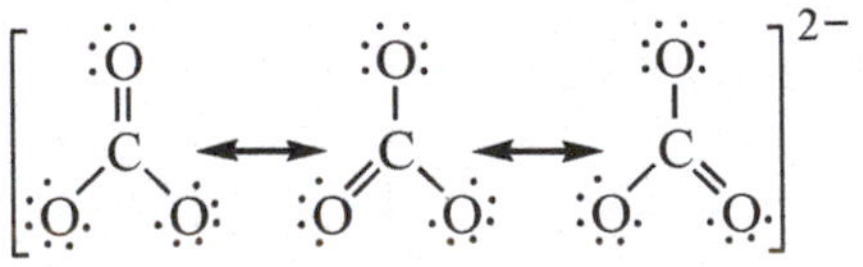

    Trigonal-planar geometry with sp^2 hybridization at C, 120° ∠ O–C–O angles, and C sp^3-O 2p single bonds. Remaining C 2p-orbital can overlap with O 2p-orbitals to form a π-bond in each of three ways corresponding to the resonance structures. Therefore, each carbon-oxygen bond can be viewed as one σ-bond and one-third of a π-bond.

Topic 2G: MOLECULAR ORBITAL (MO) THEORY

2G.1 Molecular Orbitals

- **Valence Bond (VB) deficiencies**

  → Cannot explain *paramagnetism* of O_2 (see text Box 2G.2 for explanation of paramagnetism and *diamagnetism*)

→ Difficulty treating *electron-deficient* compounds such as diborane, B_2H_6

→ No simple explanation for *spectroscopic properties* of compounds such as color

- **Molecular Orbital (MO) advantages**

  → Addresses *all* of these shortcomings of VB theory

  → Provides a *deeper* understanding of electron-pair bonds

  → Accounts for the *structure and properties* of metals and semiconductors

  → More facile for *computer calculations* than VB theory

- **MO theory**

  → In MO theory, electrons occupy MOs that are *delocalized* over the *entire* molecule.

  → In VB theory, bonding electrons are *localized* between the two atoms.

- **Molecular Orbitals (MO)**

  → Formed by superposition (**linear combination**) of atomic orbitals **(LCAO-MO)**

  → **Bonding orbital** (constructive interference):
  Increased amplitude or electron density between atoms

  → **Antibonding orbital** (destructive interference):
  Decreased amplitude or electron density between atoms

- **Types of Molecular Orbitals**

  → **Bonding MO:** energy *lower* than that of the constituent AOs

  → **Nonbonding MO:** energy *equal* to that of the constituent AOs

  → **Antibonding MO:** energy *greater* than that of the constituent AOs

- **Rules for combining AOs to obtain MOs**

  → N atomic orbitals (AOs) yield N molecular orbitals (MOs)

  → Conservation of orbitals

  → Little overlap of inner shell AOs, and these MOs are usually *nonbonding*.

- **MO energy-level diagrams**

  → Relative energies of original AOs and resulting MOs are shown schematically in energy-level diagrams.

  → Arrows used to show electron spin and to indicate location of the electrons in the separated atoms and the molecule

  Example: Formation of bonding and antibonding MOs by LCAO method and MO energy-level diagram for the formation of Li_2 from Li atoms.

Energy-level diagram

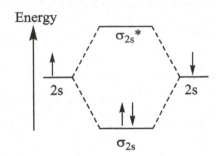

2G.2 The Electron Configurations of Diatomic Molecules

- **Procedure for determining the electronic configuration of diatomic molecules**

  → Construct all possible MOs from *valence-shell* AOs.

  → Place *valence electrons* in the lowest energy, unoccupied MOs.

  → Follow the **Pauli exclusion principle** and **Hund's rule** as for AOs: Electrons have *spins paired* in a fully occupied molecular orbital and enter unoccupied *degenerate* orbitals with *parallel spins*.

- *Valence-shell MOs* for Period 1 and Period 2 *homonuclear* diatomic molecules

Correlation Diagrams

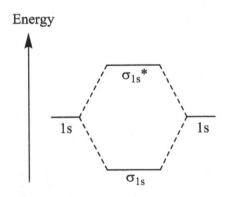

Schematic order of MO energy levels for H_2, He_2

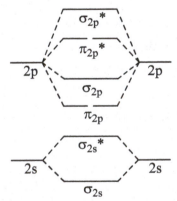

Schematic order of MO energy levels for Li_2, Be_2, B_2, C_2, N_2

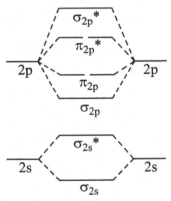

Schematic order of MO energy levels for O_2, F_2, Ne_2

- **Bond order (*b*)** → $b = \dfrac{1}{2}(N_e - N_e{}^*)$

  N_e = total number of electrons in *bonding* MOs

  $N_e{}^*$ = total number of electrons in *antibonding* MOs

  Examples: *b* in Li_2 = 1 (two valence electrons in bonding 2sσ MO). *b* in $H_2{}^-$ = 1/2 (two electrons in $1s\sigma$ bonding MO, one electron in $1s\sigma^*$-antibonding MO).

O_2^+ has 11 valence electrons, which are placed in the $2s\sigma$, $2s\sigma^*$, $2p\sigma$, two *degenerate* $2p\pi$, and one of the *degenerate* $2p\pi^*$ MOs. The valence electron configuration is $(2s\sigma)^2 (2s\sigma^*)^2 (2p\sigma)^2 (2p\pi)^4 (2p\pi^*)^1$. There is one unpaired electron. $N_e = 8$ and $N_e^* = 3$. Bond order = $(1/2) \times (8 - 3) = 5/2$.

2G.3 Bonding in Heteronuclear Diatomic Molecules

- **Bond properties**
  - → Polar in nature
  - → Atomic orbital from the *more* electronegative atom is *lower* in energy.
  - → *Bonding* orbital has a *greater* contribution from the *more* electronegative atom.
  - → *Antibonding* orbital has a *greater* contribution from the *less* electronegative atom.

- **Result of bond properties is modified, unsymmetrical correlation diagrams.**

    Example:

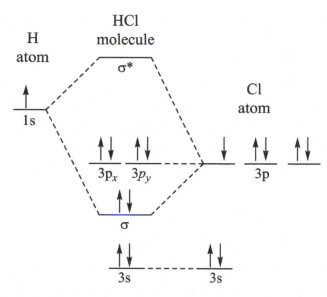

Bonding in HCl using MO theory. The correlation diagram shows occupied valence orbitals for H and Cl atoms, resulting MOs from overlap of $1s$ H and $3p_z$ Cl orbitals, lone pair nonbonding orbitals $3p_x$ and $3p_y$ of Cl, and the nonbonding Cl $3s$ orbital. The electronic configuration is $(Cl\ 3s)^2 \sigma^2 (Cl\ 3p_x)^2 (Cl\ 3p_y)^2$.

2G.4 Orbitals in Polyatomic Molecules

- **Bond properties**
  - → All atoms in the molecule are encompassed by the MOs.
  - → Electrons in MOs bond all the atoms in a molecule.
  - → Electrons are *delocalized* over all the atoms.
  - → Energies of the MOs are obtained by using spectroscopic techniques.

- **Effectively localized electrons**

  → While MO theory requires MOs to be *delocalized* over all atoms, electrons sometimes are *effectively localized* on an atom or group of atoms.

    Example: See the HCl example, shown previously, with nonbonding $Cl\,3p_x$ and $Cl\,3p_y$ orbitals effectively localized on Cl.

- **Combining VB and MO theories**

  → VB and MO theory are sometimes jointly invoked to describe the bonding in complex molecules.

    Example: In benzene, localized sp^2 hybridization (hybrids of C 2s-, C $2p_x$-, and C $2p_y$-orbitals on each C) is used to describe the σ-bonding framework of the ring, while delocalized MOs (LCAOs of six C $2p_z$-orbitals) are used to describe the π-bonding.

- **Triumphs of MO theory**

  → Accounts for paramagnetism of O_2

  → Accounts for bonding in electron-deficient molecules

    Example: In diborane (B_2H_6, with 12 valence electrons),   *six* electron pairs bond *eight* atoms using *six* MOs. VB theory with only localized covalent bonds requires *eight* electron pairs and fails to describe the bonding.   A combination of MO and VB theories has been used to describe the bonding in B_2H_6 as four B–H single bonds (8 electrons) and two, two-electron, 3-center B–H–B delocalized bonds (4 electrons).

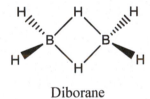

Diborane

  → Provides a framework for interpreting visible and ultraviolet (UV) spectra

    Example: For benzene (schematic energy levels shown in the following figure), the UV spectrum is interpreted as the excitation of an electron from the *highest occupied molecular orbital* (**HOMO,** *bonding* π MO) to the *lowest unoccupied molecular orbital* (**LUMO,** *antibonding* π* MO). Excitation changes the bond order of the π-system from 3 to 2.

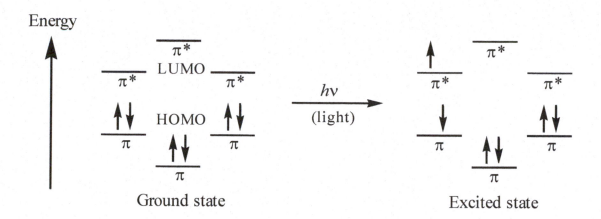

Energy

π*

π*   LUMO   π*

HOMO

π   π

π

Ground state

*h*ν
(light)

π*

π*   π*

π   π

π

Excited state

Focus 3: BULK MATTER

Topic 3A: THE NATURE OF GASES

3A.1 Observing Gases

- **Gases**
  - → Examples of *bulk matter*, forms of matter consisting of large numbers of molecules

- **Atmosphere**
  - → Not uniform in composition, temperature (T), or density (d)

- **Physical properties**
  - → Most gases behave similarly at low pressure and high temperature.

- **Compressibility**
  - → Ease with which a gas undergoes a volume (V) decrease
  - → Gas molecules are widely spaced, and gases are highly compressible. Gas molecules are in ceaseless chaotic motion.

- **Expansivity**
  - → The ability of a gas to rapidly fill the space available to it
  - → Atomic and molecular gases move rapidly and respond quickly to changes in the volume available to them.

3A.2 Pressure

$$\text{Pressure} \equiv \frac{\text{force}}{\text{area}} \quad or \quad P = \frac{F}{A}$$

Opposing force from a confined gas

- **Barometer**
  - → A glass tube, sealed at one end, filled with liquid mercury, and inverted into a beaker also containing liquid mercury (Torricelli)
  - → The height (h) of the liquid column is a measure of the atmospheric pressure (P) expressed as the length of the column for a particular liquid.

  Common pressure units: mmHg, cmHg, cmH_2O

  $$P = \frac{F}{A} = \frac{mg}{A} = \frac{(dV)g}{A} = \frac{(dhA)g}{A} = dhg$$

  where m = mass of liquid, g = standard acceleration of gravity,
  (average = 9.806 65 $m \cdot s^{-2}$ at the Earth's surface), and d = density of liquid

→ The atmospheric pressure P when the height h of a column of mercury in a barometer is *exactly* 0.76 m at 0 °C is 101 325 kg·m^{-1}·s^{-2}, a result which is obtained by using the value of the density of mercury = 13 595.1 kg·m^{-3} and the gravitational constant at the surface of the Earth.

Note: 1 Pa (pascal) = 1 kg·m^{-1}·s^{-2} = 1 N·m^{-2} (SI unit of pressure)

- **Manometer** (see **Figure 3A.5** in the text)

  → U-shaped tube filled with liquid and connected to an experimental system

  → Open-tube pressure of system shows atmospheric pressure when levels are equal.

  → Open-tube pressure of system shows pressure difference between system and atmospheric pressure when levels are unequal.

  → Closed-tube pressure (system) proportional to difference in level heights

- **Gauge pressure**

  → Difference between the pressure inside a vessel containing a compressed gas (e.g., a tire) and atmospheric pressure

3A.3 Alternative Units of Pressure

- **Atmosphere**

  → 1 atm = **760** Torr   (**bold type = exact**)
    1 atm = **101 325** Pa = **101.325** kPa
    1 bar = 10^5 Pa = **100** kPa
    1 Torr = 1 mmHg   (Not exact, but correct to better than 2×10^{-7} Torr)

Topic 3B: THE GAS LAWS

3B.1 The Experimental Observations

- **Boyle's law (fixed amount of gas n at constant temperature T)**

- **Gas behavior**

  → Boyle's law describes an isothermal system with a fixed amount of gas: $\Delta T = 0$, $\Delta n = 0$ (constant T, n).

$$\text{Pressure} \propto \frac{1}{\text{volume}} \quad \text{or} \quad P \propto \frac{1}{V} \quad \text{or} \quad PV = \text{constant}$$

The functional form is a *hyperbola* (plot of P versus V), a plot of P versus V^{-1} gives a straight line.

- **Changes between two states (1 and 2) at constant temperature**

$$P_1 V_1 = \text{constant} = P_2 V_2 \quad \text{or} \quad \frac{P_2}{P_1} = \frac{1/V_2}{1/V_1} = \frac{V_1}{V_2}$$

- **Deviations**
  - → At low temperature and/or high pressure
- **Charles's Law (fixed amount of gas n at constant pressure P)**
- **Gas behavior**
  - → Charles's law describes a constant pressure (isobaric) system with a fixed amount of gas: $\Delta P = 0$, $\Delta n = 0$ (constant P, n). The functional form is *linear*.

$$\text{Volume} \propto \text{temperature} \quad \text{or} \quad V \propto T \quad \text{or} \quad V/T = \text{constant}$$

- **Kelvin temperature scale**
  - → Two *fixed* points define the absolute temperature scale:

    $T \equiv 273.16 \text{ K}$ — The triple point of water, where ice, liquid, and vapor are in equilibrium (see **Topic 5A**).

    $T \equiv 0 \text{ K}$ — Defined from the limiting behavior of Charles's law, V approaches 0 at fixed *low* pressure (extrapolation).

- **Celsius temperature scale**
  - → $t(^\circ\text{C}) \equiv T(\text{K}) - 273.15$ *(exactly)*

    The freezing point of water at 1 atm external pressure is 0 °C or 273.15 K.

- **Changes between two states (1 and 2)**

$$V_1/T_1 = \text{constant} = V_2/T_2 \quad \text{or} \quad \frac{V_2}{V_1} = \frac{T_2}{T_1}$$

- **Deviations**
  - → At high pressure and/or low temperature
- **Another aspect of gas behavior (fixed amount of gas n at constant volume V)**
  - → $\boxed{\text{Pressure} \propto \text{temperature} \quad \text{or} \quad P \propto T \quad \text{or} \quad P/T = \text{constant}}$
  - → This relation describes an *isochoric* system: $\Delta V = 0$ (constant V). The functional form is *linear*.
- **Changes between two conditions (1 and 2)**

$$P_1/T_1 = \text{constant} = P_2/T_2 \quad \text{or} \quad \frac{P_2}{P_1} = \frac{T_2}{T_1}$$

- **Deviations**
  - → For small volume
- **Avogadro's principle (gas at constant pressure P and temperature T)**
  - → A given number of gas molecules at a particular P and T occupy the same volume regardless of chemical identity.

- **Gas behavior**

$$\boxed{\text{Volume} \propto \text{amount} \quad \text{or} \quad V \propto n \quad \text{or} \quad V/n = \text{constant}}$$

- **Molar volume**

$$\boxed{V_m = \frac{V}{n} = \text{constant} \qquad \text{for } \Delta P = 0 \text{ and } \Delta T = 0}$$

- **Changes between two conditions (1 and 2)**

$$\boxed{V_1/n_1 = \text{constant} = V_2/n_2 \quad \text{or} \quad \frac{V_2}{V_1} = \frac{n_2}{n_1}}$$

- **Deviations**
  - → At high pressure and/or low temperature
- **Another aspect of gas behavior (gas at constant volume V and temperature T)**

$$\boxed{\text{Pressure} \propto \text{amount} \quad \text{or} \quad P \propto n \quad \text{or} \quad P/n = \text{constant}}$$

- **Changes between two conditions (1 and 2)**

$$\boxed{P_1/n_1 = \text{constant} = P_2/n_2 \quad \text{or} \quad \frac{P_2}{P_1} = \frac{n_2}{n_1}}$$

- **Deviations**
  - → For small volumes
- **The ideal gas law, $PV = nRT$ (R = gas constant)**
  - → Incorporates Boyle's law, Charles's law, and Avogadro's principle
  - → R is the "universal" gas constant described below.
- **Gas behavior**
  - → As P approaches 0, all gases behave *ideally* (*limiting law*).
- **Derivation**

Boyle	**Charles**	**Avogadro**
$PV = \text{constant}_1$	$V = \text{constant}_2 \times T$	$P = \text{constant}_3 \times n$

Result: $PV = (\text{constant}_3 \times n)(\text{constant}_2 \times T) = nRT$ *equation of state*

- **Gas constant**
  - → $R = 8.205\,74 \times 10^{-2}$ L·atm·K^{-1}·mol^{-1}
  - → $R = 8.314\,46$ L·kPa·K^{-1}·mol^{-1} = $8.314\,46$ J·K^{-1}·mol^{-1}

3B.2 Applications of the Ideal Gas Law

- **Changes between two conditions (1 and 2)**

$$\frac{P_1 V_1}{n_1 T_1} = \frac{P_2 V_2}{n_2 T_2}$$

Combined gas law

  Note: The more general form of the combined gas law allows the amount of gas to change.

  Note: When using this relationship, remember that the actual units used in manipulating ratios of quantities are *not* important. Consistency in units is essential. *Absolute* temperature, however, *must* be used in all the equations.

3B.3 Molar Volume and Gas Density

- **Molar volume**

$$V_m = \frac{RT}{P}$$

- **Standard ambient temperature and pressure (SATP)**
  - → 298.15 K and 1 bar
- **Standard temperature and pressure (STP)**
  - → 0 °C and 1 atm   (273.150 K and 1.013 25 bar)
- **Molar volume at STP**

$$\rightarrow V_m = \frac{RT}{P} = \frac{(8.20574 \times 10^{-2}\ \text{L·atm·K}^{-1}\cdot\text{mol}^{-1})(273.150\ \text{K})}{(1\ \text{atm})} = 22.4140\ \text{L·mol}^{-1}$$

- **Molar concentration**

$$\frac{n}{V} = \frac{1}{V_m} = \frac{P}{RT}$$

- **Molar concentration at STP**

$$\rightarrow \frac{n}{V} = \frac{1}{V_m} = \frac{1}{22.4140\ \text{L·mol}^{-1}} = 4.46150 \times 10^{-2}\ \text{mol·L}^{-1}$$

- **Density of a gas**

$$d = \frac{m}{V} = \frac{nM}{V} = \frac{MP}{RT}$$

- **Density of a gas at STP**

$$\rightarrow \quad d = \frac{m}{V} = \frac{nM}{V} = (4.46150 \times 10^{-2} \text{ mol} \cdot \text{L}^{-1}) \times M$$

Topic 3C: GASES IN MIXTURES AND REACTIONS

3C.1 Mixtures of Gases

- **Behavior**

  $\rightarrow$ Mixtures of nonreacting ideal gases behave as if only a single component were present.

- **Partial pressure**

  $\rightarrow$ Pressure a gas would exert if it occupied the container alone

- **Law of partial pressures**

  $\rightarrow$ Total pressure of a gas mixture is the sum of the partial pressures of its components (Dalton's law).

  $$P = P_A + P_B + \dots \text{ for mixture containing A, B, } \dots$$

- **Components at the same T and V**

  $\rightarrow$ $n = n_A + n_B + \dots = $ total amount of mixture

  $$P = \frac{nRT}{V} \qquad \text{and} \qquad P_A = \frac{n_A RT}{V}$$

  $$x_A = \frac{n_A}{n} = \frac{P_A}{P} = \text{mole fraction of A in the mixture.}$$

  Thus, $P_A = x_A P$.

- **Humid gas**

  $\rightarrow$ For air, $P = P_{\text{dry air}} + P_{\text{water vapor}}$, and $P_{\text{water vapor}} = 47$ Torr at body temperature (37 °C).

  $\rightarrow$ Values for the vapor pressure of water at selected temperatures is given in **Table 5A**.2.

3C.2   The Stoichiometry of Reacting Gases

- **Accounting for reaction volumes**

  $\rightarrow$ Use mole-to-mole calculations as described in **Fundamentals Topics L** and **M**. Follow this flow chart and then convert to gas volume by using the ideal gas law.

  > *Mass of A → molar mass → moles of A → stoichiometry*
  > *→ moles of B → molar volume → volume of B*

- **Balanced chemical equation**

  → See **Example 3C.2** and the accompanying self-test exercises in the text.

  Example: $NH_4NO_3(s) \rightarrow N_2O(g) + 2\,H_2O(g)$

  A total of three moles of gas products is produced from one mole of solid.

- **Molar volumes of gases**

  → Generally more than 1000 times those of liquids and solids (condensed phases)

  → Reactions that produce gases from condensed phases can be explosive.

Topic 3D: MOLECULAR MOTION

3D.1 Diffusion and Effusion

- **Diffusion**

  → Gradual dispersal of one substance through another substance

- **Effusion**

  → Escape of a gas through a small hole (or assembly of microscopic holes) into a vacuum

  → Depends on molar mass and temperature of gas

- **Graham's law**

  → Rate of effusion of a gas *at constant temperature* is *inversely* proportional to the square root of its molar mass M.

  For two gases A and B
  $$\frac{\text{Rate of effusion of A}}{\text{Rate of effusion of B}} = \sqrt{\frac{M_B}{M_A}}$$

  → Rate of effusion is proportional to the average speed of the molecules in a gas *at constant temperature*.

  $$\frac{\text{Average speed of A molecules}}{\text{Average speed of B molecules}} = \sqrt{\frac{M_B}{M_A}}$$

  → Rate of effusion is *inversely* proportional to the time taken for given volumes or amounts of gas to effuse *at constant temperature*.

  $$\frac{\text{Rate of effusion of A}}{\text{Rate of effusion of B}} = \frac{\text{Time for B to effuse}}{\text{Time for A to effuse}} = \sqrt{\frac{M_B}{M_A}}$$

- **Effusion at different temperatures**
  - → The rate of effusion and the average speed increase as the square root of the absolute temperature increases.

$$\frac{\text{Rate of effusion at } T_2}{\text{Rate of effusion at } T_1} = \frac{\text{Average speed of molecules at } T_2}{\text{Average speed of molecules at } T_1} = \sqrt{\frac{T_2}{T_1}}$$

- **Combined relationship**
  - → Average speed of molecules is *directly* proportional to the square root of the temperature and *inversely* proportional to the square root of the molar mass.

$$\text{Average speed of molecules in a gas } \propto \sqrt{\frac{T}{M}}$$

3D.2 The Kinetic Model of Gases

- **Assumptions**
  1. Gas molecules are in continuous random motion.
  2. Gas molecules are infinitesimally small particles.
  3. Particles move in straight lines until they collide.
  4. Molecules do not influence one another except during collisions.

- **Collisions with walls**
  - → Consider molecules traveling only in one dimension x with an average of square velocity of $\langle v_x^2 \rangle$.

- **Pressure on wall**
  - → $P = \dfrac{Nm\langle v_x^2 \rangle}{V}$, where N is the number of molecules, m is the mass of each molecule, and V is the volume of the container.

- **Mean square speed**
  - → $v_{rms}^2 = \langle v^2 \rangle = \langle v_x^2 + v_y^2 + v_z^2 \rangle = \langle v_x^2 \rangle + \langle v_y^2 \rangle + \langle v_z^2 \rangle = 3\langle v_x^2 \rangle$

    rms = root mean square

- **Pressure on wall**
  - → $P = \dfrac{Nmv_{rms}^2}{3V}$ and $N = nN_A$, where N_A is the Avogadro constant.

    $P = \dfrac{nN_A m v_{rms}^2}{3V} = \dfrac{nMv_{rms}^2}{3V}$, where $M = mN_A$ is the molar mass.

$$PV = \frac{nMv_{rms}^2}{3} = nRT$$

The term nRT is from the ideal gas law.

- **Solve for v_{rms}**

$$\rightarrow \quad v_{rms} = \sqrt{\frac{3RT}{M}}$$

v_{rms} is the **root mean square** speed of the particle.

Note: *Beware of units.* Use R in units of $J \cdot K^{-1} \cdot mol^{-1}$ and convert M to $kg \cdot mol^{-1}$. The resulting units of speed will be $m \cdot s^{-1}$.

$$\text{Molar kinetic energy} = \left(\frac{1}{2}mv_{rms}^2\right)N_A = \left(\frac{1}{2}m\right)\left(\frac{3RT}{M}\right)N_A = \frac{3}{2}RT, \text{ because } mN_A = M$$

- **Kinetic model of gases**

  $\rightarrow$ Consistent with the ideal gas law

  $\rightarrow$ Suggests that the root mean square (rms) speed of gas molecules is proportional to the square root of the gas temperature

3D.3 The Maxwell Distribution of Speeds

- **Symbols**

  $\rightarrow$ Let v represent a particle's speed, N the total number of particles, $f(v)$ the Maxwell distribution of speeds, Δ a *finite* change, and d an *infinitesimal* change (*not* density) in the following relationships.

- **For a *finite* range of speeds**

$$\Delta N = Nf(v)\Delta v, \text{ where } f(v) = 4\pi\left(\frac{M}{2\pi RT}\right)^{3/2} v^2 e^{-Mv^2/2RT}$$

- **Temperature dependence of $f(v)$**

  $\rightarrow$ As T increases, the fraction of molecules with speeds greater than a specific speed increases. See the lower panel in **Figure 3D.7**. (Hot molecules move faster than cool ones on average.) We will use this concept again in **Focus 7**, when we study the rates of chemical reactions.

- **Dependence of $f(v)$ on molar mass M**

  $\rightarrow$ As M decreases, the fraction of molecules with speeds greater than a specific speed increases. See the upper panel in **Figure 3D.7**. (Light molecules move faster than heavy ones on average.)

- **For an *infinitesimal* range of speeds**

$$\frac{dN}{N} = f(v)dv$$

- **Other speed relationships**

  → Calculus can be used with the Maxwell distribution of speeds to obtain the following properties, which are introduced in a homework exercise in the text (**Focus 3 Exercise 3.16**).

Average speed

$$\langle v \rangle = \int_0^\infty v f(v)\, dv = \sqrt{\frac{8RT}{\pi M}}$$

Most probable speed

$$v_{mp} = \sqrt{\frac{2RT}{M}} \;, \text{ where } \frac{df(v)}{dv} = 0 \quad \text{(maximum in distribution)}$$

Topic 3E: REAL GASES

3E.1 Deviations from Ideality

- **Definition**

  → Attractions and repulsions among atoms and molecules cause deviations from ideality in gases.

  → Attractions have a longer range than repulsions.

- **Evidence**

  → Gases condense to liquids when cooled or compressed (attraction).

  → Liquids are difficult to compress (repulsion).

- **Compression factor (Z)**

  → A measure of the effect of intermolecular forces

  → The ratio of the observed molar volume of a gas to that calculated for an ideal gas

$$Z \equiv \frac{PV_m}{RT} = \frac{V_m}{V_m^{ideal}}$$

For an ideal gas, $Z = 1$.
For H_2, $Z > 1$ at all P (repulsions dominate).
For NH_3, $Z < 1$ at low P (attractions dominate).

Example (Figure 3E.1): For many gases, for example methane, attractions dominate at low pressures ($Z < 1$), while repulsive interactions dominate at high pressures ($Z > 1$). For hydrogen, $Z > 1$ under all conditions (repulsive forces always dominate).

3E.2 Equations of State of Real Gases

- **Virial equation**

$$\rightarrow \quad \boxed{PV = nRT \left(1 + \frac{B}{V_m} + \frac{C}{V_m^2} + \cdots \right)}$$

B = second virial coefficient

C = third virial coefficient, etc.

Note: **Virial coefficients** depend on temperature and are found by fitting experimental data to the virial equation. This equation is both more general than the van der Waals equation and more difficult to use to make predictions.

- **van der Waals equation**

$$\rightarrow \quad \boxed{\left(P + a\frac{n^2}{V^2}\right)(V - nb) = nRT}$$

where a and b are the **van der Waals parameters**

(determined experimentally for each gas)

- **Rearranged form**

$$\rightarrow \quad \boxed{P_{vdW} = \frac{nRT}{V - nb} - a\frac{n^2}{V^2}}$$

where a accounts for the attractive effects and b accounts for the repulsive ones

- **van der Waals Z**

$$\rightarrow \quad \boxed{Z \equiv \frac{P_{vdW}V_m}{RT} = \frac{V}{V - nb} - \frac{an}{RTV} = \frac{1}{1 - (nb/V)} - \frac{an}{RTV}}$$

- **Repulsion**

  → Repulsive forces imply that molecules cannot overlap.

  → Other molecules are *excluded* from the volume they occupy.

  → The *effective* volume, then, is $V - nb$, not V.

  → Thus, b is a constant corresponding to the volume excluded per mole of molecules (units: $L \cdot mol^{-1}$).

  → The potential energy describing this situation is that of an *impenetrable* hard sphere.

- **Attraction**

  → Attractive forces lead to clustering of molecules in the gas phase.

  → Clustering reduces the total number of gas phase species.

  → The rate of collisions with the wall (pressure) is thereby reduced.

→ Because this effect arises from attractions between *pairs* of molecules, it should be proportional to the *square* of the number of molecules per unit volume $(N/V)^2$, or equivalently, to the *square* of the molar concentration $(n/V)^2$.

→ The pressure is predicted to be reduced by an amount $a(n/V)^2$, where a is a positive constant that depends on the strength of the attractive forces.

→ The potential energy describing this situation is that of long-range attraction with the form $-(1/r^6)$, where r is the distance between a pair of gas-phase molecules. This type of attraction primarily arises from *dispersion* or *London* forces (see **Topic 3F.4**).

→ The units of a are $L^2 \cdot atm \cdot mol^{-2}$.

Summary

- **Virial equation**

  → Describes real gases

- **Van der Waal's equation**

  → An *approximate* equation of state

  → Takes molecular volume and intermolecular forces into account

- **Ideal gas equation**

  → Least accurate of the gas-phase equations of state

  → Ignores molecular volume and intermolecular forces

  → Easiest to use

3E.3 The Liquefaction of Gases

- **Joule–Thomson effect**

  → When attractive forces dominate, a real gas *cools* as it expands.

  → In this case expansion requires energy that comes from the kinetic energy of the gas, lowering the temperature. (The gas to be liquefied is compressed and then allowed to expand through a small hole.) Exceptions are He and H_2, for which repulsion dominates.

  → Expansion cooling is used in some refrigerators and to effect the condensation of gases such as oxygen, nitrogen, and argon.

Topic 3F: INTERMOLECULAR FORCES

3F.1 The Origin of Intermolecular Forces

- **Physical states**

  → Solid, liquid, and gas

- **Phase**
    - → Form of matter uniform in both chemical composition and physical state (solid, liquid, or gas)
    - → See **Table 3F.2** for melting and boiling points of substances.
- **Condensed phase**
    - → Solid or liquid phase

      Examples: Ag(s); Sn/Pb(s) alloy, composition variable; $H_2O(s)$; $H_2O(l)$; 1% NaCl(s) in $H_2O(l)$
    - → Molecules (or atoms or ions) are close to each other all the time, and intermolecular forces are of major importance.
- **Coulomb potential energy E_P**
    - → The interaction between two charges, Q_1 and Q_2, separated by distance r is $\Rightarrow$

    $$E_P \propto \frac{Q_1 Q_2}{r}$$

    - → See **Topic 1D.1** and **Fundamentals Topic A**.
    - → Almost all intermolecular interactions can be traced back to this fundamental expression.

      Note: The term *intermolecular* is used in a general way to include atoms and ions.

3F.2 Ion-Dipole Forces

- **Hydration**
    - → Attachment of water molecules to ions (cations and anions)
    - → Water molecules are polar and have an electric dipole moment $\mu(H_2O)$.
    - → A small positive charge on each H atom attracts anions, and a small negative charge on the O atom attracts cations.
- **Ion-dipole interaction**
    - → The potential energy (interaction) between an ion with charge $|z|$ and a polar molecule with dipole moment μ at a distance r is $\Rightarrow$

    $$E_P \propto -\frac{|z|\mu}{r^2}$$

    - → For proper *alignment* of the ion and dipole, the interaction is *attractive*: Cations attract the *partial* negative charges, and anions attract the *partial* positive charges on the polar molecule.
    - → Shorter range (r^{-2}) interaction than the Coulomb potential (r^{-1})
    - → Polar molecule needs to be *almost* in contact with ion for substantial ion-dipole interaction.
- **Hydrated compounds**
    - → Ion-dipole interactions are much *weaker* than ion-ion interaction but are relatively strong for *small, highly charged* cations (large $|z|$, small r).
    - → Accounts for the formation of salt hydrates such as $CuSO_4 \cdot 5\,H_2O$ and $CrCl_3 \cdot 6\,H_2O$
- **Size effects**
    - → Li^+ and Na^+ (small) tend to form hydrated compounds.
    - → K^+, Rb^+, and Cs^+ (larger) tend not to form hydrates.

→ Effective radius of NH_4^+ (143 pm) is similar in radius to Rb^+ (149 pm). NH_4^+ forms *anhydrous* compounds.

- **Charge effects**

  → Ba^{2+} and K^+ are similar in size, yet Ba^{2+} (larger charge) forms *hydrates*.

3F.3 Dipole-Dipole Forces

- **Dipole alignment**

  → In solids, molecules with dipole moments tend to *align* with partial positive charge on one molecule near the partial negative charge on another.

- **Dipole-dipole interaction in solids**

  → The interaction between two polar molecules with dipole moments μ_1 and μ_2 *aligned* and separated by a distance r is ⇒

  $$E_P \propto -\frac{\mu_1 \mu_2}{r^3}$$

  → *Attractive* interaction in solids for head-to-tail *alignment* of dipoles

  → Shorter range (r^{-3}) interaction than ion-dipole (r^{-2}) or Coulomb potential (r^{-1})

  → For significant interaction, polar molecules must be *almost* in contact with each other.

- **Dipole-dipole interactions in gas-phase molecules**

  → Dipole-dipole interactions are much *weaker* in gases than in solids. The potential energy of interaction is ⇒

  $$E_P \propto -\frac{\mu_1^2 \mu_2^2}{r^6}$$

  → Because gas molecules are in motion (rotating as well), they are subject to only a *weak* net attraction because of *occasional* alignment.

- **Dipole-dipole interactions in liquid-phase molecules**

  → Same potential energy relationship as in the gas phase, but the interaction is somewhat *stronger* because the molecules are closer.

  → The liquid *boiling point* is a measure of the *strength* of the intermolecular forces in a liquid. The boiling point of isomers is *often* related to the strength of their dipole-dipole interactions.

  → Typically, the *larger* the dipole moment, the *higher* the boiling point (often a relatively small effect).

  Example: *cis*-dibromoethene (dipole moment ≈ 2.4 D, bp 112.5 °C) versus
  trans-dibromoethene (zero dipole moment, bp 108 °C)

3F.4 London Forces

- **Nonpolar molecules**

  → Condensation of *nonpolar* molecules to form liquids implies the existence of a type of intermolecular interaction other than those described earlier.

- **London force**

  → Occurs in addition to any dipole-dipole interactions

  → Is universal; applies to all molecules

→ Accounts for the attraction between any pair of ground-state molecules (*polar* or *nonpolar*)

→ Arises from *instantaneous* partial charges (*instantaneous* dipole moment) in one molecule *inducing* partial charges (dipole moment) in a neighboring one

→ Exists between atoms and *rotating* molecules as well

→ Strength depends on polarizability and shape of molecule.

Example: Other things equal, rod-shaped molecules tend to have stronger London forces than spherical molecules because they can approach each other more closely.

- **Polarizability, α**

  → Of a molecule is related to the ease of deformation of its electron cloud

  → Is proportional to the total number of electrons in the molecule

  → Because the number of electrons correlates with molar mass (generally increases), polarizability does as well.

  → The potential energy between two molecules (polar or nonpolar) with polarizability α_1 and α_2, separated by a distance r, is $\Rightarrow$

  $$E_P \propto -\frac{\alpha_1 \alpha_2}{r^6}$$

  Note: Same (r^{-6}) dependence as dipole-dipole (r^{-6}) with *rotating* molecules, but the London interaction is *usually* stronger at normal temperatures.

- **Liquid boiling point**

  → A measure of the strength of intermolecular forces in a liquid

  → In comparing the interactions discussed in **Topics 3F.3** and **3F.4**, the *major* influence on boiling point in both *polar* and *nonpolar* molecules is the London force.

  Example: We can predict the relative boiling points of the nonpolar molecules: F_2, Cl_2, Br_2, and I_2. Boiling point correlates with polarizability, which depends on the number of electrons in the molecule. The boiling points are expected to increase in the order of F_2 (18 electrons), Cl_2 (34), Br_2 (70), and I_2 (106). The experimental boiling temperatures are F_2 (−188 °C), Cl_2 (−34 °C), Br_2 (59 °C), and I_2 (184 °C).

- **Dipole-induced dipole interaction**

  → The potential energy between a polar molecule with dipole moment μ_1 and a nonpolar molecule with polarizability α_2 at a distance r is $\Rightarrow$

  $$E_P \propto -\frac{\mu_1^2 \alpha_2}{r^6}$$

  → Also applies to molecules that are both polar: each one can induce a dipole in the other.

  → Weaker interaction than dipole-dipole

  Example: Carbon dioxide ($\mu = 0$) dissolved in water ($\mu > 0$)

3F.5 Hydrogen Bonding

- **Hydrogen bonds**

  → Interaction *specific* to certain types of molecules (with strong attractive forces)

  → Account for unusually high boiling points in ammonia (NH_3, −33 °C), water (H_2O, 100 °C), and hydrogen fluoride (HF, 20 °C)

→ Arise from an H atom, *covalently* bonded to a N, O, or F atom in *one* molecule, strongly attracted to a *lone pair* of electrons on a N, O, or F atom in *another* molecule

→ Are *strongest* when the three atoms are in a *straight line* and the distance between terminal atoms is within a given range

→ Common symbol for the H bond is three dots: $\cdots$

→ *Strongest* intermolecular interaction between *neutral* molecules

Examples: N–H$\cdots$:N hydrogen bonds between NH_3 molecules in pure NH_3

O–H$\cdots$:O hydrogen bonds between H_2O molecules in pure H_2O

- **Hydrogen bonding in gas-phase molecules**

→ Aggregation of some molecules persists in the vapor phase.

→ In HF, fragments of *zigzag* chains and $(HF)_6$ rings are formed. In CH_3COOH (acetic acid), hydrogen-bonded *dimers* are formed. The abbreviated (no C–H bonds shown) Lewis structure of the dimer $(CH_3COOH)_2$ is shown on the right.

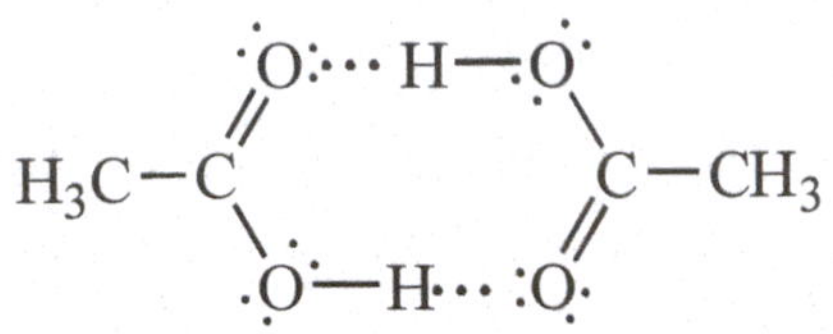

$$H_3C-C \overset{\text{O}\cdots\text{H}-\text{O}}{\underset{\text{O}-\text{H}\cdots\text{O}}{}} C-CH_3$$

- **Importance of hydrogen bonding**

→ Accounts for the open structure of solid water

→ Maintains the shape of biological molecules

→ Binds the two strands of DNA together

3F.6 Repulsions

- **Nature of repulsions**

→ Molecules (or atoms that do not form bonds) that are very close together repel one another.

→ The Pauli exclusion principle forms the basis of understanding this repulsion.

Example: As two He atoms approach each other, at short separations, their $1s$ atomic orbitals overlap and form a bonding and an antibonding molecular orbital.

The bonding molecular orbital is filled with two electrons, and the remaining two electrons are required by the exclusion principle to fill the antibonding orbital.

An antibonding orbital is more antibonding in character than a bonding orbital is bonding; the result is an increase in energy as the two atoms merge into each other.

The same effect occurs for all molecules whose atoms have filled shells (e.g., H_2), even though the details of the bonding and antibonding orbitals they form may be much more complicated.

The result is that all molecules repel each other when they come into contact and their orbitals overlap.

- **Electron density**
  - → Electron density in all atomic orbitals (AOs) and the molecular orbitals (MOs) they form decreases exponentially toward zero at large distances from the nucleus.
  - → The overlap between orbitals on neighboring molecules will also depend exponentially on their separation.
  - → As a result, repulsions between molecules usually depend exponentially on separation (see text **Figure 3F.1**). Repulsions are effective only when the two molecules are very close together. Once they are close, the energy of repulsion increases rapidly as the distance between the two molecules decreases.
  - → The strong dependence of repulsion on separation is the underlying reason objects around us have definite, well-defined shapes.

Topic 3G: LIQUIDS

3G.1 Order in Liquids

- **Liquid phase**
  - → Mobile molecules with restricted motion
  - → Between the extremes of gas and solid phases
- **Long-range order**
  - → Characteristic of a *crystalline solid*
  - → Atoms or molecules are arranged in orderly patterns that are repeated over long distances.
- **Short-range order**
  - → Characteristic of the *liquid phase*
  - → Atoms or molecules are positioned in orderly patterns at nearest-neighbor distances only.
  - → Local order is maintained by a continual process of forming and breaking nearest-neighbor interactions.

3G.2 Viscosity and Surface Tension

- **Viscosity**
  - → Resistance of a substance to flow: *Greater* viscosity yields *slower* flow.
  - → Viscous liquids include those with hydrogen bonding between molecules.
    - **Examples of liquids with high viscosity:**
    - H_3PO_4 (phosphoric acid) and $C_3H_8O_3$ (glycerol) (each forms many H bonds); liquid phases of metals; and long-chain molecules that can be entangled, such as hydrocarbon oil and greases

- **Viscosity and temperature**
  - → *Usually* viscosity *decreases* with *increasing T*
    Exception: Unusual behavior of sulfur, for which viscosity initially increases with increasing *T* as S_8 rings break, forming chains that tangle together

- **Surface tension (several definitions)**
  - → Tendency of surface molecules in a liquid to be pulled into the interior of the liquid by an imbalance in intermolecular forces
  - → *Inward pull* that determines the resistance of a liquid to an increase in surface area
  - → A measure of the *force* that must be applied to *surface molecules* so that they undergo the same *force* as molecules in the *interior* of the liquid
  - → A measure of the tightness of the surface layer (Symbol: γ (gamma)   Units: $N \cdot m^{-1}$ or $J \cdot m^{-2}$)

- **Capillary action**
  - → Rise of liquids up narrow tubes when the *adhesion* forces are greater than *cohesion* forces
  - → **adhesion:** Forces that bind a *substance* to a *surface*
  - → **cohesion:** Forces that bind *molecules* of a substance together to form a *bulk material*

- **Meniscus (curved surface that a liquid forms in a tube)**
  - → *Adhesive* forces greater than *cohesive* forces (forms a ⌣ shape)
  - → *Cohesive* forces greater than *adhesive* forces (forms a ⌢ shape)
  - → Glass surfaces have exposed O atoms and O–H groups to which hydrogen-bonded liquids such as H_2O can bind. In this case, the *adhesive* forces are greater than the *cohesive* ones. Water *wets* glass, forms a ⌣ shape at the surface, and undergoes a capillary rise in a glass tube.
  - → Mercury liquid does not bind to glass surfaces. The *cohesive* forces are greater than the *adhesive* ones. Mercury does *not* wet glass, forms a ⌢ shape at the surface, and undergoes a capillary *lowering* in a glass tube.

3G.3 Liquid Crystals

  - → Materials that cannot be characterized as solid, liquid, or gas
  - → Substances that flow like viscous liquids but whose molecules form a moderately ordered array similar to that in a crystal.
  - → Examples of a *mesophase,* an intermediate state of matter with the fluid properties of a liquid and some molecular ordering similar to that of a crystal

- **Isotropic material**
  - → Properties independent of the direction of measurement
    Ordinary liquids are isotropic, with viscosity values equal in every direction.

- **Anisotropic material**
  - → Properties depend on the direction of measurement

Certain rod-shaped molecules form liquid crystals, in which molecules are free to slide past one another along their axes but resist motion perpendicular to that direction.

- **Classes of liquid crystals**
    → Differ in the arrangement of the molecules
      See text **Figures 3G.7, 3G.8,** and **3G.9**.

    Nematic phase: Molecules lie together in the same direction but are *staggered*.

    Smectic phase: Molecules lie together in the same direction in *layers*, but not staggered.

    Cholesteric phase: Molecules form *nematiclike layers*, but the molecules of neighboring layers are *rotated* with respect to each other. The resulting liquid crystal has a *helical* arrangement of molecules.

- **Thermotropic liquid crystals**
    → Made by melting solid-phase material
    → Exist over small temperature range between solid and liquid
      Example: *p*-azoxyanisole

- **Lyotropic liquid crystals**
    → Layered structures produced by the action of a solvent on a solid or a liquid.
    → Examples include cell membranes and aqueous solutions of detergents and lipids (fats).

3G.4 Ionic Liquids

→ Molecular substances tend to have low melting points.

→ Ionic, network, and metallic substances tend to have high melting points.

→ Ionic liquids are exceptions.

→ Ionic liquids are ionic substances with low melting points. They can be liquid at room temperature and below.

→ Ionic liquids have remarkably low vapor pressures; they constitute a new class of solvents developed to have low vapor pressures but also to dissolve organic compounds.

→ Compounds in which one of the ions (usually the cation) is a large, organic ion that prevents the liquid from crystallizing at ordinary temperatures

Example: Cation: 1-Butyl-3-methylimidazolium (**Margin Figure 4** on **page 199** in the text)
             Anion: Tetrafluoroborate: BF_4^-

→ One formulation is used to dissolve rubber in old tires for recycling.

→ Another is used to extract radioactive waste from groundwater.

Topic 3H: SOLIDS

3H.1 Classification of Solids

- **Amorphous solid**

  → Atoms or molecules (neutral or charged) lie in random positions.

- **Crystalline solid**

  → Atoms, ions, or molecules are associated with points in a **lattice**, which is an *orderly* array of equivalent points in three dimensions.

  → Structure has long-range order.

- **Crystal faces**

  → Flat, well-defined planar surfaces with definite interplanar angles

- **Classification of crystalline solids**

Metallic solids:	Cations in a sea of electrons **Examples:** $Fe(s)$, $Li(s)$
Ionic solids:	Mutual attractions of cations and anions **Examples:** $NaCl(s)$, $Ca(SO_4)_2(s)$
Molecular solids:	Discrete molecules held together by the *intermolecular forces* discussed earlier **Examples:** sucrose, $C_{12}H_{22}O_{11}(s)$; ice, $H_2O(s)$; benzene(s)
Network solids:	Atoms bonded *covalently* to their neighbors throughout the entire solid **Examples:** diamond, $C(d)$; graphite planes, $C(gr)$

3H.2 Molecular Solids

→ Solid structures that reflect the nonspherical nature of their molecules and the relatively weak intermolecular forces that hold them together

→ Characterized by low melting temperatures and less hardness than ionic solids

→ May be amorphous (wax) or crystalline (sucrose)

3H.3 Network Solids

- **Crystals**

  → Atoms joined to neighbors by *strong* covalent bonds that form a network extending throughout the solid

  → Characterized by high melting and boiling temperatures

  → Tend to be hard and rigid structures (diamond)

- **Elemental network solids**

  → Network solids formed from one element only, such as the allotropes graphite and diamond, in which the carbon atoms are connected differently

  → *Allotropes* are forms of an element with different solid-state structures.

- **Graphite**
  - → The thermodynamically stable allotrope of carbon (Soot contains small crystals of graphite.)
  - → Produced pure commercially by heating C rods in an electric furnace for several days
  - → Contains sp^2-hybridized C atoms
  - → Consists structurally of large sheets of fused benzene-like hexagonal units. A π-bonding network of delocalized electrons accounts for the high electrical conductivity of graphite.
  - → When certain impurities are present, graphitic sheets can slip past one another, and graphite becomes an excellent dry lubricant.
  - → Graphite is soluble only in a few liquid metals.
  - → Soot and carbon black have commercial applications in rubber and inks. Activated charcoal is an important, versatile purifier. Unwanted compounds are adsorbed onto its microcrystalline surface.
  - → A chicken-wire-like sheet of carbon atoms in graphite is called a *graphene* sheet. Graphene, a single monolayer sheet of graphite, is a new material with exceptional promise in the electronics industry.
  - → Very pure sheets of graphene can be prepared and stacked together with water molecules that act as a kind of glue between them.
  - → The result is a very strong, flexible, but very thin paperlike material that conducts electricity but is tougher than diamond.

- **Diamond**
  - → Found naturally embedded in *kimberlite*, a soft rock
  - → Hardest substance known and an excellent conductor of heat
  - → An excellent abrasive, and the heat generated by friction is rapidly conducted away
  - → Natural diamonds are mined extensively. Synthetic diamonds are produced from graphite at high pressures (> 80 kbar) and temperatures (> 1500 °C) and by thermal decomposition of methane (the preferred method).
  - → Diamond is soluble in liquid metals, such as Cr and Fe, but less soluble than graphite. This solubility difference is utilized in synthesizing diamond at high pressure and high temperature.
  - → Carbon in diamond is sp^3-hybridized, and each C atom in a diamond crystal is bonded directly to four other C atoms and indirectly interconnected to all of the other C atoms in the crystal through C–C single σ-bonds.
  - → The σ-bonding network of localized electrons accounts for the electrical insulating properties of diamond.

- **Ceramics**
  - → Usually oxides with a network structure having great strength and stability because covalent bonds must break to deform the crystal
  - → Tend to shatter rather than bend under stress

    Examples: Quartz, silicates, and high-temperature superconductors

3H.4 Metallic Solids

- **Close-packed structures**

  → Atoms occupy smallest total *volume* with the *least* amount of empty space.

  → Metal atoms are treated as *spheres* with radii *r*.

  → Type of metal determines whether a close-packed structure is assumed.

- **Close-packed layer**

  → Atoms (*spheres*) in a planar arrangement with the *least* amount of empty space; contour of the layer has *dips* or *depressions.*

  → Each atom in a layer has *six* nearest neighbors (hexagonal pattern).

  → Layers stack such that bottoms of spheres in one layer fit in *dips* of the layer below.

- **Stacked layers**

  → In a close-packed structure, each atom has *three* nearest neighbors in the layer above, *six* in its original layer, and *three* in the layer below, for a total of 12 nearest neighbors.

- **Coordination number (CN)**

  → Number of nearest neighbors of each atom in the solid

  → Impossible to pack identical spheres (metal atoms) with CN > 12

- **Two arrangements of stacking close-packed layers**

    Pattern: ABABABAB…   hcp ≡ *hexagonal close-packing* (repeat after two layers)

    ABCABCABC…   ccp ≡ *cubic close-packing* (repeat after three layers)

- **Occupied space in hcp/ccp structures**

  → 74% of space is occupied by the spheres and 26% is *empty.*

- **Tetrahedral hole**

  → Formed when a *dip* between *three* atoms in one layer is *covered* by *another* atom in an adjacent layer

  → Two tetrahedral holes per atom in a close-packed structure (hcp or ccp)

- **Octahedral hole**

  → Space between *six* atoms, *four* of which are at the corners of a square plane, forms an octahedral hole. The square plane is oriented 45° with respect to two close-packed layers. *Two* atoms in the square plane are in layer **A** and the other *two* in layer **B**. There is *one* additional atom in layer **A** that is *above* the square plane and *another* in layer **B** that is *below* the square plane. The result is an octahedral arrangement of six atoms with a hole in the center.

→ One octahedral hole per atom in a close-packed structure (hcp or ccp)

*Figure of an octahedron and an octahedron rotated
to show the relation to two close-packed layers*

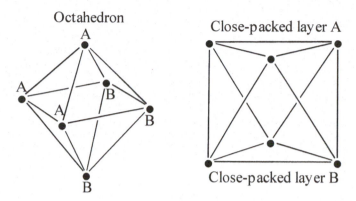

- **Body-centered cubic (bcc) structure**
  - → Coordination number of 8 (*not* close-packed)
  - → Spheres touch along the *body* diagonal of a cube.
  - → Can be converted into close-packed structures under high pressure
- **Primitive cubic (pc) structure**
  - → Coordination number of 6 (*not* close-packed)
  - → Spheres touch along the *edge* of the cube. Only one known example: Po (polonium)

Crystal Structure	Coordination Number	Occupied Space	Examples
ccp (fcc)	12	74%	Ca, Sr, Ni, Pd, Pt, Cu, Ag, Au, Al, Pb Group 18: noble gases at low T, except He
hcp	12	74%	Be, Mg, Ti, Co, Zn, Cd, Tl
bcc	8	68%	Group 1: alkali metals, Ba, Cr, Mo, W, Fe
pc	6	52%	Po (covalent character)

3H.5 Unit Cells

- **Lattice**
  - → A regular array of equivalent points in three dimensions (or any other dimensions)
- **Unit cell**
  - → Smallest repeating unit that generates the full array of points by translation
- **Crystal**
  - → Constructed by associating atoms, ions, or molecules with each lattice point

→ **Metallic crystals:** One metal atom per lattice point (ccp (fcc), bcc, pc)

　　　　　　　　　　　　 Two metal atoms per lattice point (hcp)

- **Unit cells for metals**

  → Cubic system: pc, bcc, and fcc (ccp)

  → Hexagonal system: *primitive unit cell* with *two* atoms per lattice point (hcp)

- **Unit cells in three dimensions**

  → Edge lengths: *a*, *b*, *c*

  → Angles between two edges: α (between edges *b* and *c*); β (between *a* and *c*); γ (between *a* and *b*)

- **Seven crystal systems (unit cell shapes)**

  → Cubic, hexagonal, tetragonal, orthorhombic, rhombohedral, monoclinic, and triclinic

- **Bravais lattices**

  → 14 basic patterns of arranging points in three dimensions

  → Each pattern has a different unit cell (figure).

The 14 Bravais Lattices ⇒

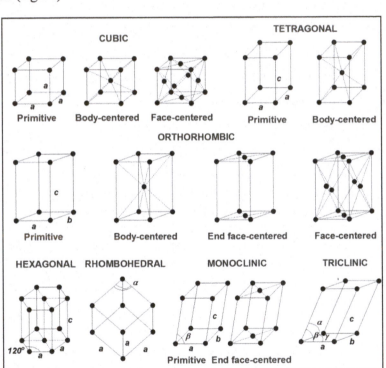

- **Volume of several types of unit cells**

  Cubic: $V = a \times a \times a = a^3$

  　　 $(\alpha = \beta = \gamma = 90°)$

  Tetragonal: $V = a \times a \times c = a^2 c$

  　　 $(\alpha = \beta = \gamma = 90°)$

  Orthorhombic: $V = a \times b \times c = abc$

  　　 $(\alpha = \beta = \gamma = 90°)$

- **Properties of unit cells**

  → Each *corner point* shared among 8 cells　(1/8 per cell)

  　　8 corner points × 1/8 per cell = 1 corner point per cell　(*all cells*)

  → Each *face-centered point* shared between 2 cells　(1/2 per cell)

  　　6 face points × 1/2 per cell = 3 face points per cell　(*face-centered cells*)

  　　2 face points × 1/2 per cell = 1 face point per cell　(*end face-centered cells*)

  → Each *body-centered point* unshared　(1 per cell)

  　　1 body point × 1 per cell = 1 body point per cell　(*body-centered cells*)

- **Properties of *cubic system* unit cells**
  - → *One* metal atom occupies each *lattice point*

    Primitive

    One metal atom (*sphere*) at each *corner point* in a unit cell
    (*one* atom (radius r) per cell)
    Spheres *touch* along an *edge:* $a = 2r$
    Volume of the unit cell: $V = a^3 = (2r)^3 = 8r^3$
    Density of a pc metal: (M = molar mass):

    $$d = \frac{(1 \text{ atom per cell})(\text{mass of one atom})}{(\text{volume of one unit cell})} = \frac{M/N_A}{a^3} = \frac{M/N_A}{8r^3}$$

    Body-centered

    One metal atom (*sphere*) at each *corner point* in a unit cell
    One metal atom (*sphere*) at the *center* of the unit cell
    Total atoms in unit cell = 8 (1/8) + 1 (1) = 2
    (*two* atoms (radius r) per unit cell (edge length a))

    Spheres *touch* along the *body diagonal:* $\sqrt{3}a = 4r$ or $a = \dfrac{4}{\sqrt{3}}r$

    Volume of the unit cell: $V = a^3 = \dfrac{4^3}{3^{3/2}}r^3 \approx 12.3168r^3$

    Density of a bcc metal:

    $$d = \frac{(2 \text{ atoms per cell})(\text{mass of one atom})}{(\text{volume of one unit cell})} = \frac{(2)(M/N_A)}{a^3} = \frac{(2)(M/N_A)}{\left(4^3/3^{3/2}\right)r^3}$$

    Face-centered

    One metal atom (*sphere*) at each *corner point* in a unit cell
    One metal atom (*sphere*) at the *center* of each *face* in the unit cell
    Total atoms in unit cell = 8 (1/8) + 6 (1/2) = 4
    (*four* atoms (radius r) per unit cell (edge length a))

    Spheres *touch* along the *face diagonal:* $\sqrt{2}a = 4r$ or $a = \dfrac{4}{\sqrt{2}}r = \sqrt{8}r$

    Volume of the unit cell: $V = a^3 = 8^{3/2}r^3$
    Density of a fcc (ccp) metal:

    $$d = \frac{(4 \text{ atoms per cell})(\text{mass of one atom})}{(\text{volume of one unit cell})} = \frac{(4)(M/N_A)}{a^3} = \frac{(4)(M/N_A)}{\left(8^{3/2}\right)r^3}$$

- **Calculations**
  - → Determining the *unit cell* type from the *measured density* and *atomic radius* of metals in the *cubic* system; the *atomic radius* of a metal atom is determined from the crystal type and the edge length

of a unit cell, which are both obtained by x-ray diffraction techniques (see **Major Technique 3** available online).

Example: The *density* of lead is 11.34 g·cm^{-3} and the *atomic radius* is 175 pm. Assuming the unit cell is *cubic,* we can determine the type of Bravais lattice. Using the equations given above, calculate the density of Pb assuming pc, bcc, or fcc Bravais lattices. Results are pc, 8.02 g·cm^{-3}; bcc, 10.4 g·cm^{-3}; fcc,11.3 g·cm^{-3}. The Bravais lattice is most likely fcc.

3H.6 Ionic Solids

- **Model   (r(tetrahedral hole) < r(octahedral hole) < r(cubic hole))**

  → Spheres of *different* radii and *opposite* charges represent cations and anions

  → Larger spheres (*usually* anions) *typically* occupy the unit cell lattice points. Smaller spheres (*usually* cations) fill *holes* in the unit cell.

- **Radius ratio   [Use rule with caution; there are many exceptions.]**

  → Ratio of radius of the *smaller* sphere to the radius of the *larger* sphere, symbol ρ

  → Defined for ions of the *same* charge number ($z_{cation} = |z_{anion}|$)

Radius Ratio–Crystal Structure Correlations

radius ratio ≤ 0.414:	tetrahedral *holes* fill (fcc for *larger* spheres) (*zinc-blende structure* (one *form* of ZnS))
0.414 < radius ratio < 0.732:	octahedral *holes* fill (fcc for *larger* spheres) (*rock-salt structure* (NaCl))
radius ratio ≥ 0.732:	cubic *holes* fill (pc for *larger* spheres) (*cesium-chloride structure* (CsCl))

- **Coordination number**

  → Number of nearest-neighbor ions of *opposite* charge

- **Coordination of ionic solid**

  → Represented as (cation coordination number, anion coordination number)

- **Properties of the *ionic structure* of unit cells**

  Zinc blende (see **Figure 3H.32** in the text)

  → Radius ratio (ZnS) = $r(Zn^{2+})/r(S^{2-})$ = (74 pm)/(184 pm) = 0.40

  → S^{2-} ions at the *corners* and *faces* of an fcc unit cell
  (*four* anions (atomic radius r_{anion}) per unit cell (edge length a))

  → Zn^{2+} ions in *four* of the *eight* tetrahedral holes in the unit cell
  (*four* cations (atomic radius r_{cation}) per unit cell (edge length a))

  → Cation and anion spheres *touch* in the tetrahedral locations.
  Each cation has *four* nearest-neighbor anions, and each anion has *four* nearest-neighbor cations: *(4,4)-coordination.*

Rock salt

→ Radius ratio (NaCl) = $r(Na^+)/r(Cl^-)$ = (102 pm)/(181 pm) = 0.564

→ Cl^- ions at the *corners* and *faces* of an fcc unit cell
  (*four* anions (atomic radius r_{anion}) per cell)

→ Na^+ ions in all *four* octahedral holes in the unit cell
  (*four* cations (atomic radius r_{cation}) per cell)

→ Cation and anion spheres *touch* along the cell edges (length a).

  Each cation has *six* nearest-neighbor anions, and each anion has *six* nearest-neighbor cations: *(6,6)-coordination*.

→ Spheres *touch* along the *edge:* $a = 2r_{cation} + 2r_{anion}$ = 566 pm

→ Volume of the unit cell: $V = a^3$ = 1.81×10^8 pm^3 = 1.81×10^{-22} cm^3

→ Density of NaCl: M(58.44 g·mol^{-1}) and d(experimental) = 2.17 g·cm^{-3}

$$d = \frac{(4 \text{ NaCl per cell})(M/N_A)}{a^3} = \frac{4(58.44/N_A)}{(1.81 \times 10^{-22})} = 2.14 \text{ g·cm}^{-3}$$

Cesium chloride

→ Radius ratio (CsCl) = $r(Cs^+)/r(Cl^-)$ = (170 pm)/(181 pm) = 0.939

→ Cl^- ions at the *corners* of a pc (*primitive cubic*) unit cell
  (*one* anion (radius r_{anion}) per cell)

→ Cs^+ ion at the *center* of the *cube* in the unit cell
  (*one* cation (radius r_{cation}) per cell)

→ Cation and anion spheres *touch* along the *body diagonal* of the cell.

  Each cation has *eight* nearest-neighbor anions and each anion has *eight* nearest-neighbor cations: *(8,8)-coordination*.

→ Spheres *touch* along the *body diagonal*: $\sqrt{3}\,a = 2r_{cation} + 2r_{anion}$ = 702 pm and a = 405 pm

→ Volume of the unit cell: $V = a^3$ = 6.64×10^7 pm^3 = 6.64×10^{-23} cm^3

→ Density of CsCl: M(168.36 g·mol^{-1}) and d(experimental) = 3.99 g·cm^{-3}

$$d = \frac{(1 \text{ CsCl per cell})(M/N_A)}{a^3} = \frac{1(168.36/N_A)}{(6.64 \times 10^{-23})} = 4.21 \text{ g·cm}^{-3}$$

• **Calculations**

  → Determining the *ionic structure* in the *cubic* system from *ionic radii* using the *radius ratio*

  → Determining the *density* of the *ionic solid* from the *density* of the *unit cell*

  Note: Usually within 10% of the experimental value

Topic 3I: INORGANIC MATERIALS

3I.1 Alloys

- **Properties**

  → Alloys are metallic materials made by mixing two or more molten metals.

  → May be homogeneous or heterogeneous

  → Alloys are often softer than the component metals and melt at a lower temperature.

- **Heterogeneous Alloys**

  → Mixture of crystalline phases; various samples have different compositions; for example, solders and mercury amalgams

  → Melt and freeze over a range of temperatures

- **Homogeneous Alloys**

  → Atoms of different elements are distributed uniformly; for example, brass, bronze, and coinage metals.

  → See **Table 3I.1** in the text for the composition of typical homogeneous alloys.

- **Substitutional**

  → Atoms nearly the same size (< 15% difference) can substitute for each other more or less freely (mainly d-block elements).

  → Lattice is distorted and electron flow is hindered.

  → Alloy has lower thermal and electrical conductivity than pure element but is harder and stronger.

  → Homogeneous examples include the Cu-Zn alloy, brass (up to 40% Zn in Cu), and bronze (metal other than Zn or Ni in copper: modern casting bronze contains 90% Cu and 10% Sn).

- **Interstitial Alloys**

  → Small atoms (> 60% smaller than host atoms) can occupy holes or *interstices* in a lattice (see text **Figure 3I.4**).

  → Interstitial atoms interfere with electrical conductivity and the movement of atoms.

  → Restricted motion makes the alloy harder and stronger than the host metal.

  → Examples include low-carbon steel (C in Fe, which is soft enough to be stamped), high-carbon steel (hard and brittle unless subjected to heat treatment), and stainless steel (a mixture of Fe with metals such as Cr and Ni, which aid in its resistance to corrosion).

- **Nonferrous alloys**

  → Nonferrous alloys are prepared by mixing molten elements, followed by molding or by *sintering*, which consists of mixing powdered metals, followed by molding the mixture by pressing it while hot.

  → Can be *homogeneous* or *heterogeneous* solutions

→ *Homogeneous alloys* include brass; bronze; coins containing gold or silver; and alnico, a magnetic alloy of aluminum, nickel, and cobalt used in speakers and mobile phones.

→ *Heterogeneous alloys* include tin–lead solder; mercury amalgams (once used for dental fillings); and Bi/Cd, which melts over a range of temperatures, except for one composition (eutectic), which melts at a fixed temperature like a pure compound.

- **Substitutional alloys**

  → Homogeneous alloys with metals of similar radius. **Example:** Brass, consisting of Cu and Zn

  → Typically harder than the pure metals but with lower electrical and thermal conductivity

3I.2 Silicates

- **Silica, or silicon dioxide, SiO_2**

  → Derives its strength from its covalently bonded network structure

  → Occurs naturally in pure form only as quartz and sand

  → Quartz contains helical chains of SiO_4 units that share O atoms.

  → Above 1500 °C, quartz transforms to the mineral *cristobalite*, which has a diamond-like structure with C–C bonds replaced by Si–O–Si units.

  → Sandstone and granite are based on silica and silicates.

- **Silicates, compounds containing SiO_4^{4-} units**

  → View as tetrahedral silicon oxoanions with considerably covalent Si–O bonds

  → Properties depend on charge on each tetrahedron, number of corner-shared O atoms, and the way in which linked tetrahedra lie together.

  → Materials with silicates include gemstones, fibers, and glasses (see below).

  → *Orthosilicates,* the simplest silicates, contain SiO_4^{4-} ions. **Examples:** Na_4SiO_4 and $ZrSiO_4$, *zircon*, an inexpensive diamond substitute.

  → *Pyroxenes* contain chains of tetrahedral SiO_4^{4-} *units* in which one O atom bridges two Si atoms (Si–O–Si) from adjacent tetrahedra, such that the "average" unit is SiO_3^{2-} (see text **Figure 3I.7**). Cations placed along the chains provide electrical neutrality. **Examples:** $Al(SiO_3)_2$ and $CaMg(SiO_3)_2$

  → Chains of silicate ions can form cross-linked ladder-like structures containing $Si_4O_{11}^{6-}$ units. One example is the fibrous mineral *tremolite*, $Ca_2Mg_5(Si_4O_{11})_2(OH)_2$, one form of *asbestos*, a material that withstands extreme heat but is a known carcinogen.

  → Other types of silicates have sheets containing $Si_2O_5^{2-}$ units, with cations lying between the sheets and linking them together. One example is *talc*, $Mg_3(Si_2O_5)_2(OH)_2$. Additional structural types of silicon oxides exist.

- **Aluminosilicates**

  → *Aluminosilicates* form when a Si^{4+} ion is replaced by Al^{3+} plus additional cations for charge balance. One example is *mica*, $KMg_3(Si_3AlO_{10})_2(OH)_2$.

→ The extra cations hold the sheets of tetrahedra together, accounting for the hardness of these materials. *Feldspar* is a silicate material in which up to half the silicon (Si(IV)) is replaced by aluminum (Al(III)). **Example:** $KAlSi_3O_8$, a typical feldspar.

3I.3 Calcium Carbonate

- **Calcium compounds**
  - → Used as structural materials in organisms, building, and engineering applications
  - → Small, doubly charged Ca^{2+} cation and CO_3^{2-} anion give $CaCO_3(s)$ a high lattice energy, consistent with its applications.
  - → $CaCO_3$ occurs naturally as chalk; limestone; and marble, a very dense form with colored impurities, most commonly Fe cations.
  - → The most common forms of pure calcium carbonate are *calcite* and *aragonite* (hard, more dense, less abundant than *calcite*).
  - → All carbonates are the fossilized remains of marine life and are soluble in acids.
  - → Acid rain erodes limestone buildings and statuary.
  - → In nature, calcium salts are used as structural materials in seashells ($CaCO_3$) and in bones [$Ca_3(PO_4)_2$].

3I.4 Cement and Concrete

- **Composite materials**
  - → Two or more substances combined into a heterogeneous material while retaining their individual characteristics
  - → Combine the advantages of the component materials
  - → Can exhibit properties superior to those of the component materials
  - → Fiberglass is strong and flexible; it consists of inorganic materials in a polymer matrix.
  - → Seashells owe their strength to a tough organic matrix and their hardness to the calcium carbonate crystals embedded in the matrix.
  - → Bones have low density but are strong because of their composite nature. In bone tissue, crystals of phosphate salts are embedded in fibers of a natural polymer called *collagen* that hold them in place but allow a small amount of flexibility.
  - → Artificial bone and cartilage mimic bone in that they have inorganic solids mixed into a crack-resistant polymer matrix. The result is composite material with flexibility and great strength.
  - → Some lightweight composites can have three times the strength-to-density ratio of steel.

    Examples: The graphite composite used for tennis rackets is composed of graphite fibers embedded in a polymer matrix.

    The fibrous refractory insulation tiles used on the body of the space shuttle contained alumina-borosilicate fibers in a silica ceramic matrix.

- **Cements**
  - → Building blocks in construction are usually held together by binders called *cements*.
  - → Most common type is *Portand cement*, which is made by heating a mixture of crushed limestone, clay or shale, sand, and oxides such as iron ore in a kiln (see **Table 3I.2** in the text).
  - → Different types of Portland cement are used to meet different requirements.
  - → When heated, cements lose water and harden to form clinkers.
  - → See the text for chemical reactions describing the hardening process.
- **Concrete**
  - → Consists of a binder and a filler
  - → Filler is usually gravel, but sometimes polymer or vermiculite pellets are added to lower the density. *Vermiculite* is a low-density, clay-like silicate material with a structure that has been expanded by heating.
  - → Binder is cement, usually Portland cement. *Mortar* is a cement-like substance used to hold bricks and other building materials together.

Topic 3J: MATERIALS FOR NEW TECHNOLGIES

3J.1   Electrical Conduction in Solids

- **Mobility of electrons**
  - → A metal can be defined as an array of cations bonded by a sea of electrons.
  - → Responsible for characteristic luster and light reflectivity of a metal
  - → Accounts for *malleability, ductility,* and *electrical conductivity* of metals
  - → Produces a relatively strong metallic bond that leads to high melting points for most metals
- **Ability to conduct electricity**
  - → Desirable property of many solids
  - → There are two types of such conduction.
  - → Electronic conduction: Charge carried by electrons as in metals and graphite.
  - → Ionic conduction: Charge carried by ions as in molten salts and electrolyte solutions.
- **Molecular Orbital (MO) theory**
  - → Accounts for electrical properties of metals and semiconductors
  - → Treats solid as a large molecule with delocalized electrons spread over the entire solid
- **Bands**
  - → Groups of MOs having closely spaced (essentially continuous) energy levels (see **Figure 3J.2** in the text)

→ Formed by the spatial overlap of a very large number of *atomic orbitals*, AOs

→ Separate into two groups (one mostly *bonding*, the other mostly *antibonding*)

- **Conduction band**
    → Empty or partially filled band
    Example: Formation of a conduction band in Na metal by the overlap of 3*s*-orbitals

- **Valence band**
    → Completely filled band of MOs (insulators such as molecular solids)

- **Band gap**
    → An energy range between bands in which there are *no* orbitals

- **Electronic conductors**
    → Metals, semiconductors, and superconductors

Insulator:	Does not conduct electricity: substance with a full *valence band* far in energy from an empty *conduction band* (large *band gap*)
Metallic conductor:	Current carried by *delocalized* electrons in *bands* Conductivity *decreases* with *increasing* temperature Substance with a partially filled *conduction band* (gap *not* relevant)
Semiconductor:	Current carried by *delocalized* electrons in *bands* Conductivity *increases* with *increasing* temperature Substance with a full *valence band* close in energy to an empty *conduction band* (small *band gap*): Excitation of some electrons from the *valence* to the *conduction band* yields conductivity.
Superconductor:	*Zero* resistance (infinite conductivity) to an electric current below a definite transition temperature

- **Electrical conductivity**
    → Symbol is κ (kappa). Electrical conductivity of a sample of length *l* and cross-sectional area *A* is defined in terms of its resistance *R*.

    → The equation is $\kappa = \dfrac{l}{A \times R}$

    → Units are Siemens per meter (S m^{-1}, where 1 S = 1 Ω^{-1}, with Ω, omega, the symbol for ohm).

    → The conductivity of a metal is usually about 10^7 S m^{-1}; that of a semiconductor about 10^{-4} S m^{-1}.

    → A metallic conductor has a much higher electrical conductivity than a semiconductor, but it is the temperature dependence of the conductivity that distinguishes the two types of conductors.

3J.2 Semiconductors

- **n-type semiconductor**
    → Adding some *valence electrons* to the *conduction band* (n = electrons added)

> **Example:** A small amount of As (Group 15) added to Si (Group 14): Extra electrons are transferred from As into the previously empty *conduction band* of Si.

- **p-type semiconductor**

  → Removing some *valence electrons* from the *valence band* (p = positive "holes" in the *valence band*)

  > **Example:** A small amount of In (Group 13) added to Si (Group 14): Valence electron deficit the previously full *valence band* of Si yields a *conduction band*.

- **p-n junctions**

  → p-type semiconductor in contact with an n-type semiconductor

  → Solid-state electronic devices

  > **Example:** One type of transistor is a p-n sandwich with two separate wires (*source* and *emitter*) connected to the n side. It acts as a *switch* in the following way: When a positive charge is applied to a polysilicon contact (*base* wire) positioned between the *source* and *emitter* wires, a current flows from the *source* to the *emitter*. Other examples include diodes and integrated circuits.

- **Enhanced (extrinsic) semiconductors**

  → Prepared by *doping*, that is, adding a small amount of impurity to the material

3J.3 Superconductors

- **Superconductor**

  → Loss of electrical resistance when a substance is cooled below a characteristic transition temperature (T_s)

  → Conventional superconductivity arises from the ability of electrons to pair by making use of lattice vibrations.

  → The electron pairs, called **Cooper pairs** after the scientist who first proposed the mechanism, can travel almost freely through the lattice, rather as oxen yoked together are less easily deflected by obstacles than individual oxen.

  → A Cooper pair forms when one electron distorts cations in its vicinity, and those cations attract a second electron. Electrons are weakly attracted in this way; the Cooper pair (and the resulting superconductivity) survives only if the temperature is low enough to avoid having the pair eliminated by lattice vibrations.

  → Superconductivity was first observed in 1911 in mercury, for which $T_s = 4$ K. Over the years, many other *metallic* superconductors were identified, some having transition temperatures as high as 23 K.

  → Such **low-temperature superconductors** must be cooled with liquid helium, which is very expensive. As a result, they have not found widespread use.

- **High-Temperature Superconductors (HTSCs)**

  → If T_s is > 77 K, liquid nitrogen can be used as the coolant (0.2% of the cost of liquid helium).

  → Almost all HTSCs are hard, brittle ceramic oxides that have sheets of copper and oxygen atoms sandwiched between layers of either cations or a combination of cations and oxide ions, and all are derived from their respective parent insulators by doping.

  → The layered structures result in strongly anisotropic electrical and magnetic properties. Electric current flows easily along the planes of the copper–oxygen sheets but only weakly perpendicular to them.

  → From 1986 to 1988, the record high transition temperature rose from 35 K for a ceramic lanthanum–copper oxide material doped with barium to 93 K for an yttrium–barium–copper oxide material (1987), to 125 K for a thallium–barium–calcium–copper material (1988).

  → By 2006, the highest transition temperature attained was 130 K. Brittle ceramic materials are problematic for use as electrical wires. One possible solution has been to deposit the superconducting material on the surface of wire or tape made of a metal such as silver.

  → Twenty-five kilograms of superconducting wire can carry as much current as 1800 kg of today's copper-based electrical cable.

  → Potential applications of superconductors include electrical transmission (little or no heat loss in wires) and energy-efficient transportation, such as magnetic levitation (maglev) trains.

3J.4 Luminescent Materials

- **Light emission from materials**

  → *Incandescence* – light emission from a heated object, such as a lamp filament or hot soot in a candle flame. See also *black-body radiation* in **Topic 1B.1**.

  → *Luminescence* – light emission caused by other processes including light absorption, chemical reaction, impact with electrons, radioactivity, or mechanical shock.

  → *Chemiluminescence* – light emission caused by chemical reaction. The reaction products are formed in energetically excited states that decay by light emission.

    Example: The reaction of hydrogen atoms and fluorine molecules produces *infrared chemiluminescence*. Hydrogen fluoride, HF, is produced with considerable *vibrational* energy; it decays by *emitting infrared* radiation (see **Online Major Technique 1** for a discussion of infrared and microwave spectroscopy).

  → *Visible chemiluminescence* is usually produced by *electronic* excitation of product molecules.

  → *Bioluminescence* – a form of chemiluminescence produced by living organisms, such as fireflies and certain bacteria.

- **Light emission from molecules**

  → *Fluorescence* is *prompt* emission of light from molecules excited by radiation of higher frequency. Normally, absorption of ultraviolet radiation by a molecule leads to emission in the visible region; higher frequency *absorption* leads to lower frequency *emission*.

Note: With high-power lasers, simultaneous absorption of two or more photons by a molecule may lead to emission of a single photon with a *greater* frequency than that of the individual photons absorbed.

→ *Phosphorescence* is *slow* or *delayed* emission of light from molecules excited by radiation of higher frequency. In this case, the initially excited molecule in a singlet state (spins paired, ↑↓) undergoes a transition to a triplet state (spins parallel, ↑↑) that decays to a singlet state slowly because spin pairing is a relatively slow process.

→ *Triboluminescence* (from the Greek word, *tribos,* a rubbing) is luminescence produced by a mechanical shock to a crystal. It is readily observed in striking or grinding sugar crystals in the dark. Trapped nitrogen gas escapes while in an excited state and produces the radiation.

- **Phosphors**

  → *Phosphorescent materials* that glow when activated by the impact of fast electrons, as well as high-frequency radiation such as ultraviolet or x-ray

  → Clusters of three *phosphors* are used for each dot in a color TV or computer screen. Commonly used phosphors are europium-activated yttrium orthovanadate, YVO_4, for the red color; silver-activated zinc sulfide for blue; and copper-activated zinc sulfide for green.

  → *Fluorescent lamps* utilize fluorescent materials activated by ultraviolet light. Mercury atoms in the lamp are excited by electrons in a discharge and emit light at 254 and 185 nm, which is absorbed by a phosphor coated on the lamp's surface. A commonly used phosphor is *calcium halophosphate,* $Ca_5(PO_4)_3F_{1-x}Cl_x$, doped with manganese(II) and antimony(III) ions. An Sb(III)-activated phosphor emits blue light and a Mn(II)-activated phosphor emits yellow light. The net result is light with an approximately white spectral range.

  → *Fluorescent materials* have important applications in medical research. Dyes such as *fluorescein* are attached to protein molecules to probe biological reactions. *Fluorescent materials,* such as sodium iodide and zinc sulfide, can be activated by radioactivity and are used in scintillation counters to measure radiation (see **Focus 10**). Light-emitting diode (LED) displays also use luminescent materials.

3J.5 Magnetic Materials

- **Paramagnetism**

  → Tendency of a substance to move into a magnetic field. It arises when an atom or molecule has at least one unpaired electron, which aligns with the applied field.

  → Because spins on neighboring atoms or molecules are aligned almost randomly, *paramagnetism* is very weak, and alignment of electron spins is lost when the magnetic field is removed.

- **Ferromagnetism**

  → In some *d*-metals, when the unpaired electrons of many neighboring atoms align with one another in an applied magnetic field, the much stronger effect of **ferromagnetism** arises. The regions of aligned spins are called **domains**, and they survive even after the applied field is turned off.

  → Much stronger than paramagnetism, ferromagnetic materials are used to make permanent magnets.

→ Ferromagnetic materials are also used to coat computer hard drives. Electromagnetic recording heads align spins to establish domains, which remain for years.

→ **Antiferromagnetic** material – neighboring spins are locked into an *antiparallel* arrangement, such that the magnetic moments cancel. Manganese is antiferromagnetic.

→ **Ferrimagnetic** material – the spins on neighboring atoms are different, and although they are locked together in an antiparallel arrangement, the two magnetic moments do not completely cancel.

> **Examples:** *Ferromagnetism* occurs in alloys such as alnico and some *d*-metal compounds, such as the oxides of iron and chromium. Ceramic magnets, used in refrigerator magnets, are made of barium ferrite ($BaO \cdot nFe_2O_3$) or strontium ferrite ($SrO \cdot nFe_2O_3$) by compressing the powdered ferrite in a magnetic field and heating it until it hardens. Because these magnets are ceramic, they are hard and brittle, with low densities. However, they are the most widely used magnets because of their low cost.

- **Ferrofluids**

  → Ferromagnetic liquids, suspensions of finely powdered magnetite, Fe_3O_4, in a viscous, oily liquid (such as mineral oil) that contains a detergent (such as oleic acid, a long-chain carboxylic acid)

  → The iron oxide particles do not settle out because they are attracted to the polar ends of the detergent molecules, which form clusters called *micelles* around the particles. The nonpolar ends of the detergent molecules point outward, allowing the micelles to form a colloidal suspension in the oil.

  → When a magnet is brought near a ferrofluid, the particles in the liquid try to align with the magnetic field but are kept in place by the oil. As a result, it is possible to control the flow and position of the ferrofluid by means of an applied magnetic field.

  → One application of ferrofluids is in the braking systems of exercise machines. The stronger the magnetic field, the greater the resistance to motion.

- **Molecular magnets**

  → Single-molecule magnets contain several *d*-block metal atoms bonded to groups of nonmetal atoms such as carbon, hydrogen, and oxygen

  → Respond to a magnetic field as if they were nano-size compass needles and have a great potential for miniaturizing electronic storage media.

  → Typically the d-block metal atoms are imbedded in cage-like structures, and some of the nano-size molecular magnets have interesting shapes, such as rings, tubes, or spheres.

3J.6 Nanomaterials

- **Nanoparticles**

  → Particles ranging in size from 1 to 100 nm, larger than molecules but too small to exhibit bulk properties

  → Can be manufactured and manipulated at the *molecular level* using nanotechnology

- **Nanomaterials**
    - → Materials composed of *nanoparticles* or *regular arrays* of molecules or atoms such as *nanotubes*
    - → Properties differ from the properties of atoms and from those of bulk materials
    - → Used to make miniature circuits and drug delivery systems

- **Quantum Dots**
    - → Three-dimensional clusters of semiconducting materials, such as cadmium selenide (CdSe), containing 10 to 10^5 atoms
    - → Can be made in solution or by depositing atoms on a surface, with the size of the nanocrystal being determined by the details of the synthesis
    - → Color varies with the radius of the quantum dot, and it is readily observed in suspensions of CdSe quantum dots of different sizes (see text **Figure 3J.15**).
    - → Some quantum dots emit light when an excited electron drops to a lower level in the dot. An application of this phenomenon is the monitoring of processes taking place in biological cells.

        Example: A CdSe quantum dot can be attached to the surface of a cell by an organic linking molecule. The spatial distribution of emission intensity (labeled molecule) can then be viewed with a microscope.

- **Understanding Quantum Dots**
    - → The particle-in-a-box energy levels can be used to predict the qualitative behavior of an electron trapped in a *spherical* cavity of radius *r*. The relevant equation from **Topic 1C.1** is now

        $$E_n = \frac{n^2 h^2}{8 m_e r^2}$$

    - → As the radius *r* of the quantum dot increases, the energy level separations decrease and radiation of lower frequency (lower energy) is needed to effect transitions. Or, as noted earlier, color varies with the radius of the quantum dot.

3J.7 Nanotubes

- **Carbon nanotubes**
    - → Cylinders a few nanometers in diameter made from rolled sheets of graphite-like arrays of atoms
    - → Graphene sheets containing millions of carbon atoms rolled into a cylinder only 1–3 nm in diameter
    - → First identified in 1991, when two graphite rods sealed inside a container of He gas were subjected to an electric discharge
    - → Tensile strength parallel to the axis of the tube is the greatest of any material measured.
    - → Strength-to-mass ratio is 40 times that of steel because of the very low density of nanotubes.
    - → Carbon nanotubes conduct electricity because of an extended network of delocalized π-bonds that extends from one end of the tube to the other.
    - → The conductivity of nanotubes is a function of the way the tubes are formed.

→ With hexagon points aligned along the long axis of the tube, conductivity is similar to that of metals. Electrons travel through the π-orbitals with little resistance.

→ With points aligned perpendicular to the long axis, nanotubes behave as semiconductors.

→ Like diamonds, carbon nanotubes conduct heat well.

→ Carbon nanotubes are good candidates for the developing miniature integrated circuits because of their electrical and thermal conductivity properties.

→ Tiny 60-atom carbon spheres were also observed. This allotrope was discovered and named buckminsterfullerene in 1985.

- **Other nanotubes**

  → Some 50 types of nanotubes and nanospheres have been synthesized.

  → MoS_2 and WS_2 nanotubes and nanospheres serve as solid lubricants.

  → TiO_2 nanotubes show promise as a means of storing hydrogen gas.

  → Nanospheres of ruthenium coated with platinum serve as a catalyst for the purification of hydrogen fuel.

  → Boron nitride nanotubes are also of great interest. Unlike carbon nanotubes, boron nitride nanotubes have electrical characteristics independent of the diameter of the tube or the way the sheets are rolled. They have a moderately large band gap and so are considered to be only weakly semiconducting.

  → Carbon nanotubes grow until they are capped by bonds to another element or by five-member rings, which pucker the structure, forming bends along the length of tubes and domes at their ends.

  → For BN nanotubes, five-member rings cannot form because they would require boron–boron or nitrogen–nitrogen bonds, which are too weak to maintain the structure.

  → Instead, BN nanotubes are capped by the atom of a metal such as tungsten.

  → Nanotubes can be used as templates, or tiny molds, for nanostructures of other elements. As very small test tubes, nanotubes filled with biomolecules such as cytochrome *c* may act as nanosensors for medical applications.

  → Nanotubes carrying hydrogen molecules could become a medium for hydrogen-powered vehicles.

- **Nanotechnology**

  → Two major methods, physical and chemical (top-down and bottom-up), are used to produce nanomaterials.

  → **Physical** methods such as lithography are top-down approaches: materials are molded into shape by use of physical nanotechnology methods.

  → **Chemical** methods such as the solution and vapor-phase synthesis are bottom-up approaches: molecules are allowed to assemble themselves into specific patterns by use of intermolecular interactions. In template synthesis, metal ions are plated electrochemically on porous polymer membranes, which are then dissolved to leave metal nanotubes.

INTERLUDE

Ceramics and Glasses

Ceramics

- **Preparation and properties**

Ceramics

→ Inorganic materials, such as clays, hardened by heating to a high temperature

→ Typically very hard, insoluble in water, and stable to corrosion and high temperatures

→ Can be used at high temperatures without failing and resist deformation

→ Tend to be brittle, however

→ Often are oxides of elements on the border between metals and nonmetals

→ Used in many automobile parts, including spark plugs, pressure and vibration sensors, brake linings, and catalytic converters

→ Some d-metal oxides and compounds of B and Si with C and N are also ceramic.

→ Most ceramics are electrical insulators, but some are semiconductors or superconductors.

Aluminosilicate ceramics

→ Formed in many cases by heating clays to expel water trapped between sheets of tetrahedral aluminosilicate units

→ The result is a rigid heterogeneous mass of small interlocking crystals bound together by glassy silica.

→ One form is white *china clay,* used to make porcelain and china. It is free of iron impurities that tend to make clays reddish brown.

→ Methods developed to make ceramics less brittle are the *sol-gel process* and the *composite material technique.* These methods have produced *sol-gels* and *aerogels.*

→ An *aerogel* is a ceramic foam of great strength that has low density and low thermal conductivity. It is used to insulate skylights and in the Mars Rover.

Aluminum Oxide

→ A ceramic found in several forms

→ *Corundum*, α-alumina, a hard, stable, crystalline substance

→ *Emery*, impure microcrystalline corundum, an abrasive

→ *Sapphires*, large, single crystals of alumina doped with iron and titanium impurities

Glasses

- **Preparation and properties**
  - → A *glass* is an ionic solid with an amorphous structure, similar to a liquid.
  - → Glasses have short-range order but lack long-range order.
  - → Glass is characterized by a network structure based on a nonmetal oxide, commonly silica, SiO_2, melted with metal oxides that function as *network modifiers*.
  - → The development of optical fibers led to considerable recent advances in glassmaking.
  - → Heating SiO_2 ruptures Si–O bonds, enabling metal ions to form ionic bonds with some of the O atoms. The nature of the glass depends on the metal added.
  - → *Soda-lime glass* (about 90% of all glass), used for windows and bottles, is made by adding Na^+ (12% Na_2O) and Ca^{2+} (12% CaO). Heating Na_2CO_3 (soda) produces the Na_2O; heating $CaCO_3$ (lime) produces the CaO.
  - → *Borosilicate glass,* such as Pyrex, is produced using less soda and lime and adding 16% boron trioxide, B_2O_3. These glasses expand very little on heating and are used for ovenware and laboratory glassware.
  - → In glass, tiny crystallites do not scatter light, so glass is transparent.

- **Optical fibers**
  - → Made by drawing fiber from heated glass
  - → Coated with plastic
  - → Transmit information much faster than wires
  - → Vary in composition in cross-section to avoid loss of light

- **Reaction with acids and bases**
  - → Glass is resistant to attack by most chemicals.
  - → But the silica in glass can react with HF:

$$SiO_2(s) + 6\,HF(aq) \rightarrow SiF_6^{2-}(aq) + 2\,H_3O^+(aq)$$

  - → Removal of silica from glass by the ions F^- (from HF), OH^-, and CO_3^{2-} is called *etching*.
  - → Silica also reacts with the Lewis base OH^- in molten NaOH and the carbonate anion in molten Na_2CO_3 at 1400 °C:

$$SiO_2(s) + Na_2CO_3(l) \rightarrow Na_2SiO_3(s) + CO_2(g)$$

Focus 4: THERMODYNAMICS

Overview

- **Thermodynamics**
  - → Branch of science concerned with the relationship between heat and other forms of energy
- **Energy**
  - → The capacity to do work
- **Laws of thermodynamics**
  - → Generalizations based on experience with *bulk* matter
  - → Not derivable, but understandable without knowledge of the behavior of atoms or molecules
- **First law of thermodynamics**
  - → Consequence of the **law of conservation of energy**
  - → Quantitative description of energy changes in physical (e.g., phase changes) and chemical (*e.g.*, reactions) processes
  - → Energy changes in relation to heat and work:

    A paddlewheel stirrer immersed in a liquid produces heat from mechanical work yields a temperature rise in the liquid.

    A hot gas expanding in an insulated cylinder coupled to a flywheel produces mechanical work from heat and results in a temperature drop in the gas.
- **Second law of thermodynamics**
  - → Criterion for spontaneity
  - → Explains why some chemical reactions and other processes occur spontaneously whereas others do not
- **Statistical thermodynamics**
  - → Laws of thermodynamics reflected in the behavior of large numbers of atoms or molecules in a sample
  - → Links between behavior at atomic and bulk amounts of matter
- **Focus goal**
  - → Gain insight into heat and work and their relationship to energy changes in physical and chemical processes

Topic 4A: WORK AND HEAT

4A.1 Systems and Surroundings

- **The "universe" or "world"**
  - → Composed of a system and its surroundings

- **System**
  - → Portion of the universe of interest
    Examples: A beaker of ethanol, a frozen pond, 20 g of $CaCl_2(s)$, or the Earth itself.

- **Surroundings**
  - → Remainder of the universe (everything that is *not* in the system)
  - → Where observations and measurements of the system are made
    Example: The observation of heat transferred to or from a system is made in the surroundings.

- **Boundary**
  - → Dividing surface between the system and the surroundings
  - → Can be fixed (a closed, rigid container) or variable (a piston containing an expanding gas)

- **Types of systems: open, closed, or isolated**

  Open system
  - → Both *matter* and *energy* can be exchanged between the system and surroundings.
    Example: A half-liter of water in an open beaker. Matter can cross the boundary; water evaporates. Energy can cross the boundary; heat from the surroundings may enter the system.

  Closed system
  - → *Energy* can be exchanged between the system and surroundings, but *matter* cannot.
    Example: The refrigerant in an air-conditioning system. As the refrigerant expands, energy (heat) is withdrawn from the space to be cooled. As the refrigerant is compressed, energy (heat) is supplied to another space that is warmed. The mass of the refrigerant is unchanged.

  Isolated system
  - → Neither *matter* nor *energy* can be exchanged between the system and surroundings.
    Example: An *approximately* isolated system is a substance in a well-insulated container, such as ice in a plastic foam ice chest or coffee in a covered plastic foam cup.

4A.2 Work

- **Work, *w***
  - → Process of achieving motion against an opposing force
    Examples: Raising a weight, compressing a gas, winding a spring, charging a battery
  - → Fundamental thermodynamic property (see text **Table 4A.1**: *Varieties of Work*)

→ Definition: work ≡ opposing force × distance moved

→ Units: (force) *newton*, N; $1\ N = 1\ kg \cdot m \cdot s^{-2}$

(work) *joule*, J; $1\ J = 1\ N \cdot m = 1\ kg \cdot m^2 \cdot s^{-2}$

- **Internal energy, U**

    → Internal energy is the *total* energy stored in a system.

    → Includes energy of all atoms, electrons, and nuclei in a system

    → U cannot be measured.

    → *Only* a change in U ($\Delta U = U_{final} - U_{initial}$) is measurable.

    → An extensive property; varies with the size of the system (see **Fundamentals A** in the text)

    → For energy transferred to a system by doing work on the system, $\Delta U = w$.

    → Work done on a system is positive ($\Delta U > 0$); work done by a system is negative ($\Delta U < 0$).

4A.3 Expansion Work

- **Expansion work**

    → Change in volume of the system against an external pressure
    Example: Inflating a tire or a balloon

- **Nonexpansion work**

    → Does not involve a change in volume
    Examples: Lifting a mass, discharging a battery

- **Calculating expansion work with a constant external pressure, P_{ex}**

    → Consider a gas confined in a volume, V, by a piston with surface area, A.

    → Work ≡ force × distance $= F \times d = (P_{ex} \times A) \times d = P_{ex} \Delta V$, where d is the distance the piston is displaced (*not* the density as in **Topic 3B.3**).

    → $\boxed{w = -P_{ex} \Delta V}$ By convention, work done *on* the system is positive (+) in sign.

    Units: $1\ Pa \cdot m^3 = 1\ kg \cdot m^{-1} \cdot s^{-2} \times 1\ m^3 = 1\ kg \cdot m^2 \cdot s^{-2} = 1\ J$

    $1\ L \cdot atm = 10^{-3}\ m^3 \times 101\ 325\ Pa = 101.325\ Pa \cdot m^3 = 101.325\ J$ (exactly)

- **Free expansion**

    → Expansion into a vacuum ($P_{ex} = 0$ and $w = 0$). No work is done because there is no opposing force.

- **Reversible process**

    → A process that can be reversed by an *infinitesimal* change in a variable

    → For a change in pressure, the external pressure must always be infinitesimally different from the pressure of the system. A net change is effected by a series of infinitesimal changes in the external pressure followed by infinitesimal adjustments of the system.

    → A change in temperature of the system must occur as a series of infinitesimal steps either of work done or heat flow.

    → Processes in which change occurs by *finite* amounts are *irreversible* in nature.

- **Reversible isothermal expansion or compression of an ideal gas**

  → $$w = -nRT \ln \frac{V_{\text{final}}}{V_{\text{initial}}} = -nRT \ln \frac{P_{\text{initial}}}{P_{\text{final}}}$$

  → The change in pressure is carried out in infinitesimal steps (text derivation in **Topic 4A.3**).

4A.4 Heat

- **Heat, q**

  → Fundamental thermodynamic property

  → Energy that is transferred as a result of a temperature difference, $\Delta T = T_2 - T_1 > 0$

  → Energy flows as *heat* from a region of high temperature, T_2, to one of low temperature, T_1.

  → SI unit: *joule*, $1\text{ J} = 1\text{ N·m} = 1\text{ kg·m}^2\text{·s}^{-2}$

   Common unit: The *calorie*, $1\text{ cal} \equiv 4.184\text{ J}$, is widely used in biochemistry, organic chemistry, and related fields. The *nutritional calorie*, Cal, is 1 kcal.

- **Sign of q defined in terms of the system**

  → If heat flows from the surroundings into the system, $q > 0$ (*endothermic* process).

  → If heat flows from the system into the surroundings, $q < 0$ (*exothermic* process).

  → If no heat flows between the system and surroundings, $q = 0$ (*adiabatic* process).

   Example: A thermally insulating (adiabatic) wall (plastic foam or vacuum flask) prevents the passage of energy as heat.

  → *Diathermic* (nonadiabatic) walls permit the transfer of energy as heat.

- **Internal energy, U**

  → For energy transferred to a system only by the flow of heat, $\Delta U = q$ ($w = 0$).

  → If the system is initially at temperature T_1 and $\Delta T = T_2 - T_1 > 0$, then $q > 0$ and $\Delta U > 0$.

  → **Thermal energy:** Sum of the potential and kinetic energies of all atoms, molecules, and ions in a system.

4A.5 The Measurement of Heat

- **Heat capacity, C, of a pure substance**

  → is an *extensive* property.

  → is the amount of heat absorbed *by a sample of the substance* per degree Celsius rise in temperature.

- **Specific heat capacity, C_s, of a pure substance**

  → is an *intensive* property.

  → is the amount of heat absorbed *by one gram* of a substance per degree Celsius rise in temperature. (Values for common materials are given in **Table 4A.2** in the text.)

  → $C_s = C/m$ Units: $\text{J·(°C)}^{-1}\text{·g}^{-1}$ or $\text{J·K}^{-1}\text{·g}^{-1}$ (commonly, the *former*)

  → $$q = C\Delta T = mC_s\Delta T$$

- **Molar heat capacity, C_m, of a pure substance**
  - → is an *intensive* property.
  - → is the amount of heat absorbed *by one mole* of a substance per degree Celsius rise in temperature. (Values for some substances are also given in **Table 4A.2** in the text.)
  - → $C_m = C/n$ Units: $J \cdot (°C)^{-1} \cdot mol^{-1}$ or $J \cdot K^{-1} \cdot mol^{-1}$ (commonly, the *latter*)
  - → $\boxed{q = C\Delta T = nC_m \Delta T}$
  - → Values of molar heat capacities increase with increasing molecular complexity. They depend on the temperature and the state of the substance. $C_m(\text{liquid}) > C_m(\text{solid})$

- **Calorimeter**
  - → Device that monitors heat transfer resulting from a process (chemical reaction) taking place within it
  - → Heat lost by a reaction equals heat gained by the calorimeter: $-q = q_{cal}$.
  - → For an *exothermic* process, $q_{cal} > 0$; the temperature of the calorimeter increases, $\Delta T > 0$.
  - → For an *endothermic* process, $q_{cal} < 0$; the temperature of the calorimeter decreases, $\Delta T < 0$.
  - → The heat capacity of a calorimeter, C_{cal}, is the amount of heat absorbed *by the calorimeter* per degree Celsius rise in temperature.
  - → $\boxed{q_{cal} = C_{cal}\Delta T}$
  - → A calorimeter's heat capacity is determined by adding a known amount of heat, supplied electrically, and measuring the temperature rise in the calorimeter (see **Example 4A.4** in the text). Units of heat capacity: $J \cdot (°C)^{-1}$ or $J \cdot K^{-1}$

Topic 4B: INTERNAL ENERGY

4B.1 The First Law

- **Combining heat and work**
  - → $\boxed{\Delta U = q + w}$
  - → Change in internal energy of a closed system is the *sum* of the heat *added* ($q > 0$) *to* or *removed* ($q < 0$) *from* the system *plus* the work done *on* ($w > 0$) or *by* ($w < 0$) the system.

- **First law of thermodynamics**
  - → states that the internal energy of an *isolated* system is constant.
  - → is an extension of the law of conservation of energy.
  - → is a generalization based on experience and cannot be proven.
    - **Note:** If the first law were false, one could construct a perpetual motion machine by starting with an isolated system, removing it from isolation, and letting it do

work on the surroundings. It could then be placed in isolation again and its internal energy allowed to return to the initial value. This feat has never been accomplished despite many attempts.

- **Constant volume process, $\Delta V = 0$**
  - → No expansion work is done.
  - → $\Delta U = q$ No work of any kind (expansion, electrical, etc.) is done.
  - → Heat absorbed or released by a system equals the change in internal energy for a constant-volume process if no other forms of work are done: $\Delta U = q$.

4B.2 State Functions

- **State of a system**
  - → Defined when all of its properties are fixed

- **State function**
  - → Property that depends only on the current state of the system
  - → Independent of the manner in which the state was prepared

    Examples of state functions include temperature, T; pressure, P; volume, V; and internal energy, U. (Heat, q, and work, w, are *not* state functions.)

- **Path**
  - → Sequence of intermediate steps linking an initial and a final state
  - → Consider the transition from state 1 to state 2. The changes in the state functions are independent of path. The quantities ΔU, ΔP, ΔV, and ΔT are uniquely determined. The amount of heat transferred or the work done depends on the sequence of steps followed. Heat and work are path functions.

4B.3 A Molecular Interlude

- **Internal energy, U**
  - → Energy stored as potential and kinetic energy of molecules

- **Potential energy**
  - → Energy an object has by virtue of its position

- **Kinetic energy**
  - → Energy an object has by virtue of its motion
  - → For atoms, kinetic energy is *translational* in nature.
  - → For molecules, kinetic energy has *translational, rotational,* and *vibrational* components.
  - → *Translation* is the motion of the molecule as a *whole*.
  - → See **Topic 1C.2** in the text for the particle-in-a-box translational energy levels.
  - → *Rotation* and *vibration* are motions *within* the molecule; they do not change the center of gravity (mass) of the molecule.
  - → See **Online Major Technique 1** for the rigid-rotor rotational energy levels.

→ See the **Online Major Technique 1** for a description of vibrational motion.

Translational, Rotational, and Vibrational Motion of F_2

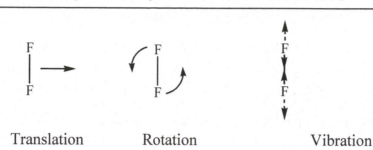

Translation   Rotation   Vibration

- **Degrees of freedom**
  - → Modes of motion (translational, rotational, and vibrational)
  - → *Mainly* translational and rotational degrees of freedom store internal energy at room temperature.
  - → Vibrational modes are largely inactive at room temperature. We will ignore them.

- **Equipartition theorem**
  - → *Average* kinetic energy of each degree of freedom of a molecule in a sample at temperature T is equal to $\frac{1}{2}kT$. The Boltzmann constant $k = 1.380\,658 \times 10^{-23}$ J·K^{-1}.

  - → A molecule moving in *three* dimensions has *three* translational degrees of freedom.
    $$U(\text{translation}) = 3 \times \frac{1}{2}kT = \frac{3}{2}kT$$

    For 1 mol of molecules, $U_m(\text{translation}) = N_A\left(\frac{3}{2}kT\right) = \frac{3}{2}RT \quad (R = N_A k)$.

  - → A *linear molecule* has *two* rotational degrees of freedom.
    $$U(\text{rotation, linear}) = 2 \times \frac{1}{2}kT = kT$$
    For 1 mol of molecules, $U_m(\text{rotation, linear}) = N_A(kT) = RT$.

  - → A *nonlinear molecule* has *three* rotational degrees of freedom.
    $$U(\text{rotation, nonlinear}) = 3 \times \frac{1}{2}kT = \frac{3}{2}kT$$

    For 1 mol of molecules, $U_m(\text{rotation, nonlinear}) = N_A\left(\frac{3}{2}kT\right) = \frac{3}{2}RT$.

  - → A molecule of an *ideal gas* does *not* interact with its neighbors, and the potential energy is 0. For an ideal gas, the internal energy is independent of volume; it depends *only* on temperature, $U_m(T)$.

  - → A molecule of a liquid, solid, or *real gas* does interact with its neighbors. The potential energy is an important component of the internal energy, which has a significant dependence on volume. For a real gas, the internal energy depends on both temperature and volume, $U_m(T,V)$.

Topic 4C: ENTHALPY

Overview

- **Heat transfer at constant volume, q_V (notation not used in the text)**

  → $\boxed{\Delta U = q_V}$ for a system in which only expansion work is possible

  → Combustion reactions are studied in a bomb calorimeter of constant volume.

  → Electric battery driving motors

- **Heat transfer at constant pressure, q_P (notation not used in the text)**

  → $\boxed{\Delta H = q_P}$ a new state function H (*enthalpy*) for a system at constant pressure

  → Most chemical reactions occur at constant pressure, largely because reaction vessels are usually open to the atmosphere.

4C.1 Heat Transfers at Constant Pressure

- **Definition of enthalpy, H**

  → $\boxed{H = U + PV}$ a state function because U, P, and V are state functions

  → Change in enthalpy of a system, $\Delta H = \Delta U + \Delta(PV)$
  At constant pressure, $\Delta H = \Delta U + P\Delta V = q_P + w + P\Delta V = q_P - P_{ex}\Delta V + P\Delta V = q_P$
  for a system open to the atmosphere and for which only expansion work may occur.
  Examples: System open to the atmosphere; gas or liquid in a piston with constant opposing pressure

- **Heat change for a constant pressure process**

  → *Exothermic* process: one in which heat is released by the system into the surroundings, $\Delta H = q_P < 0$

  → *Endothermic* process: one in which heat is absorbed by the system from the surroundings, $\Delta H = q_P > 0$

  → Because most chemical reactions take place at constant pressure, enthalpy is of great importance in chemistry.

4C.2 Heat Capacities at Constant Volume and Constant Pressure

- **General definition of heat capacity**

  → $\boxed{C = \dfrac{q}{\Delta T}}$

- **At constant volume**

  → $\boxed{C_V = \dfrac{q_V}{\Delta T} = \dfrac{\Delta U}{\Delta T}}$

- **At constant pressure**

    → $\boxed{C_P = \dfrac{q_P}{\Delta T} = \dfrac{\Delta H}{\Delta T}}$

- **Molar heat capacities**

    → $C_{V,m} = C_V/n$ and $C_{P,m} = C_P/n$

    → $C_{V,m} \approx C_{P,m}$ for *liquids and solids*, but $C_{P,m} > C_{V,m}$ for *gases*

- **Molar heat capacity relationship for ideal gases**

    → $C_P = C_V + nR$ *any amount of an ideal gas*

    → $\boxed{C_{P,m} = C_{V,m} + R}$ *one mole of an ideal gas*

    → For gases, $C_{P,m} > C_{V,m}$ because at constant pressure only some of the heat added raises the temperature; the rest is used to expand the gas (expansion work).

4C.3 The Molecular Origin of the Heat Capacities of Gases

- **Contributions to the molar heat capacity for a monatomic ideal gas**

    → Consider only translational motion.

    → $U_m = \frac{3}{2}RT$ and $\Delta U_m = \frac{3}{2}R\Delta T$ (Molar internal energy depends only on T.)

    $\boxed{C_{V,m} = \dfrac{\Delta U_m}{\Delta T} = \dfrac{\frac{3}{2}R\Delta T}{\Delta T} = \frac{3}{2}R}$ $= 12.471\ 77\ \text{J·K}^{-1}\text{·mol}^{-1}$ and $\boxed{C_{P,m} = \dfrac{\Delta H_m}{\Delta T} = \frac{5}{2}R}$

- **Contributions to the molar heat capacity for a linear-molecule ideal gas**

    → Consider translational motion and 2 rotational degrees of freedom.

    → $U_m = \frac{5}{2}RT$ and $\Delta U_m = \frac{5}{2}R\Delta T$ (Molar internal energy depends only on T.)

    $\boxed{C_{V,m} = \dfrac{\Delta U_m}{\Delta T} = \dfrac{\frac{5}{2}R\Delta T}{\Delta T} = \frac{5}{2}R}$ $= 20.786\ 28\ \text{J·K}^{-1}\text{·mol}^{-1}$ and $\boxed{C_{P,m} = \dfrac{\Delta H_m}{\Delta T} = \frac{7}{2}R}$

- **Contributions to the molar heat capacity for a nonlinear-molecule ideal gas**

    → Consider translational motion and 3 degrees of freedom in rotational motion.

    → $U_m = 3RT$ and $\Delta U_m = 3R\Delta T$ (Molar internal energy depends only on T.)

    $\boxed{C_{V,m} = \dfrac{\Delta U_m}{\Delta T} = \dfrac{3R\Delta T}{\Delta T} = 3R}$ $= 24.943\ 53\ \text{J·K}^{-1}\text{·mol}^{-1}$

$$C_{P,m} = \frac{\Delta H_m}{\Delta T} = 4R \quad = 33.258\ 04\ \text{J·K}^{-1}\text{·mol}^{-1}$$

Note: The contribution from molecular rotation is for room temperature. At lower temperatures, this contribution diminishes and the heat capacity approaches the value for a monatomic gas. At room temperature, a small contribution to the heat capacity from molecular vibrational motion leads to experimental values slightly larger than the ones given earlier.

4C.4 The Enthalpy of Physical Change

- **Physical changes (e.g., phase transitions) at constant temperature and pressure**

  → Molecules in liquids and solids are held together by intermolecular attractions.

  → Phase changes that increase the separation between molecules (vaporization) require energy and are endothermic.

  → Phase changes that decrease the separation between molecules (condensation) give off energy and are exothermic.

  → Vaporization liquid → vapor $\Delta H_{vap} = H_{vapor,m} - H_{liquid,m}$ > 0 *(endothermic)*

    Condensation vapor → liquid $\Delta H_{cond} = H_{liquid,m} - H_{vapor,m}$ < 0 *(exothermic)*

  → Fusion (melting) solid → liquid $\Delta H_{fus} = H_{liquid,m} - H_{solid,m}$ > 0 *(endothermic)*

    Freezing   liquid → solid $\Delta H_{freeze} = H_{solid,m} - H_{liquid,m}$ < 0 *(exothermic)*

  → Sublimation solid → vapor $\Delta H_{sub} = H_{vapor,m} - H_{solid,m}$ > 0 *(endothermic)*

    Deposition vapor → solid $\Delta H_{dep} = H_{solid,m} - H_{vapor,m}$ < 0 *(exothermic)*

  → $\Delta H_{vap} = -\Delta H_{cond}$ $\Delta H_{fus} = -\Delta H_{freeze}$ $\Delta H_{sub} = -\Delta H_{dep}$

  → $\Delta H_{sub} = \Delta H_{fus} + \Delta H_{vap}$ (for the same T and P)

  Values of $\Delta H_{fus}°$ and $\Delta H_{vap}°$ for selected substances are given in **Table 4C.1** in the text.

  → Liquids may evaporate at any temperature, but they boil only at the temperature at which the vapor pressure of the liquid is equal to the external pressure of the atmosphere.

4C.5 Heating Curves

- **Heating (cooling) curve**

  → is a graph showing the variation of the temperature, T, of a sample as heat is added or removed at a constant rate.

  → Two types of behavior are seen in a heating curve.

  → The temperature increase of a single phase to which heat is supplied has a positive slope that depends on the value of the heat capacity. The larger the heat capacity is, the more gentle the slope.

  → At the temperature of a phase transition, two phases are present and the slope is 0 (no temperature change as heat is added). The *greater* the heat associated with the phase change, the *longer* the length of the flat line.

→ As more heat is added to the substance, the relative amounts of the two phases change. The temperature does not rise again until only one phase remains.

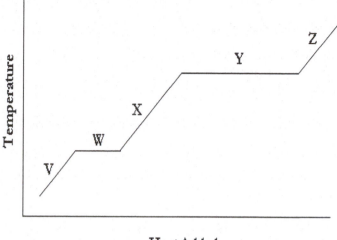

Heat Added

In the schematic heating curve shown here, line V with positive slope represents the heating of a solid phase. The flat line W shows the phase transition between solid and liquid (melting *or* fusion).

Line X represents the heating of the liquid after the solid has completely melted.

Line Y reflects the phase transition between liquid and vapor (boiling at constant external pressure).

Finally, line Z represents the heating of the vapor after the liquid disappears. Because the enthalpy of vaporization is greater than the enthalpy of fusion, line Y is longer than line W.

Topic 4D: THERMOCHEMISTRY

4D.1 Reaction Enthalpy

- **Thermochemical equation**
  → A chemical equation with a corresponding enthalpy change, ΔH, expressed in kJ for the *stoichiometric* number of moles of each reactant and product
  → Applying the principles of thermodynamics to chemical change

- **Enthalpy of reaction, ΔH**
  → The enthalpy change in a thermochemical reaction is expressed in kJ, where the reaction refers to the number of moles of each substance as determined by the stoichiometric coefficient in the balanced equation.

→ Allows calculation of the amount of heat released or required for any mass of reactant consumed or product formed in a chemical reaction

→ Combustion is reaction of a substance with oxygen, $O_2(g)$.

4D.2 The Relation Between ΔH and ΔU

• **Reactions with liquids and solids only**

→ Recall that $\Delta H = \Delta U + \Delta(PV)$

→ $\Delta H \approx \Delta U \implies q_P \approx q_V$

→ Only a small change in the volume of reactants compared to products at constant P

• **Reactions with ideal gases only**

→ $\Delta H = \Delta U + \Delta(PV) = \Delta U + \Delta(nRT) = \Delta U + (\Delta n)RT$ *constant temperature*

→ The equation is exact for ideal gases only.

• **Reactions with liquids, solids, and gases**

→ $\Delta H \approx \Delta U + (\Delta n_{gas})RT$ *constant temperature*

→ $\Delta n_{gas} = n_{final} - n_{initial}$ = change in the number of moles of gas between products and reactants

→ The equation is a very good approximation. The value of $(\Delta n_{gas})RT$ is usually much smaller than ΔU and ΔH.

4D.3 Standard Reaction Enthalpies

• **Standard conditions**

→ Values of reaction enthalpy, ΔH, depend on pressure, temperature, and the physical state of each reactant and product.

→ A standard set of conditions is used to report reaction enthalpies.

• **Standard state**

→ Standard state of a substance is its pure form at a pressure of *exactly* 1 bar.

→ For a solute in a liquid solution, the standard state is a solute concentration of 1 mol·L^{-1} at a pressure of *exactly* 1 bar.

→ Standard state of liquid water is pure water at 1 bar. Standard state of ice is pure ice at 1 bar. Standard state of water vapor is pure water vapor at 1 bar.

Note: Compare the definition of standard state in thermodynamics to the definition of standard temperature and pressure, STP, for gases (**Topic 3B.3**). For gases, standard temperature is 0 °C (273.15 K) and standard pressure is 1 atm (1.013 25 bar).

• **Standard reaction enthalpy, $\Delta H°$**

→ Reaction enthalpy when reactants in their standard states form products in their standard states

→ Degree symbol, °, is added to the reaction enthalpy.

→ Reaction enthalpy has a *weak* pressure dependence, so it is a very good approximation to use standard enthalpy values even when the pressure is *not* exactly 1 bar.

- **Temperature convention**
  - → Most thermochemical data are reported for 298.15 K (25 °C).
  - → Temperature convention is *not* part of the definition of the standard state.

- **Heat output in a reaction**
  - → Heat is treated as a reactant or product in a stoichiometric relation.
  - → In an endothermic reaction, heat is treated as a reactant.
  - → In an exothermic reaction, heat is treated as a product.

- **Standard enthalpy of combustion, $\Delta H_c°$**
  - → Change in enthalpy per mole of a substance that is burned in a combustion reaction under standard conditions (see values in **Table 4D.1** in the text)
  - → Combustion of hydrocarbons is exothermic, and heat is treated as a product.

- **Specific enthalpy**
  - → Enthalpy of combustion per *gram* ($kJ.g^{-1}$)
  - → Practical measure of a fuel's value (see values also in **Table 4D.1** in the text).

- **Energy conservation and biomass fuels**
  - → Biomass energy is being used 5 million times faster that it is being stored.
  - → Consequently, the importance of energy conservation and the use of alternative energy fuels have grown (see **Box 4D.1** in the text).

4D.4 Combining Reaction Enthalpies: Hess's Law

- **Hess's law**
  - → Overall reaction enthalpy is the sum of the reaction enthalpies of the steps into which a reaction can be divided.

    Example: Add two reactions (1 and 2) to get a third (3); add their reaction enthalpies.
    $$\Delta H(1) + \Delta H(2) = \Delta H(3)$$

  - → Consequence of the fact that the reaction enthalpy is path independent and depends only on the initial reactants and final products
  - → Used to calculate enthalpy changes for reactions difficult to carry out in the laboratory
  - → Used to calculate reaction enthalpies for unknown reactions

- **Using Hess's law**
  - → Write the thermochemical equation for the reaction whose reaction enthalpy is unknown.
  - → Write thermochemical equations for intermediate reaction steps whose reaction enthalpies *are known.*
  - → These steps must sum to the overall reaction whose enthalpy is unknown.
  - → Sum the reaction enthalpies of the intermediate reaction steps to obtain the unknown reaction enthalpy (see **Toolbox 4D.1** in the text).

4D.5 Standard Enthalpies of Formation

- **Treatment of chemical reactions**

  → Possible thermochemical equations are nearly innumerable.

  → Method to handle this problem is the use of a common reference for each substance. The reference is the *most stable* form of each element.

  → Standard enthalpies of formation of substances from their elements are tabulated and used to calculate standard reaction enthalpies for other types of reactions.

- **Standard enthalpy of formation, $\Delta H_f°$**

  → Enthalpy change for the formation of one mole of a substance from the most stable form of its elements under standard conditions

  → Elements (most stable form) → substance (one mole) $\Delta H° = \Delta H_f°$

     For example, $C(s) + 2\,H_2(g) + \frac{1}{2}O_2(g) \rightarrow CH_3OH(l)$ $\Delta H_f° = -238.86$ kJ.mol^{-1}

  → Values of $\Delta H_f°$ at 298.15 K for many substances are listed in **Appendix 2A**.

  → Most stable form of the elements at 1 bar (standard state) and 298.15 K (thermodynamic temperature convention):

     Metals and semimetals are solids with atoms arranged in a lattice (see **Topic 3H.1**). The exception is Hg, which is a liquid at room temperature.

     Nonmetal solids are B(s), C(s, graphite), P(s, white), S(s, rhombic), Se(s, black), and $I_2(s)$.

     Monatomic gases are He(g), Ne(g), Ar(g), Kr(g), Xe(g), and Rn(g); *Group 18.*

     Diatomic gases are $H_2(g)$, $N_2(g)$, $O_2(g)$, $F_2(g)$, and $Cl_2(g)$.

     Nonmetal liquid is $Br_2(l)$.

  Note: P(s, red) is actually more stable than P(s, white), but the white form is chosen because it is easier to obtain pure. Also, some of the nonmetal solids have molecular forms, for example, P_4 and S_8.

- **Standard enthalpy of reaction, $\Delta H°$**

  → Use the most stable form of the elements as a reference state to obtain enthalpy changes for any reaction if $\Delta H_f°$ values are known for each reactant and product.

  → $$\boxed{\Delta H° = \sum n\Delta H_f°(\text{products}) - \sum n\Delta H_f°(\text{reactants})}$$

     In this expression, values of n are the stoichiometric coefficients, and the symbol Σ (sigma) means a summation.

  → Procedure to obtain $\Delta H°$: (1) convert the reactants into the most stable form of their elements $[-\sum n\Delta H_f°(\text{reactants})]$; and (2) recombine the elements into products $[+\sum n\Delta H_f°(\text{products})]$. This is an application of Hess's law. Use $\Delta H_f°$ values in **Appendix 2A** to obtain $\Delta H°$.

4D.6 The Variation of Reaction Enthalpy with Temperature

- **Temperature dependence of reaction enthalpy**

  → $\Delta H°$ must be measured at the temperature of interest.

→ Approximation method is possible from the value of $\Delta H°$ measured at one temperature if heat capacity data for the reactants and products are available.

- **Kirchoff's law**

  → The difference in molar heat capacities of the products and reactants in a chemical reaction is

$$\Delta C_P = \sum nC_{P,\mathrm{m}} (\text{products}) - \sum nC_{P,\mathrm{m}} (\text{reactants})$$

  → An estimation of the reaction enthalpy at a temperature of T_2 can be obtained if the value at a temperature of T_1 and the difference in molar heat capacities are known.

$$\Delta H_2° = \Delta H_1° + \Delta C_P (T_2 - T_1)$$ 　　**Kirchoff's law**

  → A major assumption is that the molar heat capacities are independent of temperature.

  → Kirchoff's law in this form does not account for phase changes, for which separate steps must be added.

  → The difference in molar heat capacities of the products and reactants in a chemical reaction is

$$\Delta C_P = \sum nC_{P,\mathrm{m}} (\text{products}) - \sum nC_{P,\mathrm{m}} (\text{reactants})$$

  → An estimation of the reaction enthalpy at a temperature of T_2 can be obtained if the value at a temperature of T_1 and the difference in molar heat capacities are known.

$$\Delta H_2° = \Delta H_1° + \Delta C_P (T_2 - T_1)$$ 　　**Kirchoff's law**

  → A major assumption is that the molar heat capacities are independent of temperature.

  → Kirchoff's law in this form does not account for phase changes, for which separate steps must be added.

Topic 4E: CONTRIBUTIONS TO ENTHALPY

4E.1 Ion Formation

- **Ionic solids and lattice enthalpy**

  → Cations and anions are arranged in a three-dimensional lattice (see **Topics 2A.3** and **2A.4**) and held in place mainly by coulombic interactions (see **Topics 3H.1** and **3H.6**).

  → Lattice enthalpy: enthalpy required to separate the ions in the solid into gas phase ions, $\Delta H_{\mathrm{L}} = H_{\mathrm{m}}(\text{ions, g}) - H_{\mathrm{m}}(\text{ions, s}) > 0$. See values in **Table 4E.1** in the text.

  → Lattice enthalpy cannot be measured directly.

  → Calculate lattice enthalpy using a Born–Haber cycle.

4E.2 The Born–Haber Cycle

- **Definition of a Born–Haber cycle**

  → Thermodynamic cycle constructed to evaluate the *lattice enthalpy* by using other reactions in which the enthalpy changes are known

  → Steps in the Born–Haber cycle start and end with the most stable form of the elements in amounts appropriate to form one mole of the ionic compound.

    Example: Determine the lattice enthalpy of $CaF_2(s)$. Start with one mole of $Ca(s)$ and one mole of $F_2(g)$. Two moles of F atoms are required.

  (1) *Atomize*: Atomize the metal element to form cations and the nonmetal element to form the anions. **In all cases, ΔH (atomization) > 0.**

    $Ca(s) \rightarrow Ca(g)$ $\Delta H° = \Delta H_f°[Ca(g)] = 178\ kJ·mol^{-1}$

    $F_2(g) \rightarrow 2\ F(g)$ $\Delta H° = 2\Delta H_f°[F(g)] = 2(79\ kJ·mol^{-1}) = 158\ kJ·mol^{-1}$

    ΔH (atomization) = (178 + 158) kJ·mol^{-1} = +336 kJ·mol^{-1}

  (2) *Ionize (cation)*: Ionize the gaseous metal atom to form the gaseous cation. Several steps may be required. Because electrons generated in the formation of cations in the ionization steps are consumed in the formation of anions in the electron-gain steps (3), energy and enthalpy are considered to be interchangeable when treating ionization and electron gain.

    In all cases, ΔH (ionization, cation) > 0.

    $Ca(g) \rightarrow Ca^+(g) + e^-(g)$ $\Delta H° = I_1[Ca] = 590\ kJ·mol^{-1}$

    $Ca^+(g) \rightarrow Ca^{2+}(g) + e^-(g)$ $\Delta H° = I_2[Ca] = 1145\ kJ·mol^{-1}$

    $Ca(g) \rightarrow Ca^{2+}(g) + 2\ e^-(g)$ $\Delta H° = I_1 + I_2 = (590 + 1145)\ kJ·mol^{-1} = 1735\ kJ·mol^{-1}$

    ΔH (ionization, cation) = +1735 kJ·mol^{-1}

  (3) *Ionize (anion)*: Attach electron(s) to the gaseous nonmetal atom to form the gaseous anion. Several electron affinity values may be required. Recall that for electron gain, $\Delta H = -E_{ea}$ for each electron added.

    ΔH (ionization, anion) < or > 0

    $F(g) + e^-(g) \rightarrow F^-(g)$ $\Delta H° = -E_{ea}[F] = -328\ kJ·mol^{-1}$

    $2\ F(g) + 2\ e^-(g) \rightarrow 2\ F^-(g)$ $\Delta H° = -2E_{ea} = -656\ kJ·mol^{-1}$

    ΔH (ionization, anion) = –656 kJ·mol^{-1}

  (4) *Latticize*: Form the lattice of ions in the solid from the gaseous ions. The enthalpy of lattice formation is the negative of the lattice enthalpy, ΔH_L.

    ΔH (lattice formation) < 0

    $Ca^{2+}(g) + 2\ F^-(g) \rightarrow CaF_2(s)$ $\Delta H° = -\Delta H_L$ *value unknown*

    ΔH (lattice formation) = $-\Delta H_L$

(5) *Elementize*: Produce the most stable form of the elements from the ionic solid.
   ΔH (element formation) > 0

   $CaF_2(s) \rightarrow Ca(s) + F_2(g)$ $\Delta H° = -\Delta H_f°[CaF_2(s)] = 1220 \text{ kJ·mol}^{-1}$
   ΔH (element formation) = +1220 kJ·mol^{-1}

(6) *Lattice enthalpy*: The sum of all the enthalpy changes for the complete cycle is 0.
   $0 = \Delta H$ (atomization) + ΔH (ionization, cation) + ΔH (ionization, anion)
   **   + ΔH (lattice formation) + ΔH (element formation)**

   $0 = 336 + 1735 + (-656) + (-\Delta H_L) + 1220 \text{ kJ·mol}^{-1} = 2635 \text{ kJ·mol}^{-1} - \Delta H_L$

   $\Delta H_L = +2635$ kJ·mol^{-1} for $CaF_2(s) \rightarrow Ca^{2+}(g) + 2 F^-(g)$

- **Summary**

  → Strength of interactions between ions in a solid is determined by the lattice enthalpy, which is obtained from a Born–Haber cycle.

4E.3 Bond Enthalpies

- **Bond enthalpy, ΔH_B**

  → Enthalpy change accompanying the breaking of a chemical bond in the gas phase

  → Difference between the standard molar enthalpy of the fragments of a molecule and the molecule itself in the gas phase

  → Reactants and products in their standard states (pure substance at 1 bar) at 298.15 K

  → Bond enthalpies for diatomic molecules are given in **Table 4E.2** in the text.

- **Mean (average) bond enthalpies**

  → In polyatomic molecules, the bond strength between two atoms varies from molecule to molecule.

  → Variations are not very large, and the average or mean values of bond enthalpies are a guide to the strength of a bond in any molecule containing the bond (see **Table 4E.3** in the text).

- **Using mean bond enthalpies**

  → Atoms in the gas phase are the reference states used to estimate enthalpy changes for any gaseous reaction. A value of ΔH_B is required for each bond in the reactant and product molecules.

  → $$\Delta H° \approx \sum_{\text{reactants}} n\Delta H_B \text{(bonds broken)} - \sum_{\text{products}} n\Delta H_B \text{(bonds formed)}$$

  → In this expression **(not in the text)**, values of n are the stoichiometric coefficients and the symbol Σ (sigma) means a summation.

  → The procedure is to convert the reactants into gaseous atoms $[+\sum n\Delta H_B(\text{bonds broken})]$ and then recombine the atoms into products $[-\sum n\Delta H_B(\text{bonds formed})]$. This is an application of Hess's law.

→ Compare bond enthalpy to dissociation energy in **Topics 2D.3** and **2D.4**.

Topic 4F: ENTROPY

4F.1 Spontaneous Change

- **The big question:** What is the *cause* of *spontaneous change*?

- **Spontaneous (or *natural*) change**

  → Occurs *without* an external influence; can be fast or slow

  Examples: Diamond converts to graphite (infinitesimally slow rate). Iron rusts in air (very slow, but noticeable change). $2H_2(g) + O_2(g) \rightarrow 2H_2O(g)$ (very fast reaction or explosion with an external spark).

- **Nonspontaneous change**

  → Can be effected by using an external influence (using energy from the surroundings to do work on the system)

  Examples: Liquid water can be *electrolyzed* to form hydrogen and oxygen gases. Graphite can be converted to diamond under very high pressures. Gases can be compressed by application of an external pressure.

4F.2 Entropy and Disorder

- **Spontaneous changes**

  → Any spontaneous change is accompanied by an *increase* in the *disorder* of the universe (*system* plus the *surroundings*).

- **Entropy (*S*)**

  → A measure of disorder; *increase* in disorder leads to an *increase* in *S*.

- **Second law of thermodynamics**

  → For a spontaneous change, the entropy of an *isolated* system *increases*.

  → For a spontaneous change, the entropy of the *universe increases*.
  (The universe is considered to be a (somewhat large) isolated system.)

- **Isolated system**

  → No exchange of energy or matter with the surroundings

- **Macroscopic definition of entropy, infinitesimal change**

  → $$dS = \frac{dq_{rev}}{T}$$ (See **Topic 4F.3**)

- **Finite change, isothermal process**

$\rightarrow$ $\boxed{\Delta S = \dfrac{q_{\text{rev}}}{T}, \text{ for constant } T \text{ process}}$

  $\rightarrow$ Heat transfer processes are carried out *reversibly* to evaluate ΔS.
  (Temperature of surroundings *equals* that of the system at all points along the path.)

  $\rightarrow$ Reversible path: The system is only *infinitesimally* removed from equilibrium at all points along the path.

- **Entropy is an extensive property**

  $\rightarrow$ proportional to the amount of sample.

- **Entropy is a state function**

  $\rightarrow$ Changing the path of a process does not change ΔS.

  Note: According to the definition of ΔS, we must calculate ΔS by using a *reversible path*. The result then applies to *any* path because ΔS is a state function *independent of path*.

4F.3 Entropy and Volume

- **Entropy increases**

  $\rightarrow$ if a substance is heated (increase in thermal disorder).

  $\rightarrow$ if the volume of a given amount of matter increases (increase in positional disorder).

- **Volume dependence of entropy** (*ideal gas*)

  $\rightarrow$ n = number of moles, R = gas constant ($8.314\,46\ \text{J·K}^{-1}\text{·mol}^{-1}$), V_1 = initial volume, and V_2 = final volume

  $\rightarrow$ Isothermal volume change of an *ideal gas* (constant T) $\boxed{\Delta S = \dfrac{q_{\text{rev}}}{T} = -\dfrac{w_{\text{rev}}}{T} = nR\ln\dfrac{V_2}{V_1}}$

  $\rightarrow$ Recall: $\Delta U = 0$ because the energy of an *ideal* gas depends only on its temperature.

- **Pressure dependence of entropy** (*ideal gas*)

  $\rightarrow$ n = number of moles, P_1 = initial pressure, and P_2 = final pressure

  $\rightarrow$ Isothermal pressure change (*ideal gas*, constant T), $\Delta U = 0$, (Boyle's Law) $P_1V_1 = P_2V_2$, and

$$\boxed{\Delta S = nR\ln\dfrac{V_2}{V_1} = nR\ln\dfrac{P_1}{P_2}}$$

Example: Three moles of an ideal gas expands from 20 L to 80 L at constant T.

$$\Delta S = nR\ln\frac{V_2}{V_1} = (3\text{ mol})(8.314\,46\ \text{J·K}^{-1}\text{·mol}^{-1})\ln\frac{80\text{ L}}{20\text{ L}} = 34.6\ \text{J·K}^{-1}$$

ΔS of the system increases as expected from the volume increase.

Example: Four moles of an ideal gas undergoes a pressure increase from 0.500 atm to 1.75 atm at constant T.

$$\Delta S = nR \ln \frac{P_1}{P_2} = (4 \text{ mol})(8.314\ 46 \text{ J}\cdot\text{K}^{-1}\cdot\text{mol}^{-1}) \ln \frac{0.500 \text{ atm}}{1.75 \text{ atm}} = -41.7 \text{ J}\cdot\text{K}^{-1}$$

ΔS of the system decreases as expected from the volume decrease.

4F.4 Entropy and Temperature

- **Entropy increases**
  - → if a substance is heated (increase in thermal disorder).
  - → if the volume of a given amount of matter increases (increase in positional disorder).

- **Temperature dependence of entropy** (*any substance*)
  - → Let n = number of moles of a substance, C_P = heat capacity ($\text{J}\cdot\text{K}^{-1}$) at constant P, $C_{P,m}$ = molar heat capacity ($\text{J}\cdot\text{K}^{-1}\cdot\text{mol}^{-1}$) at constant P, C_V = heat capacity at constant V, $C_{V,m}$ = molar heat capacity at constant V, T_1 = initial temperature, and T_2 = final temperature.

  - → Isobaric (constant P) heating of a substance
  $$\Delta S = C_P \ln \frac{T_2}{T_1} = nC_{P,m} \ln \frac{T_2}{T_1}$$

  - → Isochoric (constant V) heating of a substance
  $$\Delta S = C_V \ln \frac{T_2}{T_1} = nC_{V,m} \ln \frac{T_2}{T_1}$$

  - → These relationships assume that the heat capacity is *constant* over the range of T_1 to T_2.

    Example: If two moles of Fe are cooled from 300 K to 200 K at constant P, the entropy change is

    $$\Delta S = nC_{P,m} \ln \frac{T_2}{T_1} = (2 \text{ mol})(25.10 \text{ J}\cdot\text{K}^{-1}\cdot\text{mol}^{-1}) \ln \frac{200 \text{ K}}{300 \text{ K}} = -20.3 \text{ J}\cdot\text{K}^{-1}; \ \Delta S < 0.$$

    ΔS of the system *decreases* as expected from the *decrease* in temperature.

  Note: $C_{P,m}$ for iron is taken from **Appendix 2A**.

4F.5 Entropy and Physical State

- **Changes of physical state**
  - → Fusion ≡ fus (solid to liquid); vaporization ≡ vap (liquid to vapor); sublimation ≡ sub (solid to vapor)
  - → Solid-to-solid phase changes; for example, Sn(gray) → Sn(white)
  - → Normal boiling point, T_b: temperature at which a liquid boils when $P = 1$ atm.

- **For these phase changes, at the transition temperatures**
  - → temperature of substance remains *constant* during the phase change.
  - → transfer of heat is *reversible*.
  - → heat supplied is identified with *enthalpy change* because pressure is constant (for these reversible processes at constant P, $q_{rev} = q_P = \Delta H$).

- **Entropy of vaporization**

  → Normal boiling point, T_b: T at which liquid boils when $P = 1$ atm

  → $$\boxed{\Delta S_{vap} = \frac{\Delta H_{vap}}{T_b}}$$ ΔS_{vap} is the *entropy of vaporization* (units: $J \cdot K^{-1} \cdot mol^{-1}$).

  → $\Delta S_{vap}° = $ *standard entropy of vaporization* (liquid and vapor both pure and both at 1 bar)

- **Trouton's rule**

  → For many liquids, $\Delta S_{vap}°$ *(liq)* ≈ 85 $J \cdot K^{-1} \cdot mol^{-1}$

  → **Rationale:** Approximately the same increase in *positional disorder* occurs when *any* liquid is vaporized. (Gas molecules are far apart and moving rapidly.)

  → **Exceptions:** Liquids with very weak or very strong intermolecular interactions (**Examples follow.**) Liquid helium, He (20 $J \cdot K^{-1} \cdot mol^{-1}$); *hydrogen-bonded* liquids such as water, H_2O (109 $J \cdot K^{-1} \cdot mol^{-1}$), and methanol, CH_3OH (105 $J \cdot K^{-1} \cdot mol^{-1}$).

- **Entropy of fusion (melting)**

  → $$\boxed{\Delta S_{fus}° = \frac{\Delta H_{fus}°}{T_f}}$$, where T_f is the *melting* point under standard conditions.

  → $\Delta S_{fus}°$ is the *standard entropy of fusion* (solid and liquid both pure and both at 1 bar).

  → $\Delta S_{fus}°$ is smaller than $\Delta S_{vap}°$ because a liquid is only slightly more *disordered* than its solid, whereas a gas is considerably more disordered than its liquid.

  Example: Methane melts *reversibly* at 1 bar and $-182.5\ °C$: $CH_4\,(s) \rightarrow CH_4\,(l)$. Under these conditions, the enthalpy of fusion is $+0.936$ $kJ \cdot mol^{-1}$. The standard (°) molar entropy of fusion is given by $\Delta S_{fus}° = \dfrac{\Delta H_{fus}°}{T_f} = \dfrac{936\ J \cdot mol^{-1}}{90.65\ K} = 10.3$ $J \cdot K^{-1} \cdot mol^{-1}$.

  Because $\Delta S > 0$ for *melting*, the final state is more disordered than the initial state, as expected.

Topic 4G: THE MOLECULAR INTERPRETATION OF ENTROPY

4G.1 The Boltzmann Formula

- **Absolute value of the entropy**

  → If entropy S is a measure of disorder, then a perfectly ordered state of matter (perfect crystal) *should* have *zero* entropy.

- **Third law of thermodynamics**
  - → Entropies of perfect crystals are the same at $T = 0$ K.
    (Thermal motion *almost* ceases at $T = 0$ K, and by convention $S = 0$ for perfect crystals at 0 K.)

- **Boltzmann approach**
  - → Entropy increases with an increase in the number of ways that molecules or atoms can be arranged in a sample *at the same total energy*.

- **Boltzmann formula**
  - → $\boxed{S = k \ln W}$; k = Boltzmann constant = $1.380\,658 \times 10^{-23}$ J·K^{-1}

    Note: $k = R/N_A$ J·K^{-1} = the gas constant divided by the Avogadro constant, and W = number of *microstates* available to the system at a certain energy

- **Boltzmann entropy**
  - → Also called the statistical entropy

- **Microstate**
  - → *One* permissible arrangement of atoms or molecules in a sample with a given total energy
  - → W = total number of permissible arrangements corresponding to the same total energy, also called an **ensemble**
  - → If only one arrangement is possible, $W = 1$ and $S = 0$

- **Boltzmann interpretation**
  - → Spontaneous change occurs toward more probable states.

- **Scaling S**
  - → For N molecules or atoms, the number of *microstates* is related to the number of molecular *orientations* permissible: $W = (orientations)^N$ and

    $$\boxed{S = k \ln W = k \ln(orientations)^N = N\,k \ln(orientations)}$$

  - → For N_A molecules or atoms, $W = (orientations)^{N_A}$ and

    $$\boxed{S = k \ln(orientations)^{N_A} = N_A k \ln(orientations) = R \ln(orientations)}$$

- **Residual entropy**
  - → For some solids, $S > 0$ at $T = 0$. (Such solids retain some *disorder*.)
    Note: Units of S are usually J·K^{-1}·mol^{-1}.

    Example: The residual entropy of CO(s) at 0 K is 4.6 J·K^{-1}·mol^{-1}. The number of *orientations* of CO molecules in the crystal is calculated as
    $S = R \ln(orientations);\ (orientations) = e^{S/R} = e^{(4.6/8.314\,51)} = e^{(0.553\,25)} = 1.7$.

    This is less than 2, the number expected for a random orientation (*disorder*) of CO molecules at 0 K (see **Figure 4G.2** in the text), suggesting some ordering at 0 K, possibly caused by alignment of the small permanent dipoles of neighboring CO molecules.

4G.2 The Equivalence of Statistical and Thermodynamic Entropies

- Boltzmann's *molecular interpretation* compared with the *thermodynamic approach*

- Qualitative comparison

- Volume increase in an ideal gas

  → **Thermodynamic approach:** $\Delta S = nR \ln \dfrac{V_2}{V_1}$; if $V_2 > V_1$, then $\Delta S > 0$

  → **Molecular approach:** Consider the gas container to be a box. From particle-in-a-box theory, as V increases, energy levels accessible to gas molecules pack more closely together and W increases. (See **Figure 4G.4** in the text for a one-dimensional box. The same is true for a three-dimensional box.)

  So $\Delta S = S_{V_2} - S_{V_1} = k \ln \dfrac{W_2}{W_1} > 0$.

- Temperature increase in an ideal gas

  Thermodynamic approach: $\Delta S = nC_{(V \text{ or } P, \text{m})} \ln \dfrac{T_2}{T_1}$; if $T_2 > T_1$, then $\Delta S > 0$

  Molecular approach: From particle-in-a-box theory, at low T, gas molecules occupy a small number of energy levels (W = small); at higher T, more energy levels are available (W = larger). (See **Figure 4G.5** in the text for a one-dimensional box. The same is true for a three-dimensional box.)

  So $\Delta S = S_{T_2} - S_{T_1} = k \ln \dfrac{W_2}{W_1} > 0$.

- **Qualitative Comparison**

  → Assume that the number of microstates available to any molecule is proportional to the volume available to it: W = constant $\times V$.

  → For N molecules, $W = (\text{constant} \times V)^N$

  → For an expansion of N_A molecules from V_1 to V_2, $\boxed{\Delta S = N_A k \ln \left(\dfrac{\text{constant} \times V_2}{\text{constant} \times V_1} \right) = R \ln \left(\dfrac{V_2}{V_1} \right)}$

  which is *identical* to the thermodynamic expression.

- **Summary**

  → Macroscopic and microscopic approaches give the same predictions for ΔS. The Boltzmann approach provides deep insight into entropy on the molecular level in terms of the energy states available to a system.

Topic 4H: ABSOLUTE ENTROPIES

4H.1 Standard Molar Entropies

- **Use of different formalisms**
  - → Use Boltzmann formalism to *calculate* entropy (sometimes difficult to do).
  - → Use thermodynamic formalism to *measure* entropy.

- **Standard ($P = 1$ bar) molar entropies of *pure* substances**
  - → Determined experimentally from *heat capacity* data ($C_{P,m}$), *enthalpy* data (ΔH) for phase changes, and the third law of thermodynamics
  - → For a gas at final temperature T (assuming only one solid phase and heat capacities independent of temperature)

$$S_m°(T) = S_m°(0) + \int_0^{T_f} \frac{C_{P,m}(\text{solid})dT}{T} + \frac{\Delta H_{fus}°}{T_f} + \int_{T_f}^{T_b} \frac{C_{P,m}(\text{liquid})dT}{T} + \frac{\Delta H_{vap}°}{T_b} + \int_{T_b}^{T} \frac{C_{P,m}(\text{gas})dT}{T}$$

  - → If we assume that $S_m(0) = 0$ according to the third law (no residual entropy), then

$$S_m°(T) = \int_0^{T_f} \frac{C_{P,m}(\text{solid})dT}{T} + \frac{\Delta H_{fus}°}{T_f} + \int_{T_f}^{T_b} \frac{C_{P,m}(\text{liquid})dT}{T} + \frac{\Delta H_{vap}°}{T_b} + \int_{T_b}^{T} \frac{C_{P,m}(\text{gas})dT}{T}$$

  Note: For solids, only the first term in this equation is used if only *one* solid phase exists. For liquids, the first three terms are used. For gases, all five terms are required.

  Example: Plots of $C_{P,m}$ as a function of T and $C_{P,m}/T$ as a function of T for solid copper, Cu(s), are shown next. In the second graph, the area under the curve ($T = 0$ to $T = 298.15$ K) yields the experimental value of the entropy for Cu(s): For the pure substance at 1 bar, $S_m°$ (298.15 K) = 33.15 J·K^{-1}·mol^{-1}.

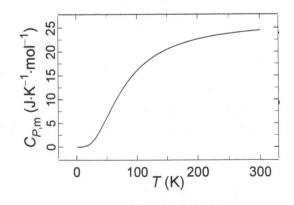
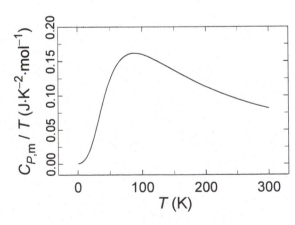

- **Appendix 2A**
  - → Lists of values of *experimental* standard molar entropies at 25 °C for many substances (*standard* ≡ *pure* substance at 1 bar)
  - → Entropies of gases tend to be larger than those of solids or liquids ($S_{gas} > S_{liquid}$ *or* S_{solid}), as expected from the association of entropy with *randomness* and *disorder*.
  - → Other things equal, molar entropy increases with molar mass. Heavier species have more energy levels available to them than lighter ones, so W and S are larger. (See **Figure 4H.3** in the text for use of particle-in-a-box model to understand this better.)

4H.2 Standard Reaction Entropies

- **Standard reaction entropies, $\Delta S°$**
  - → Determined similarly to calculation of the standard enthalpy of reaction, $\Delta H°$, for a chemical reaction

- **For any chemical reaction**
  - →
  $$\Delta S° = \sum n S_m °(\text{products}) - \sum n S_m °(\text{reactants})$$

    Example: Calculate $\Delta S°$ for the reaction $2\,\text{Li(s)} + \text{Cl}_2\text{(g)} \rightarrow 2\,\text{LiCl(s)}$.
    Using $S_m°$ values from Appendix 2A, we obtain
    $$\Delta S° = \sum n S_m °(\text{products}) - \sum n S_m °(\text{reactants})$$
    $$= 2\,S_m°\,(\text{LiCl}) - [2\,S_m°\,(\text{Li}) + S_m°\,(\text{Cl}_2)]$$
    $$= 2(59.33) - [2(29.12) + 222.96] = -162.54 \text{ J·K}^{-1}\text{·mol}^{-1}.$$

    $\Delta S° < 0$, as expected for a net decrease in the number of moles of gas when reactants form products.

- **Generalizations**
  - → In the surroundings, for an exothermic reaction, $\Delta S_{surr} > 0$; for an endothermic reaction, $\Delta S_{surr} < 0$.
  - → For reaction entropy, the entropy of gases dominates changes.
  - → $\Delta S°$ is *positive* if there is a net *production* of gas in a reaction.
  - → $\Delta S°$ is *negative* if there is a net *consumption* of gas in a reaction.

- **Properties of S and ΔS**
  - → S is an *extensive* property of state, like enthalpy (H) and ΔH.

- **For any process or change in state**

  → *Reverse* the process; *change* the sign of ΔS.

  → *Change* the amounts of all materials in a process; make a *proportional change* in the value of ΔS.

  → Add *two* reactions together to get a *third* one; *add* the ΔS values for the first *two* reactions to get ΔS for the *third* reaction.

Topic 4I: GLOBAL CHANGES IN ENTROPY

4I.1 The Surroundings

- **Second law**

  → For *spontaneous* change, the entropy of an *isolated* system *increases.*

- ***Any* spontaneous change**

  → Entropy of system ***plus*** surroundings *increases.*

  → Criterion for spontaneity *includes* the *surroundings*: $\Delta S_{tot} = \Delta S + \Delta S_{surr}.$

- **Criterion**

  → A process is *spontaneous* as written if $\Delta S_{tot} > 0$.

  → A process is *nonspontaneous* as written if $\Delta S_{tot} < 0$.
    In this instance, the *reverse* process is spontaneous.

  → For a system at *equilibrium*, $\Delta S_{tot} = 0$ (see **Topic 4I.3**).

- **Calculating ΔS_{surr} for a process at constant T and P**

  → $$\boxed{\Delta S_{surr} = \frac{q_{surr}}{T} = -\frac{\Delta H}{T}}$$

  → The heat capacity of the surroundings is vast, so its temperature remains essentially constant.

  → We assume that, at constant pressure, the *heat* associated with the process is transferred *reversibly* to the surroundings, so $\Delta S_{surr} = q_{surr,rev}/T$.

  → For a system at constant P, $q_P = \Delta H_{sys}$

  → So $\Delta S_{surr} = -\Delta H_{sys}/T.$
    Note: The process occurring in the system itself may be *reversible* or *irreversible.*

    Example: The entropy change of the surroundings when 1 mol of $H_2O(l)$ vaporizes at 25 °C
    is

    $$\Delta S_{surr} = \frac{q_{surr}}{T} = -\frac{\Delta H_{vap}}{T} = -\frac{40700\ J\cdot mol^{-1}}{(273.15+25)\,K} = -137\ J\cdot mol^{-1}\cdot K^{-1}.$$

    The entropy of the surroundings decreases while that of the system increases.

4I.2 The Overall Change in Entropy

- **Spontaneity depends on ΔS and ΔS_{surr}.**

  → Four cases arise because ΔS and ΔS_{surr} can *each* be either positive (+) or negative (−).

Case	$[\Delta S_{tot}$	$=$	ΔS	$+$	$\Delta S_{surr}]$	Spontaneity
1	+		+		+	*always* spontaneous
2	?		−		+	spontaneous if $|\Delta S_{surr}| > |\Delta S|$
3	?		+		−	spontaneous if $|\Delta S| > |\Delta S_{surr}|$
4	−		−		−	*never* spontaneous

- **Application to chemical reactions**

  → *Exothermic* reactions ($\Delta S_{surr} > 0$) correspond to cases 1 and 2, that is, *spontaneous* if $\Delta S > 0$ or if $|\Delta S_{surr}| > |\Delta S|$ when $\Delta S < 0$.

  → *Endothermic* reactions ($\Delta S_{surr} < 0$) correspond to cases 3 and 4, that is, *spontaneous* only if $\Delta S > 0$ *and* $|\Delta S| > |\Delta S_{surr}|$.

- **For a given change in state (same ΔS), the *path* can affect ΔS_{surr}, ΔS_{tot}, and *spontaneity*.**

  → Comparison of a *reversible* and an *irreversible* isothermal expansion of an *ideal gas* for the same initial and final states (see **Example 4I.3** in the text):

  (1) ΔS (*irreversible*) $= \Delta S$ (*reversible*) (entropy being a state function)

  (2) $\Delta U = q + w = 0$ and $\Delta H = 0$ because energy and enthalpy of an ideal gas depend only on T, and temperature is constant for this process ($\Delta T = 0$).

  (3) q and w are different for the two processes (q and w are *path* functions). Heat given off to the surroundings ($-q = q_{surr}$) is different for the two processes.

  (4) A *reversible* process does the *maximum* work w (see **Figure 4A.6** in the text and **Topic 4A**). Because $q + w = 0$, a *reversible* process delivers *more* heat to the surroundings.

  (5) Because $\Delta S_{surr} = q_{surr} / T$, ΔS_{surr} (*irreversible*) $< \Delta S_{surr}$ (*reversible*).

  (6) Thus, ΔS_{tot} (*irreversible*) $> \Delta S_{tot}$ (*reversible*) $= 0$.

  (7) The *reversible* process corresponds to one in which the system is only *infinitesimally* removed from equilibrium as the process proceeds, whereas for the *irreversible* process $[\Delta S_{tot}(irreversible) > 0]$ is *spontaneous*.

- **Clausius inequality and the second law**

  → From point (6), we can write in general that $\Delta S \geq q/T$. This is the Clausius inequality.

  → For an isolated system, $q = 0$ and $\Delta S \geq 0$.

- **Summary: For an isolated system, ΔS cannot decrease (another statement of the second law).**

4I.3 Equilibrium

- **System at equilibrium** → No tendency to change in forward or reverse direction without an external
  influence

- **Types of equilibrium**

  Thermal: No tendency for heat to flow into or out of the system

  Example: An aluminum rod at room temperature

  Mechanical: No tendency for any part of a system to move

  Example: An undeformed spring with no tendency to stretch or compress

  Physical: Two phases of a substance at the transition temperature with no tendency for
  either phase to increase in mass

  Example: Steam and water at the normal boiling point (100 °C)

  Chemical: A mixture of reactants and products with no *net* tendency for the formation of either
  more reactants or more products

  Example: A saturated solution of sucrose in water in contact with solid sucrose

- **Universal thermodynamic criterion for equilibrium**

  → $\Delta S_{tot} = 0$ for any system at equilibrium.

  If this were not true, ΔS_{tot} would be greater than 0 in either the forward or the reverse direction,
  and *spontaneous change* would occur until $\Delta S_{tot} = 0$.

  → Total entropy change ($\Delta S_{tot} = \Delta S + \Delta S_{surr}$) may be calculated to determine whether a system is
  at equilibrium.

Topic 4J: GIBBS FREE ENERGY

4J.1 Focusing on the System

- **ΔG, an alternative criterion for *spontaneity***

  (1) $\Delta S_{tot} = \Delta S + \Delta S_{surr}$ (always true)

  (2) If P and T are constant, $\Delta S_{surr} = -\dfrac{\Delta H}{T}$.

  (3) If P and T are constant, $\Delta S_{tot} = \Delta S - \dfrac{\Delta H}{T}$ (function of *system only*).

  (4) Definition of Gibbs free energy: $G \equiv H - TS$

  (5) If T is constant, $\Delta G = \Delta H - \Delta(TS) = \Delta H - T\Delta S$.

  (6) According to (5) and (3), $\Delta G = -T\Delta S_{tot}$.

- **Summary:** If a process at constant T and P is
  - → *spontaneous* ($\Delta S_{tot} > 0$), then $\Delta G < 0$.
  - → *nonspontaneous* ($\Delta S_{tot} < 0$), then $\Delta G > 0$.
  - → at *equilibrium* ($\Delta S_{tot} = 0$), then $\Delta G = 0$ as well.

- **Dependence of spontaneity on ΔS and ΔH** for **a system at constant T and P**
  - → Four cases arise because *both* ΔS and ΔH can be *either* positive (+) *or* negative (−).
  - → Spontaneity depends on the sign and magnitude of ΔH and ΔS, as well as the temperature T.

Case	[ΔG	=	ΔH	−	$T\Delta S$]	Spontaneity
1	−		−		+	*always* spontaneous
2	?		−		−	spontaneous, if $\lvert\Delta H\rvert > T\lvert\Delta S\rvert$
3	?		+		+	spontaneous, if $T\lvert\Delta S\rvert > \lvert\Delta H\rvert$
4	+		+		−	*never* spontaneous

Case 1

- → *Exothermic* process or reaction ($\Delta H < 0$) with $\Delta S > 0$
- → *Enthalpy* and *entropy* favor *spontaneity*.
  Enthalpy- and *entropy*-driven process or reaction
- → *Temperature* dependence: If ΔH and ΔS are T independent, the reaction is *spontaneous* at all T.

Case 2

- → *Exothermic* process or reaction ($\Delta H < 0$) with $\Delta S < 0$
- → *Only enthalpy* favors *spontaneity*.
  If the process is spontaneous, it is *enthalpy*-driven.
- → *Temperature* dependence: If ΔH and ΔS are T independent, the reaction is *spontaneous* at low T (enthalpy "wins") but *nonspontaneous* at high T. A crossover temperature exists at which $\Delta H = T\Delta S$ and the system is at equilibrium.

Case 3

- → *Endothermic* process or reaction ($\Delta H > 0$) with $\Delta S > 0$
- → *Only entropy* favors *spontaneity*.
  If the process is spontaneous, it is driven by *entropy*.
- → *Temperature* dependence: If ΔH and ΔS are independent of T, the reaction is *nonspontaneous* at low T and *spontaneous* at high T (entropy "wins"). A crossover temperature exists at which $\Delta H = T\Delta S$ and the system is at equilibrium.

Case 4

- → *Endothermic* process or reaction ($\Delta H > 0$) with $\Delta S < 0$
- → *Nonspontaneous* process or reaction ($\Delta G > 0$)

→ *Temperature* dependence: If ΔH and ΔS are independent of T, then the reaction is *nonspontaneous* at all T.

- **Temperature dependence of *G* for a *pure substance***
  → $G = H - TS$
  → The free energy of a substance decreases with increasing T.
  → But, for a given substance, $S_{m,solid} < S_{m,liq} < S_{m,gas}$.
  → So $G_{m,solid}$ decreases more slowly than $G_{m,liq}$, which decreases more slowly than $G_{m,gas}$.
  → These principles form the basis for a thermodynamic understanding of melting, vaporization, and sublimation (see **Figures 4J.3** and **4J.4** in the text).

4J.2 Gibbs Free Energy of Reaction

- **Chemical reactions** → ΔG and $\Delta G°$ are defined in a manner similar to the reaction enthalpy, ΔH, and standard reaction enthalpy, $\Delta H°$.

- **Definitions (*n* = stoichiometric coefficient)**
  → If G_m is the molar Gibbs free energy of a reactant or product, then the Gibbs free energy of reaction is

  $$\boxed{\Delta G = \sum nG_m(\text{products}) - \sum nG_m(\text{reactants})}$$

  → If $G_m°$ is the *standard* molar Gibbs free energy, the standard Gibbs free energy of reaction is

  $$\boxed{\Delta G° = \sum nG_m°(\text{products}) - \sum nG_m°(\text{reactants})}$$

  → Values of G_m or $G_m°$ cannot be determined directly. So these equations *cannot* be used to determine ΔG and $\Delta G°$.
  → Values of ΔG and $\Delta G°$ are determined from the Gibbs free energies of *formation*, ΔG_f and $\Delta G_f°$.

- **Standard Gibbs free energy of formation, $\Delta G_f°$**
  → Standard Gibbs free energy of *formation* of a compound or element
  → Gibbs free energy change for the formation of one mole of a compound from the most stable form of its elements under standard conditions ($P = 1$ bar for each reactant and product)
  → $\Delta G_f° \equiv 0$ for *all elements* in their most stable form (same convention as for enthalpy).

- **Calculating $\Delta G_f°$ for a given compound**
  (1) *Write and balance the formation reaction.*
      One mole of compound on the product side and the most stable form of the elements with appropriate coefficients on the reactant side
  (2) *Calculate $\Delta H_f°$ and $\Delta S_f°$ for the reaction by using the data in **Appendix 2A**.*
      The value of $\Delta S_f°$ is obtained from the standard molar entropy values as follows:
      $\Delta S_f° = S_m°(\text{compound}) - \sum nS_m°(\text{reactants})$.

(3) Solve for the formation of one mole of compound, using the expression

$$\Delta G_f{}^\circ = \Delta H_f{}^\circ - T\Delta S_f{}^\circ$$
$$= \Delta H_f{}^\circ(\text{compound}) - T\left[S_m{}^\circ(\text{compound}) - \sum nS_m{}^\circ(\text{reactants})\right]$$

- **Thermodynamically *stable* compound is characterized by**
  → a *negative* standard free energy of formation ($\Delta G_f{}^\circ < 0$).
  → a *thermodynamic* tendency to *form* from its elements.
    Examples: Ores such as $Al_2O_3(s)$ and $Fe_2O_3(s)$

- **Thermodynamically *unstable* compound is characterized by**
  → a *positive* standard free energy of formation ($\Delta G_f{}^\circ > 0$).
  → a *thermodynamic* tendency to *decompose* into its elements.
    Examples: Some hydrocarbons such as acetylene (C_2H_2) and benzene (C_6H_6)

- **Properties of thermodynamically *unstable* compounds**
  → *Many* decompose into their elements over a *long* time span.
  → *Thermodynamically unstable* substances (like liquid octane and diamond) that decompose into their elements *slowly* are said to be *kinetically stable*.

- **Classification of thermodynamically *unstable* compounds**
  labile: Decompose *or* react readily, for example TNT and NO
  nonlabile: Decompose *or* react slowly, for example liquid octane
  inert: Exhibit virtually no reactivity *or* decomposition, for example diamond

- **Calculating the standard Gibbs free energy of reaction $\Delta G°$ for a given chemical reaction**
  (1) *Write the balanced chemical equation for the reaction.*
  (2) *Calculate $\Delta H°$ and $\Delta S°$ for the reaction, using the data in **Appendix 2A**.*
  (3) *Solve for $\Delta G°$, using the expression*

$$\Delta G° = \Delta H° - T\Delta S°$$
$$= \left[\sum n\Delta H_f{}^\circ(\text{products}) - \sum n\Delta H_f{}^\circ(\text{reactants})\right] - T\left[\sum nS_m{}^\circ(\text{products}) - \sum nS_m{}^\circ(\text{reactants})\right]$$

  (4) *Alternatively, calculate $\Delta G°$ from the $\Delta G_f{}^\circ$ values listed in Appendix 2A.*

$$\Delta G° = \sum n\Delta G_f{}^\circ(\text{products}) - \sum n\Delta G_f{}^\circ(\text{reactants})$$

The equation yields a faster result but does *not* allow for an estimation of the *temperature* dependence of $\Delta G°$.

Example: Calculate $\Delta G°$ at 25 °C for the reaction

$$2\,Al_2O_3(s) + 3\,C(graphite,s) \rightarrow 3\,CO_2(g) + 4\,Al(s).$$

$\Delta G° = \sum n\Delta G_f°(products) - \sum n\Delta G_f°(reactants)$

$\quad = 3\,\Delta G_f°[CO_2(g)] + 4\,\Delta G_f°[Al(s)] - \{2\,\Delta G_f°[Al_2O_3(s)] + 3\,\Delta G_f°[C(graphite,s)]\}$

$\quad = 3(-394.36\;kJ·mol^{-1}) - 2(-1582.3\;kJ·mol^{-1}) = 1981.5\;kJ·mol^{-1}$

Since $\Delta G°$ is positive, the reaction is not expected to be *spontaneous* at 25 °C under standard conditions.

Note: $\Delta G_f°[Al(s)]$ and $\Delta G_f°[C(graphite,s)]$ are zero (elements in their standard states).

4J.3 The Gibbs Free Energy and Nonexpansion Work

- **Definition of free energy**
  - → The maximum nonexpansion work that can be obtained from a process at constant P and T

- **Processes at constant T and P**

  Expansion work (performed against an opposing pressure)
  - → $dw = -PdV$ (*infinitesimal* change in volume; see **Topic 4A.3**)
  - → $w = -P\Delta V$ (*finite* change in volume; constant opposing P)
  - → Expansion work is performed against an opposing pressure.
  - → The sign of w is defined in terms of the *system*:
    If work is done *on* the system *by* the surroundings, $w > 0$ (the system gains energy).
    If work is done *by* the system *on* the surroundings, $w < 0$ (the system loses energy).

  Nonexpansion work
  - → dw_e (*infinitesimal* change, subscript "e" stands for extra)
  - → w_e (*finite* change)
  - → Any other type of work, including electrical work, mechanical work, muscular contraction, neuronal signaling, and chemical synthesis (making chemical bonds)

- **Relationship between nonexpansion work and (finite) Gibbs free energy changes**
  - → For a reversible process, $\Delta G = w_{rev,e}$ (constant P and T)
  - → $w_{rev,e} = maximum$ nonexpansion work obtainable from a constant P, T process
  - → *Reversible* process: maximum amount of work is done by system on the surroundings ($w_{rev,e}$ is negative).
  - → All real systems are *irreversible* and the maximum nonexpansion work is never obtained.

Example: The maximum electrical work obtainable at 1 bar and 25 °C by burning one mole of propane in a fuel cell with *excess* oxygen according to the equation

$$C_3H_8(g) + 5\,O_2(g) \rightarrow 3\,CO_2(g) + 4\,H_2O(l) \text{ is given by}$$

$$w_{rev,e} = \Delta G° = \sum n\Delta G_f°(\text{products}) - \sum n\Delta G_f°(\text{reactants})$$

$$= 3\,\Delta G_f°[CO_2(g)] + 4\,\Delta G_f°[H_2O(l)] - \{\Delta G_f°[C_3H_8(g)] + 5\,\Delta G_f°[O_2(g)]\}$$

$$= 3(-394.36 \text{ kJ·mol}^{-1}) + 4(-237.13 \text{ kJ·mol}^{-1}) - (-23.49 \text{ kJ·mol}^{-1})$$

$$= -2108.11 \text{ kJ for 1 mol of } C_3H_8(g)$$

The negative value for $w_{rev,e}$ indicates work done on the surroundings.

4J.4 The Effect of Temperature

- **For any chemical reaction at constant P and T**

$$\boxed{\Delta G° = \Delta H° - T\Delta S° = \left[\sum n\Delta H_f°(\text{prod}) - \sum n\Delta H_f°(\text{react})\right] - T\left[\sum nS_m°(\text{prod}) - \sum nS_m°(\text{react})\right]}$$

→ Cases 1−4 from the table in **Topic 4J.1** of this study guide apply to any chemical reaction, assuming that $\Delta H°$ and $\Delta S°$ are independent of temperature.

→ In this case, the temperature dependence of $\Delta G°$ arises from the $(-T\Delta S°)$ term.

Example: To calculate the temperature range over which the reaction $H_2(g) + \frac{1}{2}O_2(g) \rightarrow H_2O(g)$ is spontaneous, use $\Delta G° = \Delta H° - T\Delta S°$. Set $\Delta G° = 0$ to determine the crossover temperature, at which point neither the forward nor reverse reaction is favored.

$$T_{crossover} = \frac{\Delta H°}{\Delta S°}$$

$$= \frac{\Delta H_f°\,[H_2O(g)]}{S_m°\,[H_2O(g)] - \{S_m°\,[H_2(g)] + (1/2)S_m°\,[O_2(g)]\}}$$

$$T_{crossover} = \frac{\Delta H°}{\Delta S°}$$

$$= \frac{-241.82 \text{ kJ·mol}^{-1}}{0.188\,83 \text{ kJ·K}^{-1}\cdot\text{mol}^{-1} - \{0.130\,68 + (1/2)(0.205\,14)\}\text{kJ·K}^{-1}\cdot\text{mol}^{-1}}$$

$$= \frac{-241.82 \text{ kJ·mol}^{-1}}{-0.044\,42 \text{ kJ·K}^{-1}\cdot\text{mol}^{-1}} = 5444 \text{ K}$$

In this case, enthalpy favors *spontaneity*; entropy does not. At 298.15 K, the reaction is spontaneous $[\Delta G° = \Delta H° - T\Delta S° = -241.82 - (298.15)(-0.044\,42)$ $= -228.58 \text{ kJ·mol}^{-1}]$. It is spontaneous until $T = 5444$ K. Above 5444 K, $H_2O(g)$ is expected to decompose *spontaneously* into its elements.

INTERLUDE

Free Energy and Life

- **Metabolism**
  - → In metabolic processes, reaction steps may have a *positive* (unfavorable) free energy of reaction. They can be coupled to spontaneous reactions to make the overall (or net) reaction spontaneous.

- **Concept of coupled chemical reactions**
  - → A way to make *nonspontaneous* reactions occur without changing temperature or pressure
  - → A prominent process in biological systems also found in nonbiological ones
  - → If reaction 1 has a *positive* reaction free energy $[\Delta G°(1) > 0]$ (*nonspontaneous*), it can be coupled to reaction 2 with a *more negative* reaction free energy $[\Delta G°(2) < 0]$, such that reaction 3, the sum of the two reactions, has a *negative* reaction free energy $[\Delta G°(3) < 0]$:

    $\Delta G°(3) = \Delta G°(1) + \Delta G°(2) < 0$ (overall reaction is *spontaneous*).

    Example: The sugar glucose (a food) is converted to pyruvate in a series of steps. The overall process may be represented as

    glucose + other reactants → 2 pyruvate + other products; $\Delta G° = -80.6 \text{ kJ·mol}^{-1}$

    The relatively large *negative* value of $\Delta G°$ indicates that the overall reaction occurs spontaneously under *biochemical* standard conditions. The *biochemical standard state* corresponds more closely to typical conditions in a cellular environment. The first step in the metabolic process is the conversion of glucose into glucose-6-phosphate:

glucose + phosphate → glucose-6-phosphate + H_2O;   (1) $\Delta G°(1) = +14.3$ kJ·mol^{-1}

Glucose   Glucose-6-phosphate

The *positive* standard free energy implies *nonspontaneity* for reaction 1. To generate glucose-6-phosphate, reaction 1 is coupled with 2, the hydrolysis of adenosine triphosphate (ATP) to yield adenosine diphosphate (ADP). Reaction 2 has a favorable $\Delta G° < 0$ value (*spontaneous*):

ATP + H_2O → ADP + inorganic phosphate;   (2) $\Delta G°(2) = -31.0$ kJ·mol^{-1}.

Adding (coupling) reactions 1 and 2 gives a net spontaneous reaction 3:

glucose + ATP → glucose-6-phosphate + ADP (3)

$\Delta G°(3) = \Delta G°(1) + \Delta G°(2) = -16.7$ kJ·mol^{-1}

Note: Reactions in metabolic pathways are *catalyzed* by *enzymes*. Step 1 is catalyzed by the enzyme hexokinase. Biological catalysts are discussed in **Topic 7E** and sugars, a class of carbohydrates, are discussed in **Focus 11**.

Focus 5: EQUILIBRIUM

Topic 5A: VAPOR RESSURE

5A.1 The Origin of Vapor Pressure

- **Phase**
  - → A specific physical state of matter such as a solid, liquid, or gas

- **Liquids**
  - → Evaporate to form a gas or vapor
    - **Example:** Puddle of rainwater evaporates. The reaction $H_2O(l) \rightarrow H_2O(g)$ goes to completion and the puddle disappears. Phase equilibrium is *not* attained here.

- **Gases**
  - → Condense to form a liquid or solid at sufficiently low temperature
    - **Examples:** Water vapor condenses to form rain. Steam condenses on cold surfaces.

- **Equilibrium**
  - → In a closed system, both liquid and gas phases exist in *equilibrium*.
  - → Represent equilibrium system using double-headed arrow: $H_2O(l) \rightleftharpoons H_2O(g)$

- **Liquid–gas phase equilibrium**
  - → *Dynamic equilibrium*
    Rate of evaporation equals rate of condensation.
    - **Example:** In a *covered* jar half-filled with water at room temperature, the liquid and gas phases are in equilibrium, $H_2O(l) \rightleftharpoons H_2O(g)$. Water vapor exhibits a characteristic equilibrium pressure, its *vapor pressure*.

- **Vapor pressure, P**
  - → Characteristic pressure of a vapor above a *confined* liquid or solid in dynamic equilibrium (*closed system*)
  - → Depends on temperature; increases rapidly with increasing temperature, T
  - → Different vapor pressure for different liquids and solids
    - **Examples:** Several solids with measurable vapor pressures are camphor, naphthalene, *para*-dichlorobenzene (mothballs), and dry ice [$CO_2(s)$].
      Virtually all liquids have measurable vapor pressures, but some (for example, $Hg(l)$) may be very low at room temperature.

- **Sublimation**
  - → Evaporation of solid to form a gas **Example:** [$CO_2(s) \rightarrow CO_2(g)$]

- **Volatility**
  - → Related to the ability to evaporate
  - → Liquids with high vapor pressures at a given temperature are *volatile* substances.

- **Characteristics of equilibrium between phases**
  - → *Dynamic equilibrium* occurs when molecules enter and leave the individual phases at the same rate and the total amount in each phase remains unchanged.
  - → At equilibrium, the molar free energies of the individual phases are equal.

    substance (phase 1) $\rightleftharpoons$ substance (phase 2), $\Delta G_m = 0$

5A.2 Volatility and Intermolecular Forces

- *Increasing* **the strength of intermolecular forces in a liquid**
  - → *decreases* the volatility.
  - → *decreases* the vapor pressure at a given temperature.
  - → *increases* the normal boiling point, T_b (defined in **Figure 5A.3** in the text and in **Topic 5A.4**).

- **London forces**
  - → More massive molecules are less volatile.
  - → More electrons in a molecule produce stronger intermolecular forces.

- **Dipolar forces**
  - → For molecules with the same number of electrons, dipolar molecules tend to be less volatile than substances with only London forces.
  - → A *greater* dipole moment yields a *lower* volatility.

- **Hydrogen bonding**
  - → Molecules that form H-bonds may produce even less-volatile substances than dipolar molecules.

- **Ionic forces**
  - → Salts are essentially nonvolatile.

  Note: As with many rules in chemistry, these qualitative guidelines must be applied cautiously.

- **Quantitative approach to vapor pressure (see derivation in Topic 5A.3 in the text)**
  - → $\ln(P/P^\circ) = \dfrac{-\Delta G_{vap}^{\circ}}{RT}$ *and* $\Delta G_{vap}^{\circ} = \Delta H_{vap}^{\circ} - T\Delta S_{vap}^{\circ}$ *lead to*

$$\ln\left(P/P^\circ\right) = \frac{-\Delta H_{vap}^{\circ}}{RT} + \frac{\Delta S_{vap}^{\circ}}{R}$$

- **Insight from the equation**
  - → Because ΔS_{vap}° is about the same for *all* liquids (Trouton's rule), the vapor pressure of a liquid depends mainly on ΔH_{vap}°, which is always a positive quantity.
  - → *Stronger* intermolecular forces result in a *larger* ΔH_{vap}° and a *decrease* in vapor pressure.
  - → For a given liquid, an *increase* in temperature results in an *increase* in vapor pressure.

5A.3 The Variation of Vapor Pressure with Temperature

• **Solve the vapor pressure equation for two temperatures** (P_2 at T_2 and P_1 at T_1)

→
$$\ln \frac{P_2}{P_1} = -\frac{\Delta H_{vap}°}{R}\left(\frac{1}{T_2} - \frac{1}{T_1}\right)$$
Clausius–Clapeyron equation

• **Uses of the Clausius-Clapeyron equation**

  → Measure the vapor pressure of a liquid at two temperatures (P_2 at T_2 and P_1 at T_1) to *estimate* the standard enthalpy of vaporization $\Delta H_{vap}°$.

$$\Delta H_{vap}° \approx \left(\frac{R}{\left(T_1^{-1} - T_2^{-1}\right)}\right) \ln \frac{P_2}{P_1}$$

  → Measure $\Delta H_{vap}°$ and the vapor pressure of a liquid at one temperature (P_1 at T_1) to *estimate* the vapor pressure *at any different temperature* (P_2 at T_2).

  → Measure $\Delta H_{vap}°$ and the vapor pressure at one temperature (P_1 at T_1). *Estimate* the normal boiling point of the liquid ($T_b = T_2$, when $P_2 = 1$ atm) or the standard boiling point of the liquid (T_2, when $P_2 = 1$ bar).

  → Determine $\Delta H_{vap}°$ and the vapor pressure at the normal boiling temperature (P_1 at T_b). *Estimate* the vapor pressure at a different temperature (P_2 at T_2).

  Note: In all cases, the term *estimate* is used because the assumption that $\Delta H_{vap}°$ is independent of temperature is not strictly true; it is an approximation.

  → Values for the vapor pressure of water at selected temperatures is given in **Table 5A.**2.

5A.4 Boiling

• **Liquid in an open container**

  → Rate of vaporization is greater than the rate of condensation, so the liquid evaporates.

• **Boiling**

  → Vapor pressure of liquid equals atmospheric pressure.
    Rapid vaporization occurs throughout the entire liquid; bubbles form.

• **Boiling point**

  → Temperature at which the liquid begins to boil

• **Normal boiling point**

  → Boiling point at 1 atm, T_b
    Within experimental error, $T_b = 99.974$ °C ≈ 100 °C for water.

• **Standard boiling point**

  → Boiling point at 1 bar

- **Effect of pressure on the boiling point**

  → An *increase* of pressure on a liquid leads to an *increase* in the boiling point. Pressure cookers make use of this fact.

  → A *decrease* of pressure on a liquid leads to a *decrease* in the boiling point. This helps explain why water boils at a lower temperature on a mountaintop than at sea level. It also accounts for vacuum distillation at low temperatures to purify liquids that decompose at higher temperatures.

Topic 5B: PHASE EQUILIBRIA IN ONE-COMPONENT SYSTEMS

5B.1 One-Component Phase Diagrams

- **Component**

  → A single substance **Examples:** Aluminum, octane, water, sodium chloride

  → A chemically independent species

- **Single-component phase diagram**

  → Map showing the most stable phase of a substance at different pressures and temperatures

- **Phase boundary**

  → Lines or curves separating regions on a phase diagram

  → Represents a set of P and T values for which *two* phases coexist in dynamic equilibrium

- **Triple point**

  → Point where *three* phase boundaries intersect

  → Corresponds to a single value of P and T for which *three* phases coexist in dynamic equilibrium (water, ice, and water vapor for H_2O).

- **Critical point**

  → High-temperature terminus of the liquid-vapor phase boundary

- **Critical temperature, T_c**

  → Temperature above which a gas cannot condense into a liquid
  Only one phase above T_c is observed.

  Note: Critical point and critical temperature are discussed in **Topic 5B.2**. These definitions are repeated later.

Phase Diagram of Water

Note the regions (bounded areas) labeled ice (solid), liquid, and vapor (gas). In each area, only a single phase (ice, water, or water vapor) is stable (points A (vapor) and B (liquid)). Within each region, pressure and/or temperature may vary independently with no accompanying phase transition. Two independent variables, or *two* degrees of freedom, characterize such a region. The three lines that separate the regions define the phase boundaries of solid–liquid, liquid– vapor, and solid–vapor. At a boundary, only *P* or *T* may be varied if the two phases are to remain in equilibrium. Only one independent variable, or *one* degree of freedom, exists for water in states corresponding to the boundary (point C). All three phases coexist at the *triple point* with its specific *P* and *T*, or *zero* degrees of freedom.

- **Slope of the solid-liquid phase boundary line**

  → Positive for most substances
  Solid sinks because it is denser than liquid.

  At constant temperature, a pressure increase yields no phase change for the solid.

  At constant temperature, a pressure increase may cause the liquid to solidify.

  → Negative for H₂O
  Because the solid is less dense than the liquid, ice floats.

  At constant temperature, a pressure increase may cause ice to melt.

  At constant temperature, a pressure increase yields no phase change for the liquid.

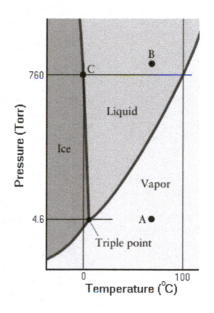

- **Increasing the pressure on graphite to make diamond and liquid carbon**

  → What occurs when the pressure on graphite at 3000 K is increased from 1 bar to 500 kbar?

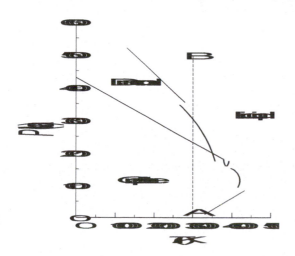

In the initial state of graphite (point A), one phase exists. Temperature is held constant (3000 K), and the pressure is allowed to vary freely. The C(graphite)-C(diamond) phase boundary is reached at approximately 230 kbar. At this point, C(diamond) forms and exists in equilibrium with C(graphite). All the C(graphite) is eventually converted to C(diamond). (This process may be quite slow.) In the diamond region, one phase exists. Additional increases in the pressure on C(diamond) result in the equilibrium of C(diamond) and liquid carbon at about 400 kbar. When all the diamond has melted, the pressure on C(l) is free to increase to attain the final state (point B).

Note: The allotropes of carbon include graphite, diamond, and buckminsterfullerene (C_{60}); the last, discovered in 1985, is composed of soccer-ball-shaped molecules. The thermodynamic stability of buckminsterfullerene has not yet been determined, and the validity of its inclusion on the C phase diagram is, therefore, uncertain. (Metastable phases, such as supercooled water, do not appear on phase diagrams.) The crystal structure is face-centered cubic, with C_{60} molecules at the corners and faces of a cubic unit cell. The unit cell is shown here:

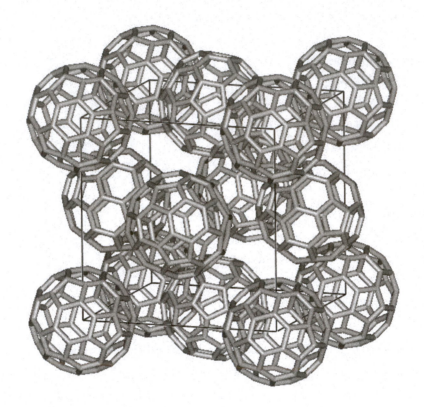

5B.2 Critical Properties

- **Critical point**

  → Terminus of the liquid—gas phase boundary at high temperature

- **Critical temperature**

  → The temperature above which a vapor cannot condense to a liquid. Only one phase, the gas phase, is observed above T_c.

- **Supercritical fluid**

  → Substance above its critical temperature, T_c

Phase Diagram of Water

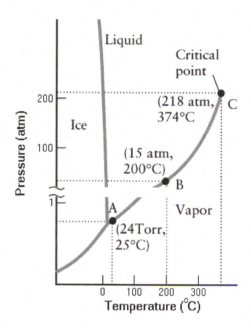

This discontinuous diagram shows both low- and high-pressure regions. Point A represents typical "room temperature" conditions. Liquid water has a vapor pressure of 23.76 Torr (0.031 26 atm) at 25 °C (298.15 K). At 200 °C (473.15 K), the vapor pressure is 15 atm (point B). As the temperature rises, the densities of the liquid and vapor in equilibrium approach one another and become nearly equal at the critical point (point C). Above the critical temperature, only one phase with the properties of a very dense vapor remains. The *critical pressure*, P_c, is the vapor pressure measured at the critical temperature. The temperature, pressure, and density values for water at the critical point are T_c = 374.1 °C (647 K), P_c = 218.3 atm, and d_c = 0.32 g·cm^{-3}, respectively.

- **Intermolecular forces**

  → Both T_c and T_b have a tendency to *increase* with the *increasing* strength of intermolecular forces.

Substance	T_c (K)	T_b (K)
He (helium)	5.2	4.3
Ar (argon)	150	88
Xe (xenon)	290	166
NH$_3$ (ammonia)	405	240
H$_2$O (water)	647	373

Note: The strength of the London forces *increases* with increasing molar mass (number of electrons) of a noble gas atom. With *stronger* intermolecular forces, *higher* critical and normal boiling temperatures are expected. Ammonia and water form strong hydrogen bonds. Water forms more hydrogen bonds per molecule than does ammonia. Therefore, water has critical and normal boiling temperatures *higher* than those of ammonia. In making these comparisons, care must be taken.

Topic 5C: PHASE EQUILIBRIA IN TWO-COMPONENT SYSTEMS

5C.1 The Vapor Pressure of Mixtures

- **Qualitative features**

  → The vapor pressure of a solvent in equilibrium with a solution containing a *nonvolatile solute* (e.g., sucrose or sodium chloride in water) is lower than that of the pure solvent.

  → For an *ideal solution* or a sufficiently dilute *real* solution, the vapor pressure of any volatile component is proportional to its mole fraction (**Topic 3C**) in solution (Raoult's law in words).

- **Quantitative features**

  → Raoult's law $\boxed{P = x_{\text{solvent}} P_{\text{pure}}}$

  → The vapor pressure P of a solvent is equal to the product of its mole fraction in solution, x_{solvent}, and its vapor pressure when pure, P_{pure}. The quantities P and P_{pure} are measured at the same temperature.

- **Ideal solution**

  → A hypothetical solution that obeys Raoult's law exactly for all concentrations of solute

  → Solute–solvent, solute–solute, and solvent–solvent interactions are all the same; therefore, the enthalpy of solution, ΔH_{sol}, is zero.

  → Entropy of solution, $\Delta S_{\text{sol}} > 0$

  → Free energy of solution, $\Delta G_{\text{sol}} < 0$, leading to a lowering of vapor pressure of the solvent (see **Topic 5A.3**)

  → Formation of solution is entropy driven.

  → Solutions approximating ideality are typically formed by mixtures of similar solute and solvent species, such as hexane and heptane.

  → Real solutions do not obey Raoult's law at all concentrations, but they do follow Raoult's law in the limit of low solute concentration (dilute solution).

  → Raoult's law is a limiting law.

- **Nonideal (real) solution**

  → Solution that does not obey Raoult's law at a certain concentration

  → Solute–solvent interactions differ from solvent–solvent interactions.

  → ΔH_{sol} is not equal to zero.

  → Real solutions *approximate* ideal behavior at concentrations below 10^{-1} mol·kg^{-1} for *nonelectrolyte* solutions and below 10^{-2} mol·kg^{-1} for *electrolyte* solutions.

  → Real solutions tend to behave ideally as the solute concentration approaches zero.

5C.2 Binary Liquid Mixtures

- **Ideal solution with two volatile components A and B; liquid and vapor phases in equilibrium**

  → Let $P_{\text{A,pure}}$ = vapor pressure of pure A and $P_{\text{B,pure}}$ = vapor pressure of pure B.

  → In an ideal solution, each component obeys Raoult's law at all concentrations.

→ $P_A = x_{A,liquid} P_{A,pure}$, where $x_{A,liquid}$ = mole fraction of A in the liquid mixture.

→ $P_B = x_{B,liquid} P_{B,pure}$, where $x_{B,liquid}$ = mole fraction of B in the liquid mixture.

→ For a binary mixture, $x_{A,liquid} + x_{B,liquid} = 1$.

→ P_A and P_B are the partial pressures of A and B, respectively, in the vapor above the solution. P_{total} is the total pressure above the solution. Dalton's law of partial pressures for ideal gases is $P_{total} = P_A + P_B$ (see **Topic 3C.1**).

→ Combination of Raoult's law and Dalton's law produces the expression

$$P_{total} = P_A + P_B = x_{A,liquid} P_{A,pure} + x_{B,liquid} P_{B,pure}$$

→ Nearly ideal solutions with two volatile components are formed when the components are very similar (hexane/octane or benzene/toluene), and both components are completely soluble in each other (miscible). The composition may range from $0 < x < 1$ for a given component; the labels "solute" and "solvent" are not routinely used.

- **Composition of the vapor above a binary ideal solution**

  → Each vapor component obeys Dalton's law.

  → $P_A = x_{A,vapor} P$, where $x_{A,vapor}$ = mole fraction of A in the vapor mixture and $P = P_{total}$.

  → $P_B = x_{B,vapor} P$, where $x_{B,vapor}$ = mole fraction of B in the vapor mixture.

  → For a binary mixture, $x_{A,vapor} + x_{B,vapor} = 1$.

  → Combination of Raoult's law and Dalton's law yields an expression for $x_{A,vapor}$

$$x_{A,vapor} = \frac{P_A}{P} = \frac{P_A}{P_A + P_B} = \frac{x_{A,liquid} P_{A,pure}}{x_{A,liquid} P_{A,pure} + x_{B,liquid} P_{B,pure}}$$

  → If $P_{A,pure} \neq P_{B,pure}$, then $x_{A,vapor} \neq x_{A,liquid}$ and $x_{B,vapor} \neq x_{B,liquid}$.

  The liquid and vapor compositions differ and the vapor is always *richer* in the more volatile component.

5C.3 Distillation

- **Temperature–composition phase diagram (ideal solution)**

  → Plot of the temperature of an equilibrium mixture *vs.* composition (mole fraction of one component)

  → Two curves are plotted on the same diagram, T versus x_{liquid} and T versus x_{vapor}.

  → At a given temperature, the composition of each phase at equilibrium is given.

  → See **Figure 5C.7** in the text for a graphical illustration of a temperature–composition diagram.

- **Distillation**

  → Purification of a liquid by evaporation and condensation

- **Distillate**

  → Vapor produced during distillation that is condensed and collected in the final stage

- **Fractional distillation**
  - → Continuous separation (purification) of two or more liquids by repeated evaporation and condensation in a vertical column
  - → Each step takes place on a fractionating column in small increments.
  - → Liquid and vapor are in equilibrium at each point in the column, but their compositions vary with height, as does the temperature.
  - → At the top of the column, condensed vapor (distillate) is collected in a series of samples or *fractions*; the most volatile fraction is collected first.
  - → Progress may be displayed on a temperature–composition phase diagram.

5C.4 Azeotropes

- **Nonideal solutions with volatile components**
  - → Most mixtures of liquids are not ideal.
  - → For *nonideal solutions* with *volatile* components: Raoult's law is not obeyed: the enthalpy of mixing, $\Delta H_{mix} \neq 0$, and solute–solvent interactions are different from solvent–solvent interactions.

- **Azeotrope**
  - → A solution that, like a pure liquid, distills at a constant temperature without a change in composition
  - → At the azeotrope temperature, $x_{liquid} = x_{vapor}$ for each component

- **Positive deviation from Raoult's law**
  - → Vapor pressure of the mixture is *greater* than the value predicted by Raoult's law.
  - → Enthalpy of mixing is endothermic, $\Delta H_{mix} > 0$.
  - → Solute–solvent interactions are *weaker* than solute–solute interactions.
  - → See **Figure 5C.9(a)** for an illustration of the vapor-pressure behavior of a mixture of ethanol and benzene. To make an ethanol–benzene solution, strong hydrogen bonds in ethanol are broken and replaced by weaker ethanol–benzene London interactions. Weaker interactions *increase* the vapor pressure and *decrease* the boiling temperature.

- **Minimum-boiling azeotrope**
  - → Forms if the minimum boiling temperature is less than that of each pure liquid
  - → Distillation yields azeotrope as the distillate (see **Figure 5C.10** in text).

- **Negative deviation from Raoult's law**
  - → Vapor pressure of the mixture is *smaller* than the value predicted by Raoult's law.
  - → Enthalpy of mixing is exothermic, $\Delta H_{mix} < 0$.
  - → Solute–solvent interactions are *stronger* than solute–solute interactions
  - → See **Figure 5C.9(b)** in the text for an illustration of the vapor-pressure behavior of a mixture of acetone and chloroform. In the acetone–chloroform solution,

interactions similar to hydrogen bonding occur between a lone pair of electrons on the O atom of acetone and the H atom of chloroform. Stronger solute–solvent interactions *decrease* the vapor pressure and *increase* the boiling temperature.

- **Maximum-boiling azeotrope**
  - → Forms if the maximum boiling temperature is greater than that of each of the pure liquids
  - → Distillation yields one pure component as the distillate (see **Figure 5C.11** in the text).

Topic 5D: SOLUBILITY

5D.1 The Limits of Solubility

- **Two-component solution** → Contains one solvent and one solute species
  - **Example:** NaCl (solute) in water (solvent)

- **Interactions** → Solvent–solvent, solute–solute, and solute–solvent

- **Unsaturated solution**
  - → All solute added to solvent dissolves.
  - → Amount of dissolved solute is *less* than the equilibrium amount.

- **Saturated solution**
  - → Solubility limit of solute has been reached, with any additional solute present as a precipitate.
  - → The *equilibrium* amount of solute has been dissolved in the solvent.
  - → Dissolved and undissolved solute molecules are in *dynamic* equilibrium; in other words, the rate of dissolution equals the rate of precipitation.

- **Supersaturated solution**
  - → Under certain conditions, an amount of solute greater than the equilibrium amount (solubility limit) can be dissolved in a solvent.
  - → Because *more* than the equilibrium amount is dissolved, the system is thermodynamically unstable.
  - → A slight disturbance (seed crystal) causes precipitation and rapid return to equilibrium.

- **Solubility limit**
  - → Depends on the nature of both the solute and the solvent

- **Molar solubility**
  - → Molar concentration of a saturated solution of a substance
    Units: (moles of solute) / (liter of solution) or $mol \cdot L^{-1} \equiv M$
    Example: A saturated aqueous solution of AgCl has a molar concentration of 1.33×10^{-5} $mol \cdot L^{-1}$ or 1.33×10^{-5} M at 25 °C.

- **Gram solubility**

  → Mass concentration of a saturated solution of a substance

  Units: (grams of solute) / (liter of solution) or $g \cdot L^{-1}$

    Example: A saturated aqueous solution of AgCl has a mass concentration of 1.91×10^{-3} $g \cdot L^{-1}$ at 25 °C.

- **Molal solubility**

  → Molality of a saturated solution of a substance

  Units: (moles of solute) / (kg of solvent) or $mol \cdot kg^{-1}$

    Example: A saturated aqueous solution of AgCl has a molality of 1.33×10^{-5} $mol \cdot kg^{-1}$ at 25 °C.

    Note: For dilute solutions, the molar and molal solubilities will be essentially equal, as in the examples. This is because 1 L of water has a mass very nearly equal to 1 kg at 25 °C and the mass of the solute is very small compared with that of the solvent. In concentrated solutions, the mass of the solute becomes an appreciable fraction of the mass of the solution, and a liter of solution contains substantially less than 1 kg of water. In such a solution, molality and molarity can differ considerably.

5D.2 The Like-Dissolves-Like Rule

- **Rule**

  → If solute-solute and solvent-solvent intermolecular forces (London, dipole, hydrogen-bonding, or ionic) are similar, *larger* solubilities are predicted. *Lesser* solubilities are expected if these forces are dissimilar. **Or *like dissolves like*.**

- **Use**

  → *Qualitative* guide to predict and understand the solubility of various solute species in different solvents.

    Example: Oil is composed of long-chain hydrocarbons. Oil and water do not mix. Oil molecules are held together by London forces, whereas water associates primarily by hydrogen bonds. The like-dissolves-like rule suggests that a solvent with only cohesive London forces is needed to dissolve oil. Gasoline, a mixture of shorter-chain hydrocarbons such as heptane and octane, is a possible solvent, as is benzene, an unsaturated cyclic hydrocarbon.

    Example: Glucose, $C_6H_{12}O_6$, has five –OH groups capable of forming hydrogen bonds. It is soluble in hydrogen-bonding solvents such as water and insoluble in nonpolar solvents such as hexane.

    Example: Potassium iodide, KI, is an ionic compound. Because both water and ammonia are highly polar molecules, they are expected to be effective in solvating both K^+ and I^- ions. Ethyl alcohol molecules are less polar than water, so KI is expected to be less soluble in ethanol than in water. A similar argument can be used to explain the *slight* solubility of KI in acetone.

- **Hydrophilic**
  - → Water-attracting
- **Hydrophobic**
  - → Water-repelling
- **Soaps**
  - → Long-chain molecules, with hydrophobic and hydrophilic ends, which are soluble in both polar and nonpolar solvents
- **Surfactant**
  - → Surface-active molecule with a hydrophilic head and a hydrophobic tail
    Examples: Soaps and detergents are surfactants.
- **Micelle**
  - → Spherical aggregation of surfactant molecules with hydrophobic ends in the interior and hydrophilic ends on the surface

    Note: The term "hydrophobic" is somewhat of a misnomer. In actuality, an *attraction* between solute and solvent molecules *always* exists. In the case of water, the solute-solvent interaction disrupts the local structure of water, which is dominated by hydrogen bonding. The stronger the solute–H_2O interaction, the greater the likelihood that the local solvent structure will be disrupted by hydrated solute molecules, increasing the solubility of the solute.

5D.3 Pressure and Gas Solubility

- **Pressure dependence of gas solubility**
  Qualitative features
  - → For a gas and liquid in a container, an increase in gas pressure leads to an increase in solubility of the gas in the liquid.
  - → Gas molecules strike the liquid surface and some dissolve. An increase in gas pressure leads to an increase in the number of impacts per unit time, thereby increasing solubility.
  - → In a gas mixture, the solubility of each component depends on its partial pressure because molecules strike the surface independently of one another.

  Quantitative features
  - → Henry's law $\boxed{s = k_H P}$
  - → The solubility, s, of a gas in a liquid is directly proportional to the partial pressure, P, of the gas above the liquid.

→ $s \equiv$ molar solubility (see **Topic 5D.E**)

         $k_H \equiv$ Henry's law constant, a function of temperature , the gas, and the solvent

→   Units: s (mol·L^{-1}), k_H (mol·L^{-1}·atm^{-1}), and P (atm)

     Example: Estimate the molar solubility and gram solubility of O_2 in dry air dissolved in a liter of water open to the atmosphere at 20 °C. Assume that air is 20.95% O_2 by volume. The pressure of the atmosphere is 1 atm, so the partial pressure of O_2 is 0.2095 atm.

$$s\,(O_2) = (1.3 \times 10^{-3} \text{ mol·L}^{-1}\text{·atm}^{-1})(0.2095 \text{ atm}) = 2.7 \times 10^{-4} \text{ mol·L}^{-1}$$

$$s \times M = (2.7 \times 10^{-4} \text{ mol·L}^{-1})(32.0 \text{ g·mol}^{-1}) = 8.6 \times 10^{-3} \text{ g·L}^{-1}$$

5D.4 Temperature and Solubility

- **Dependence of molar solubility on temperature**

  →   For solid and liquid solutes, solubility *usually* increases with increasing temperature, but for some salts, for example Li_2CO_3, solubility decreases with increasing temperature (see **Figure 5D.9** in the text).

  →   For gaseous solutes, solubility *usually* decreases with increasing temperature.

  →   Despite these trends, solubility behavior can sometimes appear complex, as with sodium sulfate, Na_2SO_4, whose solubility in water increases then decreases (see **Figure 5D.9** in the text). In this instance, complex solubility behavior arises because different hydrated forms of sodium sulfate precipitate at different temperatures.

  Examples:

       The solute that precipitates from a saturated solution may be different from that of the solid initially dissolved. At room temperature, anhydrous potassium hydroxide, KOH(s), readily dissolves in water. The solid that precipitates from a saturated solution of KOH is the dihydrate, KOH·2 H_2O. Therefore, the chemical equilibrium in the saturated solution is

  $$\text{KOH·2 } H_2O(s) \rightleftharpoons K^+(aq, \text{ saturated}) + OH^-(aq, \text{ saturated}).$$

       The following diagram shows that from −50 and 160 °C, the solubility of LiCl displays three discontinuities, corresponding to temperatures at which one solid hydrated form transforms into another. The four regions separated by the three discontinuities correspond, from low to high temperature, to precipitation of LiCl·3 H_2O, LiCl·2 H_2O, LiCl·H_2O, and LiCl, respectively. Note also that saturated aqueous solutions freeze at lower temperatures and boil at higher temperatures than pure water (see **Topic 5F.1**), accounting for the large temperature range of these solutions.

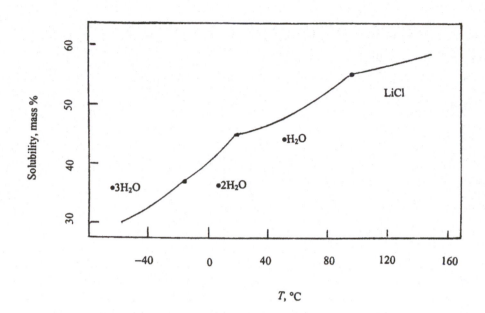

T, °C

5D.5 The Thermodynamics of Dissolving

- **Solutions of ionic substances** $\rightarrow$ $A_mB_n(s) \rightarrow mA^{n+}(aq) + nB^{m-}(aq)$

- **Enthalpy of solution, ΔH_{sol}**
  - $\rightarrow$ Enthalpy change per mole of substance dissolved
  - $\rightarrow$ Depends on *concentration* of solute

- **Limiting enthalpy of solution**
  - $\rightarrow$ Refers to the enthalpy associated with the formation of a very dilute solution
    Values are given in **Table 5D.3** of the text.
  - $\rightarrow$ Use of limiting enthalpy of solution avoids complications arising from interionic interactions that occur in more concentrated solutions because ions are far apart in very dilute solutions.

  Note: All topics that follow refer to the limiting enthalpy condition.

- **Nature of the formation of solutions of ionic substances**
  - $\rightarrow$ Conceptualized as a two-step process: sublimation of an ionic solid to form gaseous ions followed by solvation of the gaseous ions to form an ionic solution
  - $\rightarrow$ The enthalpy change for the sublimation step is designated the lattice enthalpy, ΔH_L (see **Table 4E.1** in the text).
  - $\rightarrow$ The enthalpy change for the second step (formation of hydrated ions from gas phase ions) is designated as the enthalpy of hydration, ΔH_{hyd} (see **Table 5D.4** in the text).
  - $\rightarrow$ ΔH_L *always* has a positive value, whereas ΔH_{hyd} *always* has a negative value.
  - $\rightarrow$ $\Delta H_{sol} = \Delta H_L + \Delta H_{hyd}$ is the sum of two numbers of opposite sign, both of which are typically large.
  - $\rightarrow$ As a consequence, **limiting enthalpies of solution, ΔH_{sol}, may be positive or negative.**

- **Endothermic process:** $\Delta H_L > |\Delta H_{hyd}|$ and $\Delta H_{sol} < 0$

- **Exothermic process:** $\Delta H_L < |\Delta H_{hyd}|$ and $\Delta H_{sol} > 0$

  Note: For small, highly charged ions, both ΔH_L and $|\Delta H_{hyd}|$ have large values.

- **Solutions of ionic substances**

  → $A_m B_n(s) \rightarrow mA^{n+}(aq) + nB^{m-}(aq)$

- **Free energy of solution, ΔG_{sol}**

  → Free energy change per mole of substance dissolved

  → Depends on *concentration* of solute

  → $\Delta G_{sol} = \Delta H_{sol} - T\Delta S_{sol}$ at constant T

- **Nature of solubility**

  → A substance will dissolve if $\Delta G_{sol} < 0$ and will continue to dissolve until a saturated solution, for which $\Delta G_{sol} = 0$, is obtained.

  → Both ΔH_{sol} and ΔS_{sol} change with increasing concentration to make ΔG_{sol} more positive.

  → If the solute is consumed before ΔG_{sol} reaches 0, an unsaturated solution results, with the potential to dissolve additional solute.

  → With excess solute present, ΔG_{sol} reaches 0 and a saturated solution results.

  → For endothermic enthalpies of solution, the increase in entropy of solution drives the solubility process.

  → A substance with a *large* endothermic enthalpy of solution is *usually* insoluble.

- **Temperature dependence of solubility**

  → $\Delta G_{sol} = \Delta H_{sol} - T\Delta S_{sol}$. Focus on entropy changes.

  → So for most ionic solutes, solubility *increases* with increasing temperature.

  → When gases dissolve, their motion is more constricted in the liquid phase, $\Delta S_{sol} < 0$, and gas solubility *decreases* with increasing T.

5D.6 Colloids

- **Colloids**

  → Dispersion of large particles in a solvent

  → Particles (*dispersed phase*) with lengths or diameters between 1 nm and 1 μm *dispersed* or *suspended* in a *dispersion medium* (gas, liquid, or solid solvent)

  → *Classification of colloids* is given in **Table 5D.5** in the text.

  → *Aerosols* are solids dispersed in a gas (smoke) *or* liquids dispersed in a gas (hairspray, mist, fog).

  → *Sols* or *gels* are solids dispersed in a liquid (printing ink, paint).

  → *Emulsions* are liquids dispersed in a liquid (milk, mayonnaise).

  → *Solid emulsions* are liquids dispersed in a solid (ice cream).

→ *Foams* are gases dispersed in a liquid (soapsuds or fire extinguisher foam). *Solid foams* are gases dispersed in a solid (plastic foam, such as Styrofoam).

→ *Solid dispersions* are solids dispersed in a solid (gold in glass, some alloys).

- **Aqueous colloids**

  → *Hydrophilic colloids* contain molecules with *polar groups*.
  Examples: Proteins that form gels and puddings

  → *Hydrophobic colloids* contain molecules with *nonpolar groups* only weakly attracted to water.
  Examples: Fats that form emulsions (milk and mayonnaise)

  → Rapid mixing of silver nitrate and sodium bromide solutions may produce a *hydrophobic colloidal suspension* rather than a precipitate of silver bromide. The tiny silver bromide particles are kept from further aggregation by *Brownian motion,* the motion of small particles resulting from constant collisions with solvent molecules. The *sol* is further stabilized by *adsorption* of ions on the surfaces of the particles. The adsorbed ions are hydrated by surrounding water molecules and help prevent further aggregation.

Topic 5E: MOLALITY

- **Molality, *b***

  → Moles of solute per kilogram of solvent

  → Independent of temperature (solute and solvent given as masses or mass equivalent)

  → Units of moles per kilogram ($mol \cdot kg^{-1}$)

  → Used when relative number of molecules of components is to be emphasized

  $$\text{For a binary solution,} \quad b_{solute} = \frac{n_{solute}}{\text{kilograms}_{solvent}}$$

  Note: The letter *m* is used to designate mass and sometimes molality. Care should be taken to avoid confusing the two.

- **Mole fraction, *x***

  → Ratio of moles of solute to the total number of moles of all species in a mixture

  → Independent of temperature

  → Dimensionless quantity

  → Used when relative number of molecules of components is to be emphasized

  $$\text{For a binary solution,} \quad x_{solute} = \frac{n_{solute}}{n_{solute} + n_{solvent}}$$

Topic 5F: COLLIGATIVE PROPERTIES

5F.1 Boiling-Point Elevation and Freezing-Point Depression

- **Boiling-point elevation**

  → A nonvolatile solute lowers the vapor pressure of the solvent. As a result, a solution containing a nonvolatile solute will not boil at the normal boiling point of the pure solvent. The temperature must be increased above that value to bring the vapor pressure of the solution to atmospheric pressure. Therefore, boiling is achieved at a higher temperature: the boiling point is elevated.

  → Arises from the influence of the solute on the entropy of the solvent

  → Quantitatively, $\boxed{\text{boiling-point elevation} = k_b \times \text{molality} = k_b \times b_{\text{solute}}}$

  → The boiling-point constant, k_b, depends on the solvent; it has units of K·kg·mol^{-1} (**Table 5F.1**).

  → The boiling-point elevation equation holds for nonvolatile solutes in dilute solutions that are approximately ideal.

- **Freezing-point depression**

  → Lowering the vapor-pressure of a solution decreases the triple-point temperature, the intersection of the liquid–vapor (vapor pressure) and solid–vapor phase boundaries. The solid–liquid phase boundary originating at the triple point is moved slightly to the left on the phase diagram. (The solid–vapor boundary is unchanged.) The freezing temperature of the solution is thereby lowered. (The freezing-point is depressed.)

  → Arises from the influence of the solute on the entropy of the solvent

  → Proportional to the molality of the solute

  → Quantitatively, $\boxed{\text{freezing-point depression} = k_f \times \text{molality} = k_f \times b_{\text{solute}}}$

  → The freezing-point constant, k_f, depends on the solvent; it has units of K·kg·mol^{-1} (**Table 5F.1**).

  → The freezing-point depression equation holds for nonvolatile solutes in dilute solutions that are approximately ideal.

- **Freezing-point depression corrected for ionization or aggregation of the solute**

  → $\boxed{\text{freezing-point depression} = i\,k_f \times \text{molality} = i\,k_f \times b_{\text{solute}}}$

  → The van't Hoff factor, i, determined experimentally, is an adjustment used to treat *electrolytes* (e.g., *ionic solids*, strong *acids* and *bases*) that *dissociate* and molecular solutes that *aggregate* (e.g., acetic acid dimers).

  → For electrolytes, i is the number of moles of ions formed by each mole of solute dissolved in 1 kg of solvent if all ions behave independently. In aqueous solutions, this occurs only in very dilute solutions; with increasing concentration, interionic effects reduce i below the value expected for complete ionization.

  → For molecular aggregates, i may be used to estimate the average size of the aggregates.

5F.2 Osmosis

- **Qualitative features**

  → *Osmosis*: The tendency of a solvent to flow through a membrane into a more concentrated solution.

  → Used to determine an unknown molar mass, particularly for large molecules such as polymers and proteins.

  → *Osmometry* is a technique used to determine the molar mass of a solute if the mass concentration is measured.

  → A *semipermeable membrane* allows only certain types of molecules to pass through. Typical membranes allow passage of water and small molecules, but not large molecules or ions.

  → In osmosis, the free energy of the solution is lower than that of the solvent; hence, dilution is a *spontaneous process*. The free energy of the solution can be increased by applying pressure on the solution. The increased pressure at equilibrium is called the *osmotic pressure, Π*.

  → In a static apparatus open to the atmosphere, pressure is applied by the increased height of the raised column of solution caused by the flow of solvent through the membrane into the solution. In a dynamic apparatus in a closed system, pressure is applied by the increased force of a piston confining the solution (preventing solvent flow).

  → If the external pressure $P < \Pi$, osmosis (dilution) is *spontaneous*.
  If $P = \Pi$, the system is at equilibrium (no net flow).
  If $P > \Pi$, *reverse osmosis* occurs (flow of solvent in solution to the pure solvent).
  Reverse osmosis is used to purify seawater.

- **Molarity, c**

  → Moles of solute divided by total volume of *solution*

  → Temperature dependent (volume of solution changes with temperature)

  → Units of moles per liter ($mol \cdot L^{-1}$)

$$\text{For a binary solution,} \quad c_{solute} = \frac{n_{solute}}{V_{solution}}$$

- **Quantitative features**

  → van't Hoff equation $\boxed{\Pi = iRTc_{solute}}$

  → c_{solute} = molarity of the solution = $\dfrac{\text{moles of solute}}{\text{liters of solution}}$

  i = van't Hoff factor, R = gas constant, T = temperature in kelvin
  Use units of Π in atm and of R in $L \cdot atm \cdot K^{-1} \cdot mol^{-1}$; i is dimensionless.

- **Converting from one concentration unit to another**

  → The molar mass of one component is required to convert *molality* to *mole fraction* or *mole fraction* to *molality*.

  → The density of the solution is required to convert *molarity* to *molality* or *vice versa*.

  Note: For converting from one concentration unit to another, it is convenient to assume *one* of the following: (1) The solution has a volume of one liter (*molarity* ⇔ *molality*). (2) The total amount of solvent and solute is one mole (*mole fraction* ⇔ *molality*). (3) The mass of the solvent in the solution is one kilogram (*molality* ⇔ *mole fraction*).

<hr>

Topic 5G: CHEMICAL EQUILIBRIUM

5G.1 The Reversibility of Reactions

- **Chemical reaction**

  → Reaction mixture approaches a state of dynamic equilibrium.

  → At equilibrium, the rates of the forward and reverse reactions are equal.

  Example: $H_2(g)$, $I_2(g)$, and $HI(g)$ in a closed container

Forward reaction:	$H_2(g) + I_2(g) \rightarrow 2HI(g)$
Reverse reaction:	$2HI(g) \rightarrow H_2(g) + I_2(g)$

  Dynamic equilibrium: $H_2(g) + I_2(g) \rightleftharpoons 2HI(g)$ *or* $2HI(g) \rightleftharpoons H_2(g) + I_2(g)$
  At chemical equilibrium, the reaction may be represented either way.

  Note: The approach to equilibrium may be too slow to measure. Such a system may appear to be at equilibrium, but actually is not.

  Examples: The sugar glucose, exposed to the air, is not in equilibrium with its combustion products, CO_2 and H_2O.
  Methane burning on a gas range is not at equilibrium with combustion products CO_2 and H_2O. Methane is supplied continuously and the reverse reaction does not occur.

5G.2 Equilibrium and the Law of Mass Action

- **Law of mass action**

  → The composition of a reaction mixture can be expressed in terms of an *equilibrium constant K*, which is unitless.

  → The equilibrium constant K has this form:

$$K = \left\{ \frac{\text{activities of products}}{\text{activities of reactants}} \right\}_{\text{equilibrium}}$$

- *Activity* **of substance J, a_J**
  - → Partial pressure or concentration of a substance relative to its standard value
  - → Pure number that is unitless
  - → Idealized systems

    $$a_J = \frac{P_J}{P°} \quad \text{for an ideal gas}$$

    partial pressure of the gas (bar) divided by the pressure of the gas in its standard state (1 bar)

    $$a_J = \frac{[J]}{c°} \quad \text{for a solute in a dilute solution}$$

    molarity of substance J divided its molarity in the standard state (1 mol·L^{-1})

    $a_J = 1$ for a pure solid or liquid

  - → Because the standard state values of activity are all defined as unity, they are omitted in the expression for activity and in equilibrium constant expressions.

    $$a_J = \frac{P_J}{P°} = \frac{P_J}{1} = P_J \quad \text{or} \quad \boxed{a_J = P_J \text{ for an ideal gas}}$$

    Similarly, $\boxed{a_J = [J] \text{ for a solute in a dilute solution}}$ and

    $\boxed{a_J = 1 \text{ for a pure solid or liquid}}$

- *Activity coefficient,* γ_J
  - → Pure number that is unitless
  - → Equal to 1 for ideal gases and solutes in dilute solutions
  - → Accounts for intermolecular interactions
  - → Used for concentrated solutions and gases, which do not behave ideally

    $a_J = \gamma_J P_J$ for a real gas

    $a_J = \gamma_J [J]$ for any concentration of solute in a solution

    $a_J = 1$ for a pure solid or liquid

- **Form of the equilibrium constant**
  - → For a general reaction: $a\,A + b\,B \rightarrow c\,C + d\,D$

    $$\boxed{K = \left(\frac{a_C^c a_D^d}{a_A^a a_B^b} \right)_{equilibrium}} \quad \text{law of mass action, } K, \text{ is unitless}$$

  - → K is the ratio of the activities of the products, raised to powers equal to their stoichiometric coefficient, to the corresponding expression for the reactants.
  - → Use activities given in these boxed equations to write expressions for K.
  - → Distinguish between *homogeneous* and *heterogeneous* equilibria.

- **Homogeneous equilibria**
  - → Reactants and products all in the same phase

- **Heterogeneous equilibria**

  → Reactants and products with different phases

    Remember: Activities of pure solids and liquids are set equal to 1.

  Note: For reactions in solution, expressions for equilibrium constants should be obtained by using *net ionic equations*.

    Example: Write the equilibrium constant for the reaction $2\,HI\,(g) \rightleftharpoons H_2\,(g) + I_2\,(g)$. Reactants and products are all gases whose activities are equal to their partial pressures, P_j. Therefore,

    $$K = \left(\frac{P_{H_2} P_{I_2}}{(P_{HI})^2} \right)_{\text{at equilibrium}}$$

5G.3 The Origin of Equilibrium Constants

- **ΔG as a function of the composition of the reaction mixture**

  → $G_m(J)$ is the *molar free energy* of the reactant or product J.

  → $G_m{}^\circ(J)$ is the *molar free energy* of the reactant or product J *in the standard state*.

  → $\boxed{G_m(J) = G_m{}^\circ(J) + RT \ln a_J}$ for any substance in any state

  → $\boxed{G_m(J) = G_m{}^\circ(J) + RT \ln P_J}$ for ideal gases

  → $\boxed{G_m(J) = G_m{}^\circ(J) + RT \ln[J]}$ for solutes in dilute solution

  → $\boxed{G_m(J) = G_m{}^\circ(J)}$ for pure solids and liquids ($a = 1$)

  → In *dilute aqueous* solutions, the activity of water may be approximated as $a(H_2O) = 1$.

- **ΔG_m for changes in pressure or concentration in idealized systems**

  → Standard state of a substance is its pure form at a pressure of 1 bar.

  → For a solute, the standard state is for a concentration of $1\ mol \cdot L^{-1}$.

- **ΔG for chemical reactions (see Topic 4J.2)**

  → Generalized chemical reaction: $a\,A + b\,B \rightarrow c\,C + d\,D$

  → $\boxed{\Delta G^\circ = \sum n G_m{}^\circ(\text{products}) - \sum n G_m{}^\circ(\text{reactants})}$ Units: kilojoules

  → Sometimes it is useful to interpret n as a pure number (rather than amount in moles). In the molar convention, we attach a subscript r to ΔG° and n (r for reaction).

  → $\Delta G_r{}^\circ$, the *standard Gibbs reaction free energy*, is the difference in molar free energies of the products and reactants in their standard states.

  → $\boxed{\Delta G_r{}^\circ = \sum n_r G_m{}^\circ(\text{products}) - \sum n_r G_m{}^\circ(\text{reactants})}$ Units: kilojoules per mole

- **Stoichiometric coefficients, n_r, used as pure numbers (molar convention) to obtain $\Delta G_r°$ in kJ·mol⁻¹, the same convention used for ΔG_r**

  → Evaluated by using $\boxed{\Delta G_r° = \sum n \Delta G_f°(\text{products}) - \sum n \Delta G_f°(\text{reactants})}$

  → Standard free energies of formation, $\Delta G_f°$, for a variety of substances are given in Appendix 2A.

  → $\boxed{\Delta G_r = \sum n_r G_m(\text{products}) - \sum n_r G_m(\text{reactants})}$

  → ΔG_r is the reaction free energy at any definite, fixed composition of the reaction mixture.

  → Evaluated by using $\boxed{\Delta G_r = \Delta G_r° + RT \ln Q}$

  → Q is called the **reaction quotient**.

- **Expressions for Q for a general reaction:** $a\,A + b\,B \rightarrow c\,C + d\,D$

  ➤ Real systems: $\boxed{Q = \dfrac{a_C^c\, a_D^d}{a_A^a\, a_B^b}}$

  → For ideal gases or dilute solutions, Q reduces to:

  ➤ Ideal gases: $\boxed{Q = \dfrac{P_C^c\, P_D^d}{P_A^a\, P_B^b}}$ Q is unitless.

  ➤ Dilute solutions: $\boxed{Q = \dfrac{[C]^c\,[D]^d}{[A]^a\,[B]^b}}$ Q is unitless.

- **Chemical equilibrium**

  → At equilibrium, $\Delta G = 0$ and $Q = K$ (equilibrium constant)

  → K has the same form as Q, but it is determined uniquely by the equilibrium composition of the reaction system.

  → Real systems: $\boxed{K = \left(\dfrac{a_C^c\, a_D^d}{a_A^a\, a_B^b} \right)_{\text{equilibrium}}}$ law of mass action, K is dimensionless (no units).

  → Activities can be expressed in concentration units multiplied by an activity coefficient γ.

  → For a gas, $a_J = \dfrac{\gamma_J P_J}{P°} = \gamma_J P_J$ numerically. For a solute, $a_J = \dfrac{\gamma_J [J]}{[J]°} = [J]$ numerically.

- **Evaluation of K from $\Delta G_r°$ and vice versa**

  → $\boxed{\Delta G_r° = -RT \ln K}$

5G.4 The Thermodynamic Description of Equilibrium

- **Thermodynamically,**

$$\Delta G_r^\circ = \Delta H_r^\circ - T\Delta S_r^\circ = -RT \ln K; \quad \ln K = -\frac{\Delta H_r^\circ}{RT} + \frac{\Delta S_r^\circ}{R} \qquad \textbf{Solve for } \textbf{\textit{K}}$$

$$K = e^{-\Delta H_r^\circ/RT} e^{\Delta S_r^\circ/R}$$

- **In a chemical reaction, products are favored if**

  → K is large.

  → The reaction is exothermic in the standard state (negative ΔH°, the more negative the better).

  → The reaction has a positive standard entropy of reaction ($\Delta S^\circ > 0$, final state more random).

  → Similarly, reactants are favored for endothermic reactions (positive ΔH°) with negative entropy change ($\Delta S^\circ < 0$)

- **Guidelines for attaining equilibrium**

  → If $K > 10^3$, products are favored.

  → If $10^{-3} < K < 10^3$, neither reactants nor products are strongly favored.

  → If $K < 10^{-3}$, reactants are favored.

  Example: At a certain temperature, suppose that the equilibrium concentrations of nitrogen and ammonia in the reaction

  $$N_2\,(g) + 3\,H_2\,(g) \rightleftharpoons 2\,NH_3\,(g)$$

  were found to be 0.11 M and 1.5 M, respectively. What is the equilibrium concentration of H_2 if $K_c = 4.4 \times 10^4$?

  The value of K_c is in the range that favors products. Solving the equilibrium constant expression requires the evaluation of the cube root of the concentration of H_2. Use the y^x key on your calculator.

  $$K_c = \left(\frac{[NH_3]^2}{[N_2][H_2]^3}\right)_{eq} \quad \text{and} \quad [H_2]^3 = \frac{[NH_3]^2}{K_c[N_2]}$$

  $$[H_2] = \sqrt[3]{\frac{[NH_3]^2}{K_c[N_2]}} = \left(\frac{(1.5)^2}{(4.4\times10^4)(0.11)}\right)^{1/3} = \left(4.65\times10^{-4}\right)^{1/3} = 0.077 \text{ M}$$

  The product ammonia is favored as expected.

- **For the general reaction** $a\,A\,(g) + b\,B\,(g) \rightleftharpoons c\,C\,(g) + d\,D\,(g)$

$$\boxed{K = \left(\frac{a_C^c\,a_D^d}{a_A^a\,a_B^b}\right)_{equilibrium}} \quad \text{and} \quad \boxed{Q = \left(\frac{a_C^c\,a_D^d}{a_A^a\,a_B^b}\right)_{not\ at\ equilibrium}}$$

- **Guidelines for approach to equilibrium**

  → Compare Q with K.

  → If $Q > K$, the amount of products is too high or the amount of reactants is too low. Reaction proceeds in reverse, toward reactants ($\leftarrow$).

  → If $Q < K$, the amount of reactants is too high or the amount of products is too low. Reaction proceeds forward to form products ($\rightarrow$).

  → If $Q = K$, reactants and products are at equilibrium and no observable change occurs ($\rightleftharpoons$).

Topic 5H: ALTERNATIVE FORMS OF THE EQUILIBRIUM CONSTANT

5H.1 Multiples of the Chemical Equation

- **Dependence of K on the direction of reaction and the balancing coefficients**

  → $a\,A(g) + b\,B(g) \rightleftharpoons c\,C(g) + d\,D(g)$ $\Rightarrow$ $K_1 = \left(\dfrac{a_C^c\, a_D^d}{a_A^a\, a_B^b} \right)_{\text{equilibrium}}$

  → $c\,C(g) + d\,D(g) \rightleftharpoons a\,A(g) + b\,B(g)$ $\Rightarrow$ $K_2 = \left(\dfrac{a_A^a\, a_B^b}{a_C^c\, a_D^d} \right)_{\text{equilibrium}} = \dfrac{1}{K_1} = K_1^{-1}$

  → $na\,A(g) + nb\,B(g) \rightleftharpoons nc\,C(g) + nd\,D(g)$ $\Rightarrow$ $K_3 = \left(\dfrac{a_C^{nc}\, a_D^{nd}}{a_A^{na}\, a_B^{nb}} \right)_{\text{equilibrium}} = K_1^{n}$

  → Multiply all coefficients of a reaction by n; raise the equilibrium constant to the nth power

- **Combining chemical reactions**

  → Adding reaction 1 with equilibrium constant K_1 to reaction 2 with equilibrium constant K_2 yields a combined reaction with equilibrium constant K that is the *product* of K_1 and K_2. $(\Delta G_1^\circ + \Delta G_2^\circ)$ leads to $K = K_1 \times K_2$

  → Subtracting reaction 2 with individual equilibrium constant K_2 from reaction 1 with equilibrium constant K_1 yields a combined reaction with equilibrium constant K that is the *quotient* of K_1 and K_2.

  $(\Delta G_1^\circ - \Delta G_2^\circ)$ leads to $K = \dfrac{K_1}{K_2}$

- **Summary:** Add two reactions $(1 + 2)$, multiply equilibrium constants $(K_1 \times K_2)$. Subtract two reactions $(1 - 2)$, divide equilibrium constants (K_1 / K_2).

5H.2 Composite Equation

- **Gas-phase equilibria (ideal gas reaction mixture)**

  $\rightarrow$ $a\,A(g) + b\,B(g) \rightleftharpoons c\,C(g) + d\,D(g)$ (general gas-phase reaction at equilibrium)

  $\rightarrow$ $\boxed{K = \left(\dfrac{P_C^c P_D^d}{P_A^a P_B^b}\right)_{equilibrium}}$ ideal gases (activity = partial pressure in the units of bar)

  $\rightarrow$ $\Delta n_r = (c+d) - (a+b) = \Sigma(\text{coefficients of products}) - \Sigma(\text{coefficients of reactants})$

5H.3 Molar Concentrations of Gases

- **Gas-phase equilibria (ideal gas reaction mixture)**

  $\rightarrow$ $a\,A(g) + b\,B(g) \rightleftharpoons c\,C(g) + d\,D(g)$ (general gas-phase reaction at equilibrium)

  $\rightarrow$ $\boxed{K = \left(\dfrac{P_C^c P_D^d}{P_A^a P_B^b}\right)_{equilibrium}}$ ideal gases (activity = partial pressure in the units of bar)

  $\rightarrow$ $\boxed{K_c = \left(\dfrac{[C]^c [D]^d}{[A]^a [B]^b}\right)_{equilibrium}}$ ideal gases (activity = molar concentration)

  $\rightarrow$ Both K and K_c are unitless.

  $\rightarrow$ In general, $P_J = \dfrac{n_J RT}{V} = RT\dfrac{n_J}{V} = RT\,[J]$.

  $\rightarrow$ Relationship betweer K and K_c as derived in the text: $\boxed{K = (RT)^{\Delta n_r} K_c}$

  $\rightarrow$ $\Delta n_r = (c+d) - (a+b) = \Sigma(\text{coefficients of products}) - \Sigma(\text{coefficients of reactants})$

- **Mixed-phase equilibria**

  $\rightarrow$ $a\,A(g) + b\,B(aq) \rightleftharpoons c\,C(g) + d\,D(aq)$ (some gases, some solution species at equilibrium)

  $\rightarrow$ $\boxed{K = \left(\dfrac{P_C^c [D]^d}{P_A^a [B]^b}\right)_{equilibrium}}$ $\qquad K = (RT)^{\Delta n_r} K_c$

  $\rightarrow$ $\Delta n_r = c - a = \Sigma(\text{coefficients of } gaseous \text{ products}) - \Sigma(\text{coefficients of } gaseous \text{ reactants})$
  Note: Δn_r can be positive, negative, or zero.

Topic 5I: EQUILIBRIUM CALCULATIONS

5I.1 The Extent of Reaction

- **Thermodynamically,**

$$\Delta G_r^{\circ} = \Delta H_r^{\circ} - T\Delta S_r^{\circ} = -RT \ln K; \quad \ln K = -\frac{\Delta H_r^{\circ}}{RT} + \frac{\Delta S_r^{\circ}}{R} \qquad \textbf{Solve for } K$$

$$K = e^{-\Delta H_r^{\circ}/RT} e^{\Delta S_r^{\circ}/R}$$

- **In a chemical reaction, products are favored if**

  → K is large.

  → The reaction is exothermic in the standard state (negative ΔH°, the more negative the better).

  → The reaction has a positive standard entropy of reaction ($\Delta S^{\circ} > 0$, final state more random).

  → Similarly, reactants are favored for endothermic reactions (positive ΔH°) with negative entropy change ($\Delta S^{\circ} < 0$)

- **Guidelines for attaining equilibrium**

  → If $K > 10^3$, products are favored.

  → If $10^{-3} < K < 10^3$, neither reactants nor products are strongly favored.

  → If $K < 10^{-3}$, reactants are favored.

  Example: At a certain temperature, suppose that the equilibrium concentrations of nitrogen and ammonia in the reaction

  $$N_2(g) + 3H_2(g) \rightleftharpoons 2NH_3(g)$$

  were found to be 0.11 M and 1.5 M, respectively. What is the equilibrium concentration of H_2 if $K_c = 4.4 \times 10^4$?

  The value of K_c is in the range that favors products. Solving the equilibrium constant expression requires the evaluation of the cube root of the concentration of H_2. Use the y^x key on your calculator.

  $$K_c = \left(\frac{[NH_3]^2}{[N_2][H_2]^3} \right)_{eq} \quad \text{and} \quad [H_2]^3 = \frac{[NH_3]^2}{K_c[N_2]}$$

  $$[H_2] = \sqrt[3]{\frac{[NH_3]^2}{K_c[N_2]}} = \left(\frac{(1.5)^2}{(4.4\times10^4)(0.11)} \right)^{1/3} = \left(4.65\times10^{-4} \right)^{1/3} = 0.077 \text{ M}$$

  The product ammonia is favored as expected.

5I.2 The Direction of Reaction

- **For the general reaction** $a\,A\,(g) + b\,B\,(g) \rightleftharpoons c\,C\,(g) + d\,D\,(g)$

$$K = \left(\frac{a_C^c\,a_D^d}{a_A^a\,a_B^b}\right)_{\text{equilibrium}} \quad \text{and} \quad Q = \left(\frac{a_C^c\,a_D^d}{a_A^a\,a_B^b}\right)_{\text{not at equilibrium}}$$

- **Guidelines for approach to equilibrium**

  → Compare Q with K.

  → If $Q > K$, the amount of products is too high or the amount of reactants is too low. Reaction proceeds in the reverse direction, toward reactants ($\leftarrow$).

  → If $Q < K$, the amount of reactants is too high or the amount of products is too low. Reaction proceeds in the forward direction to form products ($\rightarrow$).

  → If $Q = K$, reactants and products are at equilibrium and no observable change occurs ($\rightleftharpoons$).

5I.3 Calculations with Equilibrium Constants

- **Equilibrium table**

  → Very helpful in solving all types of equilibrium problems

- **Format and use of the equilibrium table**

  Step 1: Reactant and product species taking part in the reaction are identified.

  Step 2: The initial composition of the system is listed.

  Step 3: Changes needed to reach equilibrium are listed in terms of *one unknown quantity.*

  Step 4: The equilibrium composition in terms of one unknown is entered on the last line.

  Step 5: The equilibrium composition is entered into the equilibrium constant expression and the unknown quantity (equilibrium constant or reaction species concentration) is determined.

  Step 6: Approximations may sometimes be used to simplify the calculation of the unknown quantity in step 5.

- **Calculating equilibrium constants from pressures or concentrations**

  Example: A sample of HI(g) is placed in a flask at a pressure of 0.100 bar. After equilibrium is attained, the partial pressure of HI(g) is 0.050 bar. Evaluate K for the reaction

  $$2\,HI(g) \rightleftharpoons H_2(g) + I_2(g)$$

  (a) Set up an equilibrium table like the one that follows. In the initial system, line 1, there are no products, so some must form to reach equilibrium.

  (b) On line 2, let x be the change in pressure of I_2 required to reach equilibrium. From the stoichiometric coefficients, the change in the pressure of H_2 is x, whereas the change in the pressure of HI is $-2x$. (HI *disappears twice as fast* as iodine appears.)

  (c) Lines 1 and 2 of the table are added to give the equilibrium pressures of the reactants and products in terms of x.

(d) The value of x is determined from the change in pressure of HI $(-2x)$ from the initial to the equilibrium value.

(e) Using the calculated value of x, the equilibrium pressures from line 3 of the equilibrium table are calculated and entered into the equilibrium constant expression to obtain K.

	Species		
	HI	H_2	I_2
1. Initial pressure	0.100	0	0
2. Change in pressure	$-2x$	$+x$	$+x$
3. Equilibrium pressure	$0.100 - 2x$	x	x

$$P_{HI} = 0.100 - 2x = 0.050 \quad \text{and} \quad x = 0.025 = P_{H_2} = P_{I_2}$$

$$K = \frac{P_{H_2} P_{I_2}}{P_{HI}^2} = \frac{(0.025)^2}{(0.050)^2} = 0.25$$

- **Calculating the equilibrium composition**

  Example: A sample of ethane, $C_2H_6(g)$, is placed in a flask at a pressure of 2.00 bar at 900 K. The equilibrium constant K for the gas phase dehydrogenation of ethane to form ethene

  $$C_2H_6(g) \rightleftharpoons C_2H_4(g) + H_2(g)$$

  is 0.050 at 900 K. Determine the composition of the equilibrium mixture and the percent decomposition of C_2H_6.

  (a) Set up an equilibrium table and let x be the unknown equilibrium pressure of each product, C_2H_4 and H_2. The equilibrium pressure of C_2H_6 is then $2.00 - x$.

  (b) Solve for x by rearranging the equilibrium constant expression.

  (c) The new form is a quadratic equation, which is solved exactly by using the quadratic formula.

  (d) The equilibrium pressures can then be obtained.

  (e) The percent decomposition is the loss of reactant divided by the initial pressure times 100%, or $\left(\dfrac{x}{2.00}\right)(100)\%$.

	Species		
	C_2H_6	C_2H_4	H_2
1. Initial pressure	2.00	0	0
2. Change in pressure	$-x$	$+x$	$+x$
3. Equilibrium pressure	$2.00 - x$	x	x

$$K = \frac{P_{C_2H_4} P_{H_2}}{P_{C_2H_6}} = \frac{(x)^2}{(2.00 - x)} = 0.050$$

Rearranging yields a quadratic equation: $x^2 + (0.050)x - 0.10 = 0$

Recall: $ax^2 + bx + c = 0 \Rightarrow x = \dfrac{-b \pm \sqrt{b^2 - 4ac}}{2a}$ (exact solution)

Then, $a = 1$, $b = 0.050$, and $c = -0.10$.

$$x = \frac{-(0.050) \pm \sqrt{(0.050)^2 - 4(1)(-0.10)}}{2(1)} = \frac{-(0.050) \pm \sqrt{0.4025}}{2}$$

$$= \frac{-(0.050) \pm (0.6344)}{2} = \frac{-(0.050) + (0.6344)}{2} = \frac{0.5844}{2} = 0.29$$

The negative root is rejected because it is not physically meaningful.

$P_{C_2H_4} = P_{H_2} = x = 0.29$ bar and $P_{C_2H_6} = 2.00 - x = 1.71$ bar

$$\text{Percent decomposition} = \frac{\text{change in pressure}}{\text{initial pressure}} \times 100\% = \frac{0.29}{2.00} \times 100\% = 14.5\%$$

Note: If K is less than 10^{-3} in a problem of this type, the adjustment to equilibrium is small. The pressures of reactants then change by only a small amount. In such a case, an approximate solution that avoids the quadratic equation is possible.

Topic 5J: THE RESPONSE OF EQUILIBRIA TO CHANGES IN CONDITIONS

5J.1 Adding and Removing Reagents

- **Le Chatelier's principle**

  → When a stress is applied to a system in dynamic equilibrium, the equilibrium tends to adjust to minimize the effect of the stress.

- **Qualitative features**

  → Use Le Chatelier's principle to predict the qualitative adjustments needed to obtain a new equilibrium composition that relieves the stress.

  → Assessing the effects of the applied stress on Q relative to K is an alternative approach.

  Example: Use Le Chatelier's principle and reaction quotient arguments to describe the effect on the equilibrium

  $$2\,SO_2\,(g) + O_2\,(g) \rightleftharpoons 2\,SO_3\,(g) \qquad Q = \left(\frac{P_{SO_3}^2}{P_{SO_2}^2 \, P_{O_2}} \right)$$

  of (a) addition of SO_2, (b) removal of O_2, (c) addition of SO_3, and (d) removal of SO_3.

  Both reactant and product pressures appear in the equilibrium constant expression and in Q. Assuming no change in volume or temperature, adding reactant (or removing product) shifts the equilibrium composition to produce more product molecules, thereby helping to relieve the stress. Adding product (or removing reactant) shifts the equilibrium composition to produce more reactant molecules, thereby helping to relieve the stress.

(a) Adding reactant SO_2 results in $Q < K$. The system adjusts by producing more product SO_3 (consuming SO_2 and O_2) until $Q = K$ ($\rightarrow$).

(b) Removing reactant O_2 results in $Q > K$. The system adjusts by consuming product SO_3 (producing more SO_2 and O_2) until $Q = K$ ($\leftarrow$).

(c) Adding SO_3 results in $Q > K$. The system adjusts by producing more SO_2 and O_2 (consuming SO_3) until $Q = K$ ($\leftarrow$).

(d) Removing SO_3 results in $Q < K$. The system adjusts by consuming SO_2 and O_2 (producing more SO_3) until $Q = K$ ($\rightarrow$).

Note: In this example, the stress changes only the value of Q. The composition adjusts to return the value of Q to the original value of K.

The equilibrium constant does not change. The equilibrium constant has, however, a very weak pressure dependence that is usually safe to ignore.

- **Quantitative features**

  → Use the equilibrium table method to calculate the new composition of an equilibrium system after a stress is applied.

  Example: Assume that the reaction $2HI\,(g) \rightleftharpoons H_2(g) + I_2(g)$ has the equilibrium composition $P_{HI} = 0.050$ bar, $P_{H_2} = P_{I_2} = 0.025$ bar, and $K = 0.25$.

  If HI, H_2, and I_2 are added to the system in amounts equivalent to 0.008 bar for HI, 0.008 bar for H_2, and 0.008 bar for I_2, predict the direction of change and calculate the new equilibrium composition.

  After adding 0.008 bar to each reactant and product, the *nonequilibrium* pressures are $P_{HI} = 0.058$ bar, $P_{H_2} = 0.033$ bar, and $P_{I_2} = 0.033$ bar.

  $$Q = \frac{(0.033)(0.033)}{(0.058)^2} = 0.324 > K$$ and the adjustment to the new equilibrium is in the direction toward the reactants.

  Quantitative treatment: Set up an equilibrium table, letting $-x$ be the unknown adjustment to the initial *nonequilibrium* pressures of each product H_2 and I_2. The value of $+2x$ is then the adjustment to the initial *nonequilibrium* pressure of HI.

  A negative value of x is expected from our qualitative analysis, which suggests that reactants should form from products to relieve the stress on the system.

	Species		
	HI	H_2	I_2
1. Initial pressure	0.058	0.033	0.033
2. Change in pressure	$+2x$	$-x$	$-x$
3. Equilibrium pressure	$0.058 + 2x$	$0.033 - x$	$0.033 - x$

$$K = \frac{P_{H_2} P_{I_2}}{P_{HI}^2} = \frac{(0.033 - x)^2}{(0.058 + 2x)^2} = 0.25$$

This expression is relatively easy to solve for x by taking the square root of both sides:

$$\sqrt{K} = \frac{(0.033 - x)}{(0.058 + 2x)} = \sqrt{0.25} = 0.50$$

$0.033 - x = 0.029 + x$; $2x = 0.004$; $x = 0.002$

$P_{H_2} = P_{I_2} = 0.033 - 0.002 = 0.031$ bar *and* $P_{HI} = 0.058 + 0.004 = 0.062$ bar

Equilibrium expression check: $K = \dfrac{(0.031)^2}{(0.062)^2} = 0.25$ as expected.

5J.2 Compressing a Reaction Mixture

- **Decrease in volume of a reaction mixture at constant temperature**

  → The partial pressure of each component J of the reaction mixture increases;

  recall $P_J = \dfrac{n_J RT}{V}$ (see **Topic 3C** in this study guide and in the text).

  → To minimize the increase in pressure, the equilibrium shifts to form *fewer* gas molecules, if possible.

- **Increase in volume of a reaction mixture at constant temperature**

  → The partial pressure of each member J of the reaction mixture decreases; recall $P_J = \dfrac{n_J RT}{V}$.

  → To compensate for the decrease in pressure, the equilibrium shifts to form *more* gas molecules, if possible.

- **Increase in pressure of a reaction mixture by introducing an inert gas at constant T and V, as in a steel vessel**

  → The partial pressure of each member J of the reaction mixture does not change;

  recall $P_J = \dfrac{n_J RT}{V}$.

  → The equilibrium composition is *unaffected*. The equilibrium constant has a weak pressure dependence that is safe to ignore unless the pressure change is large.

- **Increase in volume of a reaction mixture as a result of introducing an inert gas at constant T and *total* pressure P, as in a piston**

  → increases the *volume* of the system without changing the amount of reactants or products.

  → The partial pressure of each member J of the reaction mixture decreases; recall $P_J = \dfrac{n_J RT}{V}$.

  → The return to equilibrium is equivalent to that associated with increasing the volume or decreasing the pressure of the system. The equilibrium shifts to form *more* gas molecules, if possible.

5J.3 Temperature and Equilibrium

- **Qualitative features**

  → Use Le Chatelier's principle to predict the qualitative adjustments needed to obtain a new equilibrium composition, which relieves the stress.

- **For *endothermic* reactions, heat is treated as a *reactant*.**

  → *Increasing* the temperature supplies heat to the reaction mixture, and the composition of the mixture adjusts to form more product molecules.

  → The value of the equilibrium constant, K, *increases.*

  → *Decreasing* the temperature removes heat from the reaction mixture, and the composition of the mixture adjusts to form more reactant molecules.

  → The value of the equilibrium constant, K, *decreases.*

- **For *exothermic* reactions, heat is treated as a *product*.**

  → *Increasing* the temperature supplies heat to the reaction mixture, and the composition of the mixture adjusts to form more reactant molecules.

  → The value of the equilibrium constant, K, *decreases.*

  → *Decreasing* the temperature removes heat from the reaction mixture, and the composition of the mixture adjusts to form more product molecules.

  → The value of the equilibrium constant, K, *increases.*

- **Quantitative features**

  → van't Hoff equation $\boxed{\ln \dfrac{K_2}{K_1} = \dfrac{\Delta H_r^{\circ}}{R}\left(\dfrac{1}{T_1} - \dfrac{1}{T_2}\right)}$ (see **Topic 5J.3** in the text for the derivation)

  → Gives the qualitative results shown previously (The equilibrium constant is K, not K_c.)

  → A major assumption in the derivation is that both ΔH_r° and ΔS_r° are independent of temperature between T_1 and T_2.

- **Use the van't Hoff equation to calculate**
  - → $\Delta H_r°$ if the equilibrium constant K is known at two temperatures.
  - → the equilibrium constant at any temperature T_2 if $\Delta H_r°$ and K at any other temperature are known.

INTERLUDE

Homeostasis

- **Homeostasis**
  - → Mechanism similar to chemical equilibrium, governed by Le Chatelier's principle
  - → Helps living organisms to maintain constant internal conditions
  - → Allows living organisms to control biological processes at a constant level
  - → An example is the equilibrium involving oxygen and hemoglobin, Hb.

$$Hb(aq) + O_2(aq) \rightleftharpoons HbO_2(aq)$$

  - → The following figure shows how O_2 uptake by hemoglobin and myoglobin (Mb), the oxygen storage protein, varies with the partial pressure of oxygen. The shape of the Hb saturation curve means that Hb can load O_2 almost as fully in the lungs as Mb and unload it more completely than Mb in different regions of the organisms. In the lungs, where $P(O_2) \approx 105$ Torr, 98% of the Hb molecules have taken up O_2, resulting in almost complete saturation. In resting muscular tissue, the concentration of O_2 corresponds to a partial pressure of about 40 Torr, at which value 75% of the Hb molecules are saturated with oxygen. The figure implies that sufficient oxygen is still available if a sudden surge in activity occurs. See additional information in the text.

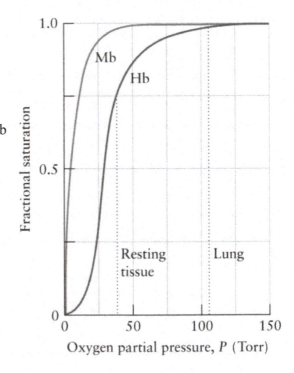

Focus 6: REACTIONS

Topic 6A: THE NATURE OF ACIDS AND BASES

6A.1 Brønsted–Lowry Acids and Bases

- **Arrhenius definitions of acids and bases in aqueous solutions**
  - → *Acid*: compound that produces hydronium ions (H_3O^+) in water
  - → *Base*: compound that produces hydroxide ions (OH^-) in water
  - → See *Fundamentals*, **Section J**
  - → Superseded by the more general Brønsted–Lowry definitions

- **Brønsted–Lowry definitions**
  - → *Acid*: proton donor; *base*: proton acceptor
  - → General theory for *any* solvent (e.g., $H_2O(l)$ or $NH_3(l)$) or *no* solvent at all
  - → Acid (*proton donor*) is *deprotonated* when reacting with a base.
  - → Base (*proton acceptor*) is *protonated* when reacting with an acid.
  - → Acids and bases react with each other in proton - transfer reactions.
  - → A strong acid is *completely deprotonated.*

    Example: $HBr(aq) + H_2O(l) \rightarrow H_3O^+(aq) + Br^-(aq)$ (*strong* acid: reaction goes *essentially* to completion; proton is transferred from HBr(aq) to $H_2O(l)$).

  - → A strong base is *completely protonated.*

    Example: $NH_2^-(aq) + H_2O(l) \rightarrow NH_3(aq) + OH^-(aq)$ (*strong* base: reaction goes *essentially* to completion; $NH_2^-(aq)$ accepts a proton from H_2O).

  - → Strong Brønsted acids and bases: a single arrow ($\rightarrow$) is used to represent their reaction, as shown.
  - → A weak acid is *partially deprotonated.*

    Example: $HF(aq) + H_2O(l) \rightleftharpoons H_3O^+(aq) + F^-(aq)$ (*Weak* acid reaction; a *chemical equilibrium* is established.)

  - → A weak base is *partially protonated.*

    Example: $NH_3(aq) + H_2O(l) \rightleftharpoons NH_4^+(aq) + OH^-(aq)$ (*Weak* base reaction; a *chemical equilibrium* is established.)

  - → Weak Brønsted acids and bases: a double arrow ($\rightleftharpoons$) is used to represent their reaction.
  - → For strong acids and bases, the equilibrium lies far to the right, justifying use of the single arrow for their reaction with the water or other solvents.

- **Conjugate base**
  - → Formed from an acid after a proton is donated
  - → Acid $\Rightarrow$ deprotonation $\Rightarrow$ conjugate base

    Examples: The conjugate base of HBr is Br^-; the conjugate base of H_3PO_4 is $H_2PO_4^-$.

- **Conjugate acid**
  - → Formed from a base after a proton is accepted
  - → Base $\Rightarrow$ protonation $\Rightarrow$ conjugate acid
    Examples: The conjugate acid of NH_3 is NH_4^+; the conjugate acid of CH_3COO^- is CH_3COOH.

- **Conjugate acid–base pair**
  - → Differ only by the presence or absence of an H^+ species:
    HCN, CN^-; H_3O^+, H_2O; H_2O, OH^-; H_2S, HS^-

- **Solvent for Brønsted–Lowry acid-base reactions**
  - → Need not be water. Other solvents include $HF(l)$, $H_2SO_4(l)$, $CH_3COOH(l)$, $C_2H_5OH(l)$, $NH_3(l)$.
  - → Acid-base reactions can also occur with *no* solvent, as in $HCl(g) + NH_3(g) \rightarrow NH_4Cl(s)$.

6A.2 Lewis Acids and Bases

- **Lewis definitions**
  - → A Lewis acid is an electron pair *acceptor*. **Examples:** $BF_3(g)$, $Fe^{2+}(aq)$, $H_2O(l)$
  - → A Lewis base is an electron pair *donor*. **Examples:** $NH_3(aq)$, $O^{2-}(aq)$
  - → Most general theory (holds for any solvent and includes metal cations)

- **Comparison of Lewis and Brønsted–Lowry Acids and Bases**
  - → Proton (H^+) transfer (Brønsted–Lowry theory) is a special type of Lewis acid-base reaction.
  - → In a proton transfer reaction, H^+ acts as a Lewis acid, accepting an electron pair from a Lewis base.
  - → Brønsted acid: supplier of one particular Lewis acid, a proton
  - → Brønsted base: a particular type of Lewis base that can use a lone pair to bond a proton

    Example: The reaction $O^{2-}(aq) + H_2O(l) \rightarrow 2\,OH^-(aq)$ can be viewed as a Brønsted–Lowry reaction (H_2O is the proton donor; O^{2-} is the proton acceptor) or as a Lewis acid-base reaction. In the latter approach, O^{2-} is the electron pair donor (*Lewis base*), H^+ on water is the electron pair acceptor (*Lewis acid*), as shown in the nearby diagram.

    Note: Curly arrow extends from a Lewis base electron pair (tail) to a Lewis acid (head); it indicate a Lewis acid-base reaction.

6A.3 Acidic, Basic, and Amphoteric Oxides

- **Oxides**
  - → React with water as Lewis acids or bases or both

- **Oxides of nonmetals**
  - → Called *acidic oxides*
  - → Tend to act as Lewis acids **Examples:** CO_2, SO_3, Cl_2O_7, N_2O_5

→ React with water to form a Brønsted acid

Examples: $CO_2(g) + H_2O(l) \rightarrow H_2CO_3(aq)$

$SO_3(g) + H_2O(l) \rightarrow H_2SO_4(aq)$
Lewis acid   Brønsted acid

→ React with bases to form a salt and water

Example: $H_2CO_3(aq) + 2\,NaOH(aq) \rightarrow Na_2CO_3(aq) + 2\,H_2O(l)$

- **Oxides of metals**

→ Called *basic oxides*

→ Tend to act as Lewis bases
Examples: Na_2O, MgO, CaO

→ React with water to form Brønsted bases

Example: $CaO(s) + H_2O(l) \rightarrow Ca(OH)_2(aq)$
Lewis base   Brønsted base

→ React with acids to form a salt and water

Example: $CaO(s) + 2\,HBr(aq) \rightarrow CaBr_2(aq) + H_2O(l)$

- **Amphoteric oxides**

→ Exhibit both acid and base character
Examples: Al_2O_3, SnO_2

→ React with both acids and bases
Example: See the text (**Topic 6A.3**) for reactions of Al_2O_3 as an acid and as a base.

- **Trends in acidity for the oxides of main-group elements**

Increasing acidity ⟶

1	2	13/III	14/IV	15/V	16/VI	17/VII
Li_2O	**BeO**	B_2O_3	CO_2	N_2O_5	(O_2)	OF_2
Na_2O	MgO	**Al₂O₃**	SiO_2	P_4O_{10}	SO_3	Cl_2O_7
K_2O	CaO	**Ga₂O₃**	**GeO₂**	**As₂O₃**	SeO_3	Br_2O_7
Rb_2O	SrO	**In₂O₃**	**SnO** **SnO₂**	**Sb₂O₃**	TeO_3	I_2O_7
Cs_2O	BaO	Tl_2O_3	**PbO** **PbO₂**	**Bi₂O₃**	PoO_3	At_2O_7

Increasing acidity ↑

Note: Acidity of oxides in water tends to increase, as shown in the table. Amphoteric oxides are in bold type, basic oxides are to their left, and acidic oxides are to their right.   Exceptions are O_2, which is neutral, and OF_2, which is only weakly acidic.

6A.4 Proton Exchange Between Water Molecules

- **Amphiprotic species**
  - → A molecule or ion that can act as either a Brønsted acid (H$^+$ *donor*) or a Brønsted base (H$^+$ *acceptor*)

    Examples: H_2O: H$^+$ *donor* (→ OH$^-$) or H$^+$ *acceptor* (→ H_3O^+)

    NH_3: H$^+$ *donor* (→ NH_2^-) or H$^+$ *acceptor* (→ NH_4^+)

- **Autoprotolysis**
  - → A reaction in which one molecule transfers a proton to another molecule of the same kind

    Examples: Autoprotolysis of water: $2 H_2O(l) \rightleftharpoons H_3O^+(aq) + OH^-(aq)$

    Autoprotolysis of ammonia: $2 NH_3(l) \rightleftharpoons NH_4^+(am) + NH_2^-(am)$

- **Equilibrium constant**
  - → Autoprotolysis of water:

    $$2 H_2O(l) \rightleftharpoons H_3O^+(aq) + OH^-(aq); \quad K_w = [H_3O^+][OH^-] = 1 \times 10^{-14} \text{ at } 25 \text{ °C}$$

    Like any equilibrium constant, $K_w = K_w(T)$.

- **Pure water at 25 °C**
  - → $[H_3O^+] = [OH^-] = 1.0 \times 10^{-7} \text{ mol·L}^{-1}$

- **Neutral solution at 25 °C**
  - → $[H_3O^+] = [OH^-] = 1.0 \times 10^{-7} \text{ mol·L}^{-1}$

- **Acidic solution at 25 °C**
  - → $[H_3O^+] > 1.0 \times 10^{-7} \text{ mol·L}^{-1}$ and $[OH^-] < 1.0 \times 10^{-7} \text{ mol·L}^{-1}$

- **Basic solution at 25 °C**
  - → $[H_3O^+] < 1.0 \times 10^{-7} \text{ mol·L}^{-1}$ and $[OH^-] > 1.0 \times 10^{-7} \text{ mol·L}^{-1}$

Topic 6B: THE pH SCALE

6B.1 The Interpretation of pH

- **Concentration variation of H_3O^+**
  - → Values higher than 1 mol·L^{-1} and less than 10^{-14} mol·L^{-1} are not commonly encountered.

- **Logarithmic scale**
  - → pH = –log $[H_3O^+]$ *and* $[H_3O^+] = 10^{-pH}$
    pH values range between 1 and 14 in most cases.
  - → Pure water at 25 °C: pH = 7.00;   ⇒ Neutral solution: pH = 7.00
  - → Acidic solution: pH < 7.00;   ⇒ Basic solution: pH > 7.00

  Note: pH has *no* units, but $[H_3O^+]$ has units of molarity.

- **Measurement of pH**

  → *Universal indicator paper* turns different colors at different pH values.

  → *pH meter* measures a difference in electrical potential (proportional to pH) across electrodes that dip into the solution (see **Topic 6L**).

  Examples: The pH of the solutions for which the $[H_3O^+] = 5.4 \times 10^{-4}$ mol·L^{-1} is
  $$pH = -\log[H_3O^+] = -(-3.27) = 3.27 \quad (acidic \text{ solution})$$

  The $[H_3O^+]$ for a solution that has a pH value of 1.70 is $[H_3O^+] = 10^{-pH} = 10^{-1.70}$
  $$= 2.0 \times 10^{-2} \text{ mol·L}^{-1}.$$

  → pH values for common materials are given in **Figure 6B.2** in the text.

6B.2 The pOH of Solutions

- **pX as a generalization of pH**

  → $pX = -\log X$ (X = anything)

  → $pOH = -\log[OH^-]$ *and* $[OH^-] = 10^{-pOH}$

  → $pK_w = -\log K_w = -\log(1.0 \times 10^{-14}) = 14.00$ at 25 °C

- **Relationship between pH and pOH of a solution**

  → $[H_3O^+][OH^-] = K_w = 1.0 \times 10^{-14}$ at 25 °C

  → $\log([H_3O^+][OH^-]) = \log[H_3O^+] + \log[OH^-] = \log K_w$

  → $-\log[H_3O^+] - \log[OH^-] = -\log K_w$

  → **pH + pOH = 14.00 at 25 °C**

Topic 6C: WEAK ACIDS AND BASES

6C.1 Acidity and Basicity Constants

- **Weak acid**

  → pH value is *larger* than that of a strong acid with the same molarity.

  → Incomplete deprotonation

- **General weak acid HA**

  → $HA(aq) + H_2O(l) \rightleftharpoons H_3O^+(aq) + A^-(aq);$ $K_a = \dfrac{a_{H_3O^+}\, a_{A^-}}{a_{HA}\, a_{H_2O}}$

  But $a_{H_2O} = 1$, $a_{HA} \approx [HA]$, $a_{H_3O^+} \approx [H_3O^+]$, and $a_{A^-} \approx [A^-]$.

  → Thus, $K_a = \dfrac{[H_3O^+][A^-]}{[HA]}$ K_a is the *acidity constant;* $pK_a = -\log K_a$ is *also commonly used.*

Some Weak Acids and Their Acidity Constants (25 °C)

	Species	Formula	K_a	pK_a
Neutral acids	Hydrofluoric acid	HF	3.5×10^{-4}	3.45
	Acetic acid	CH_3COOH	1.8×10^{-5}	4.75
	Hydrocyanic acid	HCN	4.9×10^{-10}	9.31
Cation acids	Pyridinium	$C_5H_5^+$	5.6×10^{-6}	5.24
	Ammonium	NH_4^+	5.6×10^{-10}	9.25
	Ethylammonium	$C_2H_5NH_3^+$	1.5×10^{-11}	10.82
Anion acids	Hydrogen sulfate	HSO_4^-	1.2×10^{-2}	1.92
	Dihydrogen phosphate	$H_2PO_4^-$	6.2×10^{-8}	7.21

- **In the table**
  - → Hydrogen sulfate is the strongest acid (largest K_a, *smallest* pK_a).
  - → Ethylammonium is the weakest acid (smallest K_a, *largest* pK_a).
- **Weak base**
  - → pH value is *smaller* than that of a strong base with the same molarity.
  - → **Incomplete protonation**
- **General weak base B**

  - → $B(aq) + H_2O(l) \rightleftharpoons BH^+(aq) + OH^-(aq);$ $K_b = \dfrac{a_{BH^+} a_{OH^-}}{a_B a_{H_2O}}$

    But $a_{H_2O} = 1$, $a_B \approx [B]$, $a_{BH^+} \approx [BH^+]$, and $a_{OH^-} \approx [OH^-]$.

  - → Thus, $K_b = \dfrac{[BH^+][OH^-]}{[B]}$ K_b is the *basicity constant;* $pK_b = -\log K_b$ is *also commonly used.*

Some Weak Bases and Their Basicity Constants (25 °C)

	Species	Formula	K_b	pK_b
Neutral bases	Ethylamine	$C_2H_5NH_2$	6.5×10^{-4}	3.19
	Methylamine	CH_3NH_2	3.6×10^{-4}	3.44
	Ammonia	NH_3	1.8×10^{-5}	4.75
	Hydroxylamine	NH_2OH	1.1×10^{-8}	7.97
Anion bases	Phosphate	PO_4^{3-}	4.8×10^{-2}	1.32
	Fluoride	F^-	2.9×10^{-11}	10.54

- **In the table**
  - → Phosphate, PO_4^{3-}, is the strongest base (largest K_b, *smallest* pK_b).
  - → Fluoride, F^-, is the weakest base (smallest K_b, *largest* pK_b).
- **Acid-base strength**
  - → The *larger* the value of K, the *stronger* the acid or base
  - → The *smaller* the value of pK, the *stronger* the acid or base

6C.2 The Conjugate Seesaw

- **Reciprocity of conjugate acid-base pairs**
- **Conjugate acid-base pairs of weak acids**
  - → The conjugate base of a *weak* acid is a *weak* base.
  - → The conjugate acid of a *weak* base is a *weak* acid.
  - → For our purposes, *weak* is defined by pK values between 1 and 14.
- **Conjugate seesaw (qualitative)**
  - → The *stronger* an acid, the *weaker* its conjugate base.
  - → The *stronger* a base, the *weaker* its conjugate acid.
- **Conjugate seesaw (quantitative)**
  - → Consider a general *weak* acid HA and its conjugate base A^-.

$$HA(aq) + H_2O(l) \rightleftharpoons H_3O^+(aq) + A^-(aq) \quad K_a = \frac{[H_3O^+][A^-]}{[HA]}$$

$$A^-(aq) + H_2O(l) \rightleftharpoons HA(aq) + OH^-(aq) \quad K_b = \frac{[HA][OH^-]}{[A^-]}$$

$$K_a \times K_b = \frac{[H_3O^+][A^-]}{[HA]} \times \frac{[HA][OH^-]}{[A^-]} = [H_3O^+][OH^-] = K_w$$

  - → Or, for any conjugate pair, $\boxed{K_a \times K_b = K_w}$ and $\boxed{pK_a + pK_b = pK_w}$.
  - → If $K_a > 1 \times 10^{-7}$ ($pK_a < 7$), then the acid is *stronger* than its conjugate base.
  - → If $K_b > 1 \times 10^{-7}$ ($pK_b < 7$), then the base is *stronger* than its conjugate acid.
    See **Table 6C.3** in the text.

    Example: Sulfurous acid, H_2SO_3, is a stronger acid than nitrous acid, HNO_2. Consequently, nitrite, NO_2^-, is a stronger base than hydrogen sulfite, HSO_3^-.

- **Solvent leveling**
  - → An acid is strong if it is a stronger proton donor than the conjugate acid of the solvent.
- **Strong acid in water**
  - → *Stronger* proton donor than H_3O^+ ($K_a > 1$; $pK_a < 0$)
  - → $H_3O^+(aq) + H_2O(l) \rightleftharpoons H_2O(l) + H_3O^+(aq) \qquad K_a = 1$

→   All acids with $K_a > 1$ appear equally strong in water.

→   Strong acids are *leveled* to the strength of the acid H_3O^+.

   Example: HCl, HBr, HI, HNO_3, and $HClO_4$ are strong acids in water. When they are dissolved in water, their dissociation reaction with water is essentially complete, and they behave as though they were solutions of the acid H_3O^+ with anions Cl^-, Br^-, and the like.

→   Similar arguments hold for bases.

6C.3   Molecular Structure and Acid Strength

- **Binary acids, H_nX**

   →   For elements in the *same* period, the more *polar* the H–A bond, the *stronger* the acid.
   Example: HF is a stronger acid than H_2O, which is stronger than NH_3.

   →   For elements in the *same* group, the *weaker* the H–A bond, the *stronger* the acid.
   Example: HBr is a stronger acid than HCl, which is stronger than HF.

- **Summary**

   →   For binary acids, acid strength increases from left to right across a period (polarity of H–A bond increases) and from top to bottom down a group (H–A bond enthalpy decreases).

   →   The more polar or the weaker the H–A bond, the stronger the acid.

   →   See **Topic 6C.3** in the text for a deeper thermodynamic discussion of these trends.

6C.4   The Strengths of Oxoacids and Carboxylic Acids

- **Oxoacids (also called oxyacids)**

   →   The *greater* the number of oxygen atoms attached to the *same* central atom, the *stronger* the acid.

   →   This trend also corresponds to the *increasing* oxidation number of the *same* central atom.
   Example: $HClO_3$ (Cl ox #5) is a stronger acid than $HClO_2$ (Cl ox #3), which is stronger than HClO (Cl ox #1)

   →   For the *same* number of O atoms attached to the central atom, the *greater* the electronegativity of the central atom, the *stronger* the acid.

   Example: HClO is a stronger acid than HBrO, which is stronger than HIO. The halogen is considered the central atom for comparison to other oxoacids.

- **Carboxylic acids** →   The *greater* the electronegativities of the groups attached to the carboxyl (COOH) group, the *stronger* the acid.

   Example: $CHBr_2COOH$ is a stronger acid than $CH_2BrCOOH$, which is stronger than CH_3COOH.

   Note: The preceding rules are summarized in **Tables 6C.5, 6C.6,** and **6C.7** in the text.

Topic 6D: THE pH OF AQUEOUS SOLUTIONS

6D.1 Solutions of Weak Acids

- **Initial concentration**

  → Initial concentration of an acid ($[HA]_{initial}$) or base ($[B]_{initial}$) as prepared, assuming no proton transfer has occurred.

  → Primary equilibrium for weak acid: $HA(aq) + H_2O(l) \rightleftharpoons H_3O^+(aq) + A^-(aq)$

  → Secondary equilibrium: $2H_2O(l) \rightleftharpoons H_3O^+(aq) + OH^-(aq)$

- **General result for a dilute solution of the *weak* acid HA**

Main equilibrium	$HA(aq)$ +	$H_2O(l)$ $\rightleftharpoons$	$H_3O^+(aq)$ +	$A^-(aq)$
Initial molarity	$[HA]_{initial}$	—	0	0
Change in molarity	$-x$		$+x$	$+x$
Equilibrium molarity	$[HA]_{initial} - x$		x	x

  Acidity constant expression: $K_a = \dfrac{[H_3O^+][A^-]}{[HA]} = \dfrac{x^2}{[HA]_{initial} - x}$

  Quadratic equation: $x^2 + (K_a)x - (K_a[HA]_{initial}) = 0$ *and* $x > 0$ (positive root)

  Positive root: $x = \dfrac{-K_a + \sqrt{K_a^2 + 4K_a[HA]_{initial}}}{2} = [H_3O^+] = [A^-] \geq 10^{-6}$ M   **(1)**

  $[HA] = [HA]_{initial} - x$ $[OH^-] = K_w/[H_3O^+]$ from the secondary equilibrium

  Percentage *deprotonated* $= \dfrac{x}{[HA]_{initial}} \times 100\%$ **(2)**

- **Approximation to the general result**

  Assume $[HA]_{initial} - x \approx [HA]_{initial}$, then $K_a = \dfrac{x^2}{[HA]_{initial}}$ and $x = \sqrt{K_a[HA]_{initial}}$. **(3)**

  Assumption is justified if percentage deprotonated < 5%.

  Example: For acetic acid, CH_3COOH, $K_a = 1.8 \times 10^{-5}$. The value of $[H_3O^+]$ in a 0.100 M solution of CH_3COOH is given by equation (1).

  With $[HA]_{initial} = 0.1$, the result is $x = [H_3O^+] = 1.33 \times 10^{-3}$ M.

  Using the 5% approximation method in equation (3) yields $x = [H_3O^+]$ and
  $x = \sqrt{K_a[HA]_{initial}} = [(1.8 \times 10^{-5})(0.100)]^{1/2} = 1.34 \times 10^{-3}$ M.
  The approximation is excellent in this case.

6D.2 Solutions of Weak Bases

- **General result for a *weak* base B at *dilute* concentrations**

Main equilibrium	B(aq)	+	$H_2O(l)$	$\rightleftharpoons$	$BH^+(aq)$	+	$OH^-(aq)$
Initial molarity	$[B]_{initial}$		—		0		0
Change in molarity	$-x$				$+x$		$+x$
Equilibrium molarity	$[B]_{initial} - x$				x		x

Basicity constant expression: $K_b = \dfrac{[BH^+][OH^-]}{[B]} = \dfrac{x^2}{[B]_{initial} - x}$

Quadratic equation: $x^2 + (K_b)x - (K_b[B]_{initial}) = 0$ *and* $x > 0$ (positive root)

Positive root: $x = \dfrac{-K_b + \sqrt{K_b^2 + 4K_b[B]_{initial}}}{2} = [BH^+] = [OH^-] \geq 10^{-6}$ M   **(4)**

$[B] = [B]_{initial} - x$ $[H_3O^+] = K_w/[OH^-]$ secondary equilibrium

Percentage *protonated* $= \dfrac{x}{[B]_{initial}} \times 100\%$ **(5)**

- **Approximation to the general result**

Assume $[B]_{initial} - x \approx [B]_{initial}$, then $K_b = \dfrac{x^2}{[B]_{initial}}$ and $x = \sqrt{K_b[B]_{initial}}$ **(6)**

Assumption is justified if percentage protonated is less than 5%.

Example: K_b for the weak base methylamine, CH_3NH_2, is 3.6×10^{-4}. In a 0.200 M solution of methylamine, the $[OH^-]$ is given by equation (4), with $[B]_{initial} = 0.200$ M.

Substituting the values of K_b and $[B]_{initial}$ yields $x = [OH^-] = [CH_3NH_2^+]$ $= 8.31 \times 10^{-3}$ M.

Using the approximate solution in equation (6), $x = [OH^-] = [CH_3NH_2^+] = 8.49 \times 10^{-3}$ M.

The percent difference between the exact and approximate answers is less than 5%, so the approximation is justified.

→ For basic solutions, calculate pOH $= -\log[OH^-]$ and then use pH + pOH = 14 to calculate pH and $[H_3O^+]$.

6D.3 The pH of Salt Solutions

- **Acidic cations**

  → The conjugate acids of *weak* bases
    Examples: Ammonium, NH_4^+; anilinium, $C_6H_5NH_3^+$

  → Certain small, highly charged metal cations
    Examples: Iron(III), Fe^{3+}; copper(II), Cu^{2+}

- **Neutral cations**
  - → Group 1 and 2 cations; cations with +1 charge
    Examples: Li^+, Mg^{2+}, Ag^+

- **Basic anions**
  - → Conjugate bases of weak acids
    Examples: Acetate, CH_3COO^-; fluoride, F^-; sulfide, S^{2-}

- **Neutral anions**
  - → Conjugate bases of strong acids
    Examples: Cl^-, ClO_4^-; NO_3^-

- **Acidic anions**
  - → Only a few, such as HSO_4^-, $H_2PO_4^-$

- **Acid salt**
  - → Typically, a salt with an acidic cation
    Examples: $FeCl_3$, NH_4NO_3 (In these examples, the *anions* are neutral; discussed earlier)
  - → Dissolves in water to yield an acidic solution

    Example: $NH_4NO_3(aq) \rightarrow NH_4^+(aq) + NO_3^-(aq)$ (dissociation of salt in water)
    $NH_4^+(aq) + H_2O(l) \rightleftharpoons H_3O^+(aq) + NH_3(aq)$ (Acidic cation reacts with water to give an acidic solution.)
    $NO_3^-(aq)$ is the anion of a strong acid, HNO_3, and has no basic character in water (conjugate seesaw).

- **Basic salt**
  - → A salt with a basic anion
    Examples: Sodium acetate, NaO_2CCH_3; potassium cyanide, KCN
    (In these examples, the cations are neutral; discussed earlier.)
  - → Dissolves in water to yield a basic solution

    Example: $NaO_2CCH_3(aq) \rightarrow Na^+(aq) + CH_3COO^-(aq)$ (dissociation of salt in water)

    $CH_3COO^-(aq) + H_2O(l) \rightleftharpoons OH^-(aq) + CH_3COOH(aq)$
    (Basic cation reacts with water to yield a *basic solution.*)

- **Neutral salt**
  - → A salt with neutral cations and anions
    Examples: Sodium chloride, $NaCl$; potassium nitrate, KNO_3; magnesium iodide, MgI_2
  - → Dissolves in water to yield a neutral solution

- **Method of equilibrium solution (main equilibrium)**
  - → For acid salts, use the same method as for *weak* acids, but replace HA with BH^+.
  - → For basic salts, use the same method used for *weak* bases, but replace B with A^-. (All are *anions.*)
    Example: Iron(III) chloride, $FeCl_3$, is the salt of an acid cation and a neutral anion. It dissolves in water to give an *acidic* solution:

$FeCl_3(s) + 6\,H_2O(l) \rightarrow Fe(H_2O)_6{}^{3+}(aq) + 3\,Cl^-(aq)$ (dissolution of iron(III) chloride)

Fe^{3+} is surrounded by six water molecules. The +3 charge enhances the positive charges on the H atoms, making them more acidic than those in H_2O.

$Fe(H_2O)_6{}^{3+}(aq) + H_2O(l) \rightleftharpoons Fe(H_2O)_5(OH)^{2+}(aq) + H_3O^+(aq)$;
  (reaction of a weak acid with water)

For this equation, $K_a = 3.5 \times 10^{-3}$ (see **Table 6D.1** in the text).

The cation is a weak acid and the equation is the main equilibrium; use the methodology in **Topic 6D.1** 12.11 to calculate concentration. For example, the $[H_3O^+]$ of a 0.010 M solution of $FeCl_3(s)$ is given by **equation 1**, page 165:

$$x = \frac{-K_a + \sqrt{K_a{}^2 + 4K_a[HA]_{initial}}}{2} = [H_3O^+]$$

$$= \frac{-(3.5 \times 10^{-3}) + \sqrt{(3.5 \times 10^{-3})^2 + 4(3.5 \times 10^{-3})(0.010)}}{2}$$

$x = [H_3O^+] = 4.42 \times 10^{-3}$ $pH = -\log(4.42 \times 10^{-3}) = 2.35 \approx 2.4$

The solution is quite acidic, considered corrosive.

Topic 6E: POLYPROTIC ACIDS AND BASES

6E.1 The pH of a Polyprotic Solution

• **Successive deprotonations of a polyprotic acid**

→ Polyprotic acids are a species that can donate more than one proton.

→ Stepwise equilibria must be considered if successive K_a values differ by a factor of 10^3 or more, which is normally the case. Values of K_a are given in **Table 6E.1** in the text.

→ Polyprotic bases (species that can accept more than one proton) also exist.

• **Polyprotic acids and pH**

→ Use the *main* equilibrium for a dilute solution of a polyprotic acid to calculate the pH. Neglect any secondary equilibria except for H_2SO_4, whose first deprotonation is complete.

→ The calculation is identical to that for any weak monoprotic acid.

Example: The pH of a 0.025 M solution of diprotic carbonic acid, H_2CO_3, is calculated using the first dissociation step and **equation (1)** in **Topic 6D.1**

$H_2CO_3(aq) + H_2O(l) \rightleftharpoons H_3O^+(aq) + HCO_3{}^-(aq)$ $K_{a1} = 4.3 \times 10^{-7}$ (main equilibrium)

$$x = \frac{-K_{a1} + \sqrt{K_{a1}^2 + 4K_a[HA]_{initial}}}{2} = [H_3O^+]$$

$$= \frac{-(4.3 \times 10^{-7}) + \sqrt{(4.3 \times 10^{-7})^2 + 4(4.3 \times 10^{-7})(0.025)}}{2}$$

$$x = [H_3O^+] = 1.03 \times 10^{-4}\,\text{M}; \quad pH = -\log(1.03 \times 10^{-4}) = 3.9935 \approx 4.0$$

6E.2 Solutions of Salts of Polyprotic Acids

- **pH of a solution of an amphiprotic anion HA^-**

  → Given by $pH = \frac{1}{2}(pK_{a1} + pK_{a2})$

  → Independent of the concentration of the anion

  → Equation is valid if $s \gg K_w/K_{a2}$ and $s \gg K_{a1}$. Here s = the initial concentration of the salt.

  → When these approximations are not valid, a more complex expression for the pH is needed.

- **Solution of basic anion of polyprotic acid : A^{2-} or A^{3-} Examples:** Solutions of Na_2S or K_3PO_4

  → Main equilibrium is hydrolysis of the basic anion: $A^{2-}(aq) + H_2O(aq) \rightleftharpoons OH^-(aq) + HA^-(aq)$

  → Treat using methodology for solutions for basic anions: [**Topic 6D.2, equations 4 – 6**, this guide]

6E.3 The Concentrations of Solute Species

- **Assumptions**

  → The goal is to calculate the concentrations of all solute species of a polyprotic acid in aqueous solution.

  → The polyprotic acid is the species present in largest amount.

  → Only the first deprotonation contributes significantly to $[H_3O^+]$.

  → The autoprotolysis of water doesn't contribute significantly to $[H_3O^+]$ or $[OH^-]$.

- **Determination of the concentration of all species for a solution of a weak *diprotic* acid, H_2A**

  → Solve the *main* equilibrium (K_{a1}) as for any weak acid.

  → This yields $[H_2A]$, $[H_3O^+]$, and $[HA^-]$.

  → Solve the *secondary* equilibrium (K_{a2}) for $[A^{2-}]$ using the $[H_3O^+]$ and $[HA^-]$ values previously determined.

  → Determine $[OH^-]$ using K_w and $[H_3O^+]$.

- **Determination of the concentration of all species for a solution of a weak *triprotic* acid, H_3A**

  → Solve the *main* equilibrium (K_{a1}) as for any weak acid.

  → This yields $[H_3A]$, $[H_3O^+]$, and $[H_2A^-]$.

  → Solve the *secondary* equilibrium (K_{a2}) for $[HA^{2-}]$ using the previously determined values for $[H_3O^+]$ and $[H_2A^-]$.

  → Solve the *tertiary* equilibrium (K_{a3}) for $[A^{3-}]$ using the previously determined values for $[H_3O^+]$ and $[HA^{2-}]$.

  → Determine $[OH^-]$ using K_w and $[H_3O^+]$.

6E.4 Composition and pH

- **Polyprotic acid**

  → Concentrations of the several solution species vary with the pH of the solution.

  → As the pH *rises*, the fraction of deprotonated species *increases*.

- **Composition and pH**

  → Solution to the stepwise equilibrium expressions for a polyprotic acid as a function of pH

  - **Diprotic acid (H₂A): qualitative considerations**

    (See Figure 6E.1 in the text, reproduced here)

$$H_2A(aq) + H_2O(l) \rightleftharpoons H_3O^+(aq) + HA^-(aq) \qquad K_{a1} = \frac{[H_3O^+][HA^-]}{[H_2A]} \qquad (7)$$

$$HA^-(aq) + H_2O(l) \rightleftharpoons H_3O^+(aq) + A^{2-}(aq) \qquad K_{a2} = \frac{[H_3O^+][A^{2-}]}{[HA^-]} \qquad (8)$$

  → At low pH (high $[H_3O^+]$), equilibria represented by **equations 7** and **8** shift to the left (Le Chatelier) and the solution contains a high concentration of H_2A and very little A^{2-}.

  → At high pH (low $[H_3O^+]$), equilibria represented by **equations 7** and **8** shift to the right and the solution contains a high concentration of A^{2-} and very little H_2A.

  → At intermediate pH values, the solution contains a relatively high concentration of the amphiprotic anion HA^- (see the following figure).

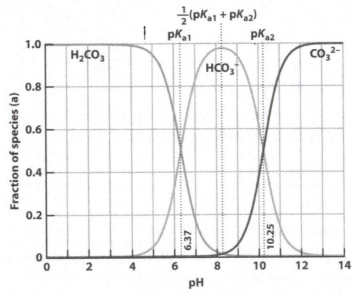

- **Diprotic acid (H₂A)** ⇒ **quantitative considerations**

  → *Fraction (f)* of H_2A, HA^-, and A^{2-} in solution as a function of $[H_3O^+]$

  → $f(H_2A) = \dfrac{[H_3O^+]^2}{H} \qquad f(HA^-) = \dfrac{[H_3O^+]K_{a1}}{H} \qquad f(A^{2-}) = \dfrac{K_{a1}K_{a2}}{H}$

→ where $H = [H_3O^+]^2 + [H_3O^+] K_{a1} + K_{a1}K_{a2}$.

→ The maximum value of (HA^-) is at $pH = \frac{1}{2} (pK_{a1} + pK_{a2})$.

→ Similar equations and graphs can be developed for triprotic acids (see text **Figure 6E.1**).

Topic 6F: AUTOPROTOLYSIS AND pH

6F.1 Very Dilute Solutions of Strong Acids and Bases

- **Very dilute solution**
  → Concentration of a strong acid or base is less than 10^{-6} M.

- **General result for a *strong* acid HA at *very dilute* concentrations**

$$HA(aq) + H_2O(l) \rightarrow H_3O^+(aq) + A^-(aq)$$

Three equations in *three* unknowns ($[H_3O^+]$, $[OH^-]$, and $[A^-]$)

Autoprotolysis equilibrium: $K_w = [H_3O^+][OH^-] = 1.0 \times 10^{-14}$

Charge balance: $[H_3O^+] = [OH^-] + [A^-]$

Material balance: $[A^-] = [HA]_{initial}$ (complete deprotonation for a strong acid)

Solution: $[OH^-] = [H_3O^+] - [HA]_{initial}$ (from charge and material balance)

$$K_w = [H_3O^+]([H_3O^+] - [HA]_{initial})$$

$$K_w = [H_3O^+]^2 - [HA]_{initial}[H_3O^+], \quad \text{let } x = [H_3O^+]$$

Quadratic equation: $x^2 - [HA]_{initial} x - K_w = 0$ *and* $x > 0$ (positive root)

Positive root: $x = \dfrac{[HA]_{initial} + \sqrt{[HA]_{initial}^2 + 4K_w}}{2} = [H_3O^+]$ **(9)**

$$\text{and } [OH^-] = K_w/[H_3O^+]$$ **(10)**

- **General result for a *strong* base MOH at *very dilute* concentrations**

$$MOH(aq) + H_2O(l) \rightarrow M^+(aq) + OH^-(aq)$$

Three equations in *three* unknowns:

Autoprotolysis equilibrium: $K_w = [H_3O^+][OH^-] = 1.0 \times 10^{-14}$

Charge balance: $[M^+] + [H_3O^+] = [OH^-]$

Material balance: $[M^+] = [MOH]_{initial}$ (complete dissociation)

Solution: $[OH^-] = [H_3O^+] + [MOH]_{initial}$

$$K_w = [H_3O^+]\{[H_3O^+] + [MOH]_{initial}\}$$

$$K_w = [H_3O^+]^2 + [MOH]_{initial}[H_3O^+], \text{ let } x = [H_3O^+]$$

Quadratic equation: $\qquad x^2 + [MOH]_{initial}\, x - K_w = 0 \ and \ x > 0 \ \text{(positive root)}$

Positive root: $\qquad x = \dfrac{-[MOH]_{initial} + \sqrt{[MOH]_{initial}^2 + 4K_w}}{2} = [H_3O^+]$ $\qquad$ **(11)**

$$and \quad [OH^-] = K_w/[H_3O^+] \qquad\qquad\qquad\qquad\text{(12)}$$

Note: Strong bases such as O^{2-} and CH_3^- produce aqueous solutions that are equivalent to MOH.

Example: The pH of a 1.5×10^{-8} M solution of the strong base KOH can be calculated from **equations 11** and **12**.

From **11**, $[H_3O^+] = \dfrac{-[MOH]_{initial} + \sqrt{[MOH]_{initial}^2 + 4K_w}}{2} = 9.28 \times 10^{-8}$ M.

$pH = -\log[H_3O^+] = -\log(9.28 \times 10^{-8}) = 7.03.$
(slightly *basic* solution as expected)

6F.2 Very Dilute Solutions of Weak Acids

- **Very dilute solution**
  → Concentration of a weak acid or base is less than about 10^{-3} M.

- **General result for a *weak* acid HA at *very dilute* concentrations**
  Four equations in *four* unknowns:

  (1) Weak acid equilibrium: $HA(aq) + H_2O(l) \rightleftharpoons H_3O^+(aq) + A^-(aq);$ $\qquad K_a = \dfrac{[H_3O^+][A^-]}{[HA]}$

  (2) Autoprotolysis equilibrium: $\quad K_w = [H_3O^+][OH^-] = 1.0 \times 10^{-14}$ at 25 °C

  (3) Charge balance: $\qquad\qquad\qquad [H_3O^+] = [OH^-] + [A^-]$

  (4) Material balance: $\qquad\qquad [HA]_{initial} = [HA] + [A^-]$ (incomplete deprotonation)

  Solution: $\qquad [A^-] = [H_3O^+] - [OH^-] = [H_3O^+] - \dfrac{K_w}{[H_3O^+]}$ (from charge balance)

  $$[HA] = [HA]_{initial} - [A^-] = [HA]_{initial} - [H_3O^+] + \dfrac{K_w}{[H_3O^+]}$$

  Substitute expressions for [HA] and [A$^-$] into K_a.

  $$K_a = \dfrac{[H_3O^+]\left([H_3O^+] - \dfrac{K_w}{[H_3O^+]}\right)}{[HA]_{initial} - [H_3O^+] + \dfrac{K_w}{[H_3O^+]}} \qquad\qquad\text{(13)}$$

Let $x = [H_3O^+]$ and then

$$K_a = \frac{x\left(x - \frac{K_w}{x}\right)}{[HA]_{initial} - x + \frac{K_w}{x}}$$ Rearrange to a cubic form.

Cubic equation: $x^3 + K_a x^2 - (K_w + K_a [HA]_{initial})x - K_a K_w = 0$

Solve by using a graphing calculator, trial and error, or the mathematical software on the text:

http://bcs.whfreeman.com/chemicalprinciples7e/.

Only one of the three roots will be physically meaningful.

Alternatively, consider the following approximations:

If $[H_3O^+] > 10^{-6}$ M, then $\dfrac{K_w}{[H_3O^+]} < 10^{-8}$, and this term can be neglected in **equation 13** to yield

$$K_a = \frac{[H_3O^+]^2}{[HA]_{initial} - [H_3O^+]} = \frac{x^2}{[HA]_{initial} - x} \qquad (14)$$

Further, if $[H_3O^+] > 10^{-6}$ M *and* $[H_3O^+] \ll [HA]_{initial}$, then **equation 14** reduces to

$$K_a = \frac{[H_3O^+]^2}{[HA]_{initial}} = \frac{x^2}{[HA]_{initial}} \qquad (15)$$

Use **equations 13, 14,** *or* **15** as appropriate to solve problems.

Topic 6G: BUFFERS

- **Common themes**
  → Many equilibria can be treated in the following way:
    1. Identify the solute species in solution.
    2. Identify the equilibrium relations among the solute species.
    3. Use the identified equilibrium along with equilibrium constants, to determine the concentrations of the solute species, often by using an equilibrium table.

6G.1 Buffer Action

- **Mixed solution**
  → A solution containing a weak acid or base and one of its salts

Example: A solution of hydrofluoric acid and sodium fluoride

- **Buffer**
  - → A mixed solution containing weak conjugate acid–base pairs that *stabilizes* the pH of a solution
  - → Provides a source and a sink for protons

- **Acid buffer**
  - → Consists of a weak acid and its conjugate base provided as a salt

    Examples: Acetic acid (CH_3COOH) and sodium acetate ($NaCH_3COO$);
    phosphorous acid (H_3PO_3) and potassium dihydrogenphosphite (KH_2PO_3)

  - → Buffers a solution on the *acid* side of neutrality (pH < 7), if $pK_a < 7$

- **Base buffer**
  - → Consists of a weak base and its conjugate acid provided as a salt

    Examples: Ammonia (NH_3) and ammonium chloride (NH_4Cl);
    sodium dihydrogenphosphate (NaH_2PO_4) and
    sodium hydrogenphosphate (Na_2HPO_4)

  - → Buffers a solution on the *base* side of neutrality (pH > 7) if $pK_b < 7$

6G.2 Designing a Buffer

- **pH of a buffer**
  - → The pH of a buffer solution is given approximately by the Henderson–Hasselbalch equation.

$$pH \approx pK_a + \log\left(\frac{[\text{base}]_{\text{initial}}}{[\text{acid}]_{\text{initial}}}\right)$$

  - → Adjustment from initial concentrations to the equilibrium concentrations of the conjugate acid–base pair is *often* negligible. If it is not, the pH can be adjusted to the desired value by adding acid or base. Use of this equation is equivalent to setting up a table of concentrations and solving an equilibrium problem if adjustment from initial concentrations is negligible.
  - → To minimize approximations and adjustments, an equilibrium table can be prepared and more exact concentrations for the conjugate pair obtained directly.

- **Preparation of an acid buffer**
  - → Use an acid (HA) with a pK_a value close to the desired pH.
  - → Add the acid (HA) and its conjugate base (A^-) in appropriate amounts to satisfy the Henderson–Hasselbalch equation.
  - → Measure the pH and make any necessary adjustments to the acid or base concentration to achieve the desired pH.

- **Preparation of a base buffer**
  - → Use a base (B) whose conjugate acid (BH^+) has a pK_a value close to the desired pH.
  - → Add the base (B) and its conjugate acid (BH^+) in appropriate amounts to satisfy the Henderson–Hasselbalch equation.

→ Measure the pH and make any necessary adjustments to the acid or base concentration to achieve the desired pH.

6G.3 Buffer Capacity

→ The amount of acid or base that can be added before the buffer loses its ability to resist a change in pH

→ Determined by its concentration and pH

→ The *greater* the concentrations of the conjugate acid-base pair in the buffer system, the *greater* the buffer capacity

- **Exceeding the capacity of a buffer**

  → .Buffers function by providing a weak acid to react with added base and a weak base to react with added acid.

  → Adding sufficient base to react with *all* of the weak acid or adding sufficient acid to react with *all* of the weak base exhausts (completely destroys) the buffer.

- **Limits of buffer action**

  → A buffer has a *practical* limit to its buffering ability that is reached before the buffer is completely exhausted.

  → As a *rule of thumb*, a buffer acts most effectively in the range

$$10 > \frac{[\text{base}]_{\text{initial}}}{[\text{acid}]_{\text{initial}}} > 0.1$$.

  → The molarity ratios on the previous page yield the following expression for the *effective pH range of a buffer*:

$$\text{pH} = \text{p}K_a \pm 1$$

Topic 6H: ACID-BASE TITRATIONS

- **Working with titrations**

  → Full chemical equation must be kept in mind to ensure the correct stoichiometry.

  Example: The neutralization of calcium hydroxide with hydrochloric acid requiring two moles of acid for each mole of base neutralized:

$$2\,\text{HCl(aq)} + \text{Ca(OH)}_2 \rightarrow \text{CaCl}_2\text{(aq)} + 2\,\text{H}_2\text{O(l)}$$

- **Analyte**

  → Unknown acid or base sample to be analyzed

- **Titrant**

  → Base or acid solution of known concentration added to neutralize the analyte

- **Stoichiometric point**
  - → Stage in a titration when exactly the correct volume of titrant has been added to the analyte to complete the reaction

6H.1 Strong Acid–Strong Base Titrations

- **pH curve**
  - → Plot of the pH of the *analyte* solution as a function of the volume of the *titrant* added during a titration

- **For all strong base–strong acid or strong acid–strong base titrations**
  - → The stoichiometric point has a pH value of 7 (neutral solution).
  - → The net ionic equation for the reaction is $H_3O^+(aq) + OH^-(aq) \rightarrow 2\ H_2O(l)$ for all strong base–strong acid or strong acid–strong base titrations.

- **Titration of a strong base (analyte) with a strong acid (titrant) (see Figure 6H.1 in the text)**
  - → The pH *changes (drops) slowly* with added titrant until the stoichiometric point is approached.
  - → The pH then *changes (drops) rapidly* near the stoichiometric point.
  - → At the stoichiometric point, the base is neutralized, at which point the titration is normally ended.
  - → In the region beyond the stoichiometric point, the pH approaches the value of the titrant solution slowly.

- **Titration of a strong acid (analyte) with a strong base (titrant) (see Figure 6H.2 in the text)**
  - → The pH *changes (rises) slowly* with added titrant until the stoichiometric point is approached.
  - → The pH then *changes (rises) rapidly* near the stoichiometric point.
  - → At the stoichiometric point, all the acid is neutralized.
  - → The pH approaches the value of the titrant solution slowly in the region well beyond the stoichiometric point.

- **pH curve calculations**
  - → As a titration proceeds, both the volume of the analyte solution and the amount of acid or base present in it change.
  - → For calculations, it is often best to calculate amounts (moles) and volume separately and to divide them to obtain concentrations.
  - → To calculate $[H_3O^+]$ in the analyte solution, the amount of acid or base remaining should be determined and divided by the total volume of the solution.
  - → Before the stoichiometric point is reached, follow the procedure outlined in **Example 6H.1** in the text.
  - → For a strong acid–strong base titration, the pH at the stoichiometric point is 7.
  - → After the stoichiometric point, follow the procedure outlined in **Toolbox 6H.1** in the text.
  - → Always assume that the volumes of the titrant and analyte solutions are additive:

  $$(V_{total} = V_{analyte} + V_{titrant\ added})$$

→ If a solid is added to a solution, its volume is usually neglected.

6H.2 Strong Acid–Weak Base and Weak Acid–Strong Base Titrations

- **Characteristics of the titrations**
    - → The titrant is the *strong* acid or base; the analyte is the *weak* base or acid.
    - → Stoichiometric point has a pH value *different* from 7.

- **Strong base–weak acid titration**
    - → A strong base *dominates* the weak acid, and the pH at the stoichiometric point is *greater* than 7.
    - → A strong base converts a weak acid, HA, into its conjugate base form, A^-.
    - → At the stoichiometric point, the pH is that of the resulting *basic* salt solution.

- **Strong acid–weak base titration**
    - → A strong acid *dominates* the weak base, and the pH at the stoichiometric point is *less* than 7.
    - → A strong acid converts a weak base, B, into its conjugate acid form, BH^+.
    - → At the stoichiometric point, the pH is that of the resulting *acidic* salt solution.

- **Shape of the pH curve**
    - → *Similar* to that of a strong acid–strong base titration, but with some important differences
    - → *Differences*
        - ⇒ A buffer region appears between the initial analyte solution and the stoichiometric point.
        - ⇒ The pH of the stoichiometric point is different from 7.
        - ⇒ There is a less abrupt pH change in the region of the stoichiometric point.

- **pH curve calculations**
    - → The pH is governed by the *main* species in solution.
    - → Calculate amounts and solution volume separately; divide to obtain concentrations.
    - → Before the titrant is added, the analyte is a solution of a weak acid or base in water. Determine the pH of the solution by following the procedures in **Topic 6D** in the text.
    - → Before the stoichiometric point is reached, the solution is a buffer with differing amounts of acid and conjugate base. To calculate the pH, follow the procedure outlined in **Example 6H.3** in the text.
    - → At the stoichiometric point, the analyte solution consists of the salt of a weak acid or of a weak base. To calculate the pH, follow the procedure outlined in **Example 6H.2** in the text.
    - → Well after the stoichiometric point is passed, the pH of the solution approaches the value of the titrant, appropriately diluted.
    - → Assume that the volumes of the added titrant solution and the analyte solution are additive:
    $$(V_{total} = V_{analyte} + V_{titrant\ added})$$
    If the sample is a solid, its volume is usually neglected.

6H.3 Acid-Base Indicators

→ Water-soluble dyes with different *characteristic colors* associated with their acid and base forms

→ Exhibit a color change over a *narrow* pH range

→ Added in low concentration to the analyte so that a titration is primarily that of the analyte, not the indicator

→ Selected to monitor the stoichiometric or *end point* of a titration

→ An indicator is a weak acid. It takes part in the following proton - transfer equilibrium:

$$HIn(aq) + H_2O(l) \rightleftharpoons H_3O^+(aq) + In^-(aq) \qquad K_{In} = \frac{[H_3O^+][In^-]}{[HIn]}$$

Values of pK_{In} are given in **Table 6H.2** in the text.

→ The color change is usually most apparent when $[HIn] = [In^-]$ and $pH = pK_{In}$. The color change begins typically within one pH unit before pK_{In} and is essentially complete about 1 pH unit after pK_{In}. The pK_{In} value of an indicator is usually chosen to be within 1 pH unit of the stoichiometric point.

$$\boxed{pK_{In} \approx pH(\text{stoichiometric point}) \pm 1}$$

- **End point of a titration**

  → Stage of a titration where the color of the indicator in the analyte is midway between its acid and base colors

  → End point of an indicator is the point at which the concentrations of its acid and base forms are equal.

  → At the end point, $pH = pK_{In}$, where "In" stands for the indicator.

  → An indicator should be chosen such that its end point is close to the stoichiometric point.

6H.4 Polyprotic Acid Titrations

- **Polyprotic acids**

  → Review the list of the common polyprotic acids in **Table 6E.1** of the text.

  → All the acids in **Table 6E.1** are *diprotic* except for phosphoric acid, which is *triprotic*.

  → Phosphorous acid, H_3PO_3, is *diprotic*. The major Lewis structures are

The proton bonded to phosphorus is *not* acidic.

- **Titration curve of a polyprotic acid**

  → Has stoichiometric points corresponding to the removal of each acidic hydrogen atom

  → Has buffer regions between successive stoichiometric points

→ At each stoichiometric point, the analyte is the salt of a conjugate base of the acid or one of its hydrogen-bearing anions.

→ See the examples of pH curves for the triprotic acid histidine $C_6H_{11}N_3O_2Cl_2$ in **Figure 6H.10** and for the diprotic acid oxalic acid $[(COOH)_2]$ in **Figure 6H.11** of the text.

- **pH calculations**

→ The pH curve can be estimated at any point by considering the primary species in solution and the *main* proton - transfer equilibrium that determines the pH.

Topic 6I: SOLUBILITY EQUILIBRIA

6I.1 The Solubility Product

- **Solubility product, K_{sp}**

→ Equilibrium constant for the equilibrium between an ionic solid and its dissolved form (saturated solution)

Example: $Sb_2S_3(s) \rightleftharpoons 2\,Sb^{3+}(aq) + 3\,S^{2-}(aq)$ $K_{sp} = \dfrac{a_{Sb^{3+}}^{2}\, a_{S^{2-}}^{3}}{a_{Sb_2S_3(s)}}$

→ The activity of a pure solid is 1, and for *dilute* solutions of sparingly soluble salts, the activity of a solute species can be replaced by its molarity.

→ To a fair approximation, $K_{sp} = [Sb^{3+}]^2[S^{2-}]^3$.   (saturated solution)

- **Molar solubility, s**

→ Maximum amount of solute that dissolves in enough water to make 1 L of solution

- **Relationship between s and K_{sp}**

→ From the example, the stoichiometric relationships are

   1 mol $Sb_2S_3 \;\hat{=}\; 2$ mol Sb^{3+} and 1 mol $Sb_2S_3 \;\hat{=}\; 3$ mol S^{2-}.

→ The molar solubility, s, of the salt $Sb_2S_3(s)$ is related to the ion concentrations at equilibrium as follows:

   $[Sb^{3+}] = 2s$ and $[S^{2-}] = 3s$.

→ The relationship between s and K_{sp} is then

$$K_{sp} = [Sb^{3+}]^2\,[S^{2-}]^3 = (2s)^2(3s)^3 = (4s^2)(27s^3) = 108s^5 \text{ and } s = \left(\frac{K_{sp}}{108}\right)^{1/5}$$

→ The relationship $s = \left(\dfrac{K_{sp}}{108}\right)^{1/5}$ holds for any salt whose general formula is X_2Y_3.

→ Similar relationships can be generated for other type of salts, for example, for an AB salt such as AgCl, $s = K_{sp}^{1/2}$.

→ Because of interionic effects and *incomplete* dissociation of some salts (e.g., PbI_2), calculations using K_{sp} are only approximate.

6I.2 The Common-Ion Effect

→ A decrease in the solubility of a salt caused by the presence of one of the ions of the salt in solution

→ Follows from Le Chatelier's principle

→ Used to help remove unwanted ions from solution

Example: Consider the equilibrium $Sb_2S_3(s) \rightleftharpoons 2\,Sb^{3+}(aq) + 3\,S^{2-}(aq)$.

Increasing the concentration of Sb^{3+} ions by addition of the soluble salt $Sb(NO_3)_3$ shifts the equilibrium to the left, causing some precipitation of $Sb_2S_3(s)$. Thus, less $Sb_2S_3(s)$ is dissolved in the presence of an external source of Sb^{3+} (or S^{2-}) than in pure water.

- **Quantitative features**

→ Use the solubility product expression to estimate the reduction in solubility resulting from the addition of a common ion.

Example: Consider the equilibrium $Sb_2S_3(s) \rightleftharpoons 2\,Sb^{3+}(aq) + 3\,S^{2-}(aq)$, $K_{sp} = [Sb^{3+}]^2\,[S^{2-}]^3$.

If $Sb_2S_3(s)$ is added to a solution already containing Sb^{3+} ions [e.g., an $Sb(NO_3)_3$ solution], then dissolution of this *sparingly soluble salt* will add a negligible amount to the initial $[Sb^{3+}]$ cation concentration. The solubility, s, and its relationship to K_{sp} is then

$$K_{sp} = [Sb^{3+}]^2\,(3s)^3 \quad \text{and} \quad s = \left(\frac{[Sb^{3+}]^2}{27} \right)^{1/3}$$

Note: Owing to interionic effects, the activities of ions differ markedly from their concentrations. As a result, simple equilibrium calculations for these systems using concentrations are at best semi-quantitative. They become more accurate as the solubility of the salt decreases.

6I.3 Complex-Ion Formation

- **Complex ions**

→ If a salt can form a complex ion with other species in solution, its solubility increases.

→ Formed by the reaction of a metal cation, acting as a Lewis acid, with a Lewis base

→ May disturb a solubility equilibrium by reducing the concentration of metal ions

→ Net result is an increase in the solubility of salts.

Example: $Fe^{2+}(aq) + 6CN^-(aq) \rightleftharpoons Fe(CN)_6^{4-}(aq)$

In this reaction, Fe^{2+} acts as a Lewis acid and CN^- acts as a Lewis base. Complexation reduces the concentration of Fe^{2+} in solution, thereby causing a shift in any solubility equilibrium in which Fe^{2+} is involved.

→ A number of formation reactions are listed in **Table 6I.2** in the text.

→ Electron pair acceptors (Lewis acids) that form complex ions include metal cations such as $Fe^{3+}(aq)$, $Cu^{2+}(aq)$, and $Ag^+(aq)$.

→ Electron pair donors (Lewis bases) that form complex ions include basic anions or neutral bases such as $Cl^-(aq)$, $NH_3(aq)$, $CN^-(aq)$, and the thiosulfate ion $S_2O_3^{2-}(aq)$.

Quantitative features

→ The equilibrium for a *formation reaction* is characterized by the *formation constant, K_f*.

Example: $Fe^{2+}(aq) + 6CN^-(aq) \rightleftharpoons Fe(CN)_6^{4-}(aq)$ $K_f = \dfrac{[Fe(CN)_6^{4-}]}{[Fe^{2+}][CN^-]^6}$

→ Values of K_f at 25 °C for several complex ion equilibria are given in **Table 6I.2** in the text.

→ Formation and solubility reactions may be combined to predict the solubility of a salt in the presence of a species that forms a complex with the metal ion.

Example: $FeS(s) \rightleftharpoons Fe^{2+}(aq) + S^{2-}(aq)$ $K_{sp} = [Fe^{2+}][S^{2-}]$ *solubility reaction*

$Fe^{2+}(aq) + 6CN^-(aq) \rightleftharpoons Fe(CN)_6^{4-}(aq)$ $K_f = \dfrac{[Fe(CN)_6^{4-}]}{[Fe^{2+}][CN^-]^6}$ *formation reaction*

Adding two reactions produces the following:

$FeS(s) + 6CN^-(aq) \rightleftharpoons Fe(CN)_6^{4-}(aq) + S^{2-}(aq)$

The equilibrium constant K for the combined reaction is

$K = K_{sp} \times K_f = \dfrac{[Fe(CN)_6^{4-}][S^{2-}]}{[CN^-]^6} = (6.3 \times 10^{-18})(7.7 \times 10^{36}) = 4.9 \times 10^{19} \gg 1$.

The equilibrium lies far to the right, indicating a large solubility for $FeS(s)$ in a cyanide solution.

Because 1 mol FeS $\approx$ 1 mol S^{2-}, the molar solubility, s, is equal numerically to the sulfide ion concentration.

Topic 6J: PRECIPITATION

6J.1 Predicting Precipitation

• If Q_{sp} is greater than K_{sp}, then a salt precipitates.

- **To predict whether precipitation occurs**
  - → Calculate Q_{sp} using actual or predicted ion concentrations.
  - → Compare Q_{sp} to K_{sp}.
  - → If $Q_{sp} > K_{sp}$, the salt will precipitate from solution.
  - → If $Q_{sp} = K_{sp}$, the solution is saturated with respect to the ions in the expression for K_{sp}. Any additional amount of these ions will cause precipitation.
  - → If $Q_{sp} < K_{sp}$, the solution is unsaturated and no precipitate will form, and additional salt could be dissolved if desired.

6J.2 Selective Precipitation

- **Separation of ions in a mixture**
  - → Separated by adding a soluble salt containing an oppositely charged ion that forms salts having very different solubilities from the original ions

- **Separation of different cations in solution**
  - → Add a soluble salt containing an *anion* with which the cations form insoluble salts.
  - → If the insoluble salts have sufficiently different solubilities, they will precipitate at different anion concentrations and can be collected separately.

    Example: Careful addition of sodium fluoride, NaF(s), to a solution containing Ca^{2+} and Pb^{2+} leads to some separation of the cations. CaF_2 ($K_{sp} = 4.0 \times 10^{-11}$) precipitates before PbF_2 ($K_{sp} = 3.7 \times 10^{-8}$). Thus the Ca^{2+} and Pb^{2+} ions are separated to some extent.

6J.3 Dissolving Precipitates

- **Ion removal**
  - → Removal from solution of *one* of the ions in the solubility equilibrium *increases* the solubility of the solid.
  - → A precipitate will dissolve completely if a sufficient number of such ions are removed.
  - → In this section, we focus on the removal of *anions*.

- **Removal of basic anions**
  - → Addition of a strong acid, such as HCl, neutralizes a basic anion.

    Example: Hydroxides react to form water:

    $$H_3O^+(aq) + OH^-(aq) \rightarrow 2\,H_2O(l)$$

  - → In some cases, gases are evolved.

    Example: Carbonates react with acid to form carbon dioxide gas:

    $$2\,H_3O^+(aq) + CO_3{}^{2-}(aq) \rightarrow CO_2(g) + 3\,H_2O(l)$$

- **Removal by oxidation of anions**

    Example: S^{2-}(aq) from a metal sulfide is oxidized to S(s) with nitric acid:

    $$3\,S^{2-}(aq) + 8\,H_3O^+(aq) + 2\,NO_3{}^-(aq) \rightarrow 3\,S(s) + 2\,NO(g) + 12\,H_2O(l)$$

- **Summary**
  - → The solubility of a salt may be increased by removal of its ions from solution. An acid can be used to dissolve a hydroxide, sulfide, sulfite, or carbonate salt. Nitric acid can be used to oxidize metal sulfides to sulfur and a soluble salt.
  - → Some solids can be dissolved by a change in temperature.

6J.4 Qualitative Analysis

- → Method used to identify and separate ions present in an unknown solution (e.g., seawater)
- → Utilizes complex formation, selective precipitation, and pH control
- → Uses solubility equilibria to remove and identify ions selectively

- **Outline of the method**
  - → The method is outlined in **Figure 6J.4** in the text and is summarized here:

Partial Qualitative Analysis Scheme

Procedure	Possible precipitate	K_{sp}
1) Add HCl(aq)	AgCl	1.6×10^{-10}
	Hg_2Cl_2	1.3×10^{-18}
	$PbCl_2$	1.6×10^{-5}
2) Add H_2S(aq)	Bi_2S_3	1.0×10^{-97}
(in acid solution, there is a	CdS	4.0×10^{-29}
low S^{2-} concentration)	CuS	7.9×10^{-45}
	HgS	1.6×10^{-52}
	Sb_2S_3	1.6×10^{-93}
3) Add base to H_2S(aq)	FeS	6.3×10^{-18}
(in basic solution, there is a	MnS	1.3×10^{-15}
higher S^{2-} concentration)	NiS	1.3×10^{-24}
	ZnS	1.6×10^{-24}

Note: In solubility calculations, the fairly strong basic character of S^{2-} must be considered.

$$S^{2-}(aq) + H_2O(l) \rightleftharpoons HS^-(aq) + OH^-(aq) \qquad K_b = 1.4$$

- → A solution may contain any or all of the cations given in the table. We wish to determine which ones are present and which are absent.

- **Procedure 1: Add HCl to the solution.**
  - → Most chlorides are soluble and will not precipitate as the $[Cl^-]$ increases.
  - → For AgCl and Hg_2Cl_2, we expect their Q_{sp} values to be exceeded and for them to precipitate.

→ $PbCl_2$ is slightly more soluble, but it should also precipitate.

→ The precipitate obtained, if any, using Procedure 1 is collected and the remaining solution tested for additional ions using Procedure 2.

- **Procedure 2: Add H₂S to the solution.**

  → Because the solution is highly acidic after Procedure 1, the concentration of sulfide will be very low:

$$H_2S(aq) + 2\,H_2O(l) \rightleftharpoons 2\,H_3O^+(aq) + S^{2-}(aq)$$

  → Thus, only sulfide salts with very low solubility products, such as CuS or HgS, will precipitate.

  → The resulting precipitate, if any, is collected and the remaining solution tested for additional ions using Procedure 3.

- **Procedure 3: Add base to the solution.**

  → Base reduces $[H_3O^+]$ and as seen from the equilibrium, increases $[S^{2-}]$.

  → At higher sulfide concentration, more soluble sulfide salts such as FeS and ZnS precipitate.

  → The resulting precipitate, if any, is collected.

- **Precipitates obtained in Procedures 1, 2, and 3 are analyzed separately for the presence of each cation in the group.**

- **Determining the presence or absence of Ag⁺, Hg₂²⁺, and Pb²⁺ in Procedure 1**

  → Of the three possible chlorides, $PbCl_2$ is the most soluble, and its solubility increases with increasing temperature.

  → Rinse the precipitate with hot water, collect the liquid, and test for Pb^{2+} by adding chromate.

  → A precipitate of insoluble lead(II) chromate indicates that Pb^{2+} is present in the initial solution:

$$Pb^{2+}(aq) + CrO_4^{2-}(aq) \rightarrow PbCrO_4(s)$$

  → Add aqueous ammonia to the remaining precipitate.

  → Silver(I) will dissolve, forming the diammine complex

$$Ag^+(aq) + 2\,NH_3(aq) \rightarrow Ag(NH_3)_2^+(aq).$$

  → In aqueous ammonia, mercury(I) will disproportionate (a single element is simultaneously oxidized and reduced) to form a gray mixture of liquid mercury and solid $HgNH_2Cl$:

$$Hg_2Cl_2(s) + 2\,NH_3(aq) \rightarrow Hg(l) + HgNH_2Cl(s) + NH_4^+(aq) + Cl^-(aq)$$

  → Formation of the gray mixture confirms the presence of mercury(I) in the original solution.

  → Silver(I) is confirmed by adding HCl to the ammine solution to precipitate white AgCl(s):

$$Ag(NH_3)_2^+(aq) + Cl^-(aq) + 2\,H_3O^+(aq) \rightarrow AgCl(s) + 2\,NH_4^+(aq) + 2\,H_2O(l)$$

  → Similar determinations can be used to identify the presence of the remaining cations.

Topic 6K: REPRESENTING REDOX REACTIONS

6K.1 Half-Reactions

→ *Conceptual* way of reporting an oxidation or reduction process

→ Important for understanding oxidation–reduction (*redox*) reactions and electrochemical cells

- **Oxidation**

  → *Removal* of electrons from a species

  → Increase in oxidation number of a species

  → Represented by an *oxidation* half-reaction

  Example: $Zn(s) \rightarrow Zn^{2+}(aq) + 2\,e^-$. Zinc metal loses two electrons to form zinc(II) ions.

- **Reduction**

  → *Gain* of electrons by a species

  → Decrease in oxidation number of a species

  → Represented by a *reduction* half-reaction

  Example: $Cu^{2+}(aq) + 2\,e^- \rightarrow Cu(s)$. A Cu(II) ion gains two electrons each to form copper atom.

- **Redox couple**

  → Ox/Red: Oxidized (Ox) and reduced (Red) species in a half-reaction with the oxidized form listed first by convention

  Example: The Zn^{2+}/Zn redox couple

  → Half-reactions express separately the oxidation and reduction contributions to a redox reaction.

  → May be added to obtain a redox reaction if the electrons cancel

  Example: The following redox reaction is obtained from the half reactions:

Oxidation half-reaction:	$Zn(s) \rightarrow Zn^{2+}(aq) + 2\,e^-$
Reduction half-reaction:	$Cu^{2+}(aq) + 2\,e^- \rightarrow Cu(s)$
Redox reaction:	$Zn(s) + Cu^{2+}(aq) \rightarrow Zn^{2+}(aq) + Cu(s)$

6K.2 Balancing Redox Equations

- **Balancing methods**

  → Two methods are commonly used to balance redox equations: the *half-reaction method* and the *oxidation number method*.

  → We will use the half-reaction method.

  → In either method, mass and charge are balanced separately.

  → In redox reactions, it is conventional to represent hydronium ions, $H_3O^+(aq)$, as $H^+(aq)$.

- **Half-reaction method**

  → A six-step procedure (see also **Toolbox 6K.1** in the text)

  1. Identify all species being oxidized and reduced from changes in their oxidation numbers.

  2. Write unbalanced skeletal equations for each half-reaction.

3. Balance all elements in each half-reaction except O and H.

4. a) For an *acidic* solution, balance O by using H_2O; then balance H by adding H^+.

   b) For a *basic* solution, balance O by using H_2O, then balance H by adding H_2O to the side of each half-reaction that needs H and adding OH^- to the other side.

Note: Adding H_2O to one side and OH^- to the other side has the *net* effect of adding H atoms to balance hydrogen if charge is ignored. Charge is balanced in the next step.

5. Balance charge by adding electrons to the appropriate side of each half-reaction.

6. a) Multiply the half-reactions by factors that give an equal number of *electrons* in each reaction.

   b) Add the two half-reactions to obtain the balanced equation.

→ An alternative procedure for *basic* solutions is to balance the redox equation for *acid* solutions first. Then, add

$$nH^+(aq) + nOH^-(aq) \rightarrow nH_2O(l) \quad or \quad nH_2O(l) \rightarrow nH^+(aq) + nOH^-(aq)$$

to the redox equation to eliminate nH^+ from the product *or* reactant side, whichever is necessary.

Topic 6L: GALVANIC CELLS

6L.1 The Structure of Galvanic Cells

- **Galvanic cell**

  → Electrochemical cell in which a spontaneous reaction produces an electric current

- **Galvanic cell components**

  → Two *electrodes* (metallic conductors and/or conducting solids) make electrical contact with the cell contents, the conducting medium.

  → An *electrolyte* (an ionic solution, paste, or crystal) is the conducting medium.

  → *Anode*, labeled (−) or $\ominus$: the electrode at which *oxidation* occurs.

  → *Cathode*, labeled (+) or $\oplus$: the electrode at which *reduction* occurs.

  → If the anode and cathode do not share an electrolyte, the oxidation reaction and reduction reaction compartments are separated to prevent direct mixing of the electrolyte solutions. Electrical contact between these solutions is maintained by a *salt bridge,* a gel containing an electrolyte such as KCl(aq) or an equivalent device.

- **Hydrogen electrode, or half-cell**

  → Inert Pt wire, used to conduct electricity, immersed in an acid solution, through which hydrogen gas is bubbled

→ Half-reaction: $2H^+(aq) + 2e^- \rightarrow H_2(g)$ (reduction) or $H_2(g) \rightarrow 2H^+(aq) + 2e^-$ (oxidation)

- **The *Daniell cell*, an example of a galvanic cell**
  → Zinc metal reacts spontaneously with Cu(II) ions in aqueous solution.
  → Copper metal precipitates and Zn(II) ions enter the solution.
  → The reaction is exothermic, as suggested by the standard enthalpy of the cell reaction:

$$Zn(s) + Cu^{2+}(aq) \rightarrow Zn^{2+}(aq) + Cu(s) \qquad \Delta H_r° = -225.56 \text{ kJ·mol}^{-1}$$

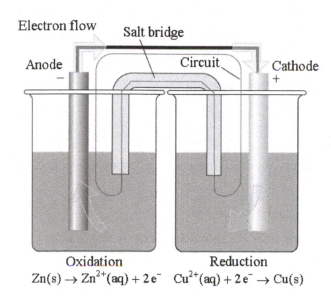

Electron flow   Salt bridge

Anode   Circuit   Cathode
  −   +

Oxidation   Reduction
$Zn(s) \rightarrow Zn^{2+}(aq) + 2e^-$ $Cu^{2+}(aq) + 2e^- \rightarrow Cu(s)$

The galvanic cell to the left contains *anode* and *cathode compartments* linked by a salt bridge to prevent the mixing of ions.

The anode compartment contains a zinc electrode in an electrolyte solution such as KCl or $ZnCl_2$.

The cathode compartment contains an electrode (not necessarily Cu) in an electrolyte solution containing $Cu^{2+}(aq)$ ions.

The salt bridge contains an electrolyte such as KCl in an aqueous gelatinous medium to control the flow of $K^+(aq)$ and $Cl^-(aq)$ ions.

When current is drawn from the cell, the electrodes are linked by an external wire to a load (examples include a light bulb and a motor).

Current (electrons) flows through the wire from anode to cathode as indicated and the spontaneous chemical *cell reaction*

$$Zn(s) + Cu^{2+}(aq) \rightarrow Cu(s) + Zn^{2+}(aq) \text{ occurs.}$$

As the reaction proceeds, Zn(s) is oxidized to $Zn^{2+}(aq)$ in the anode compartment (oxidation) and $Cu^{2+}(aq)$ is reduced to Cu(s) in the cathode compartment (reduction). The Cu(s) deposits on the cathode, and the Zn(s) anode slowly disappears. The closed - line labeled *circuit* traces the direction that current flows in the cell. Current is carried by *electrons* in the electrodes and external connections, and by *ions* in the conducting medium.

6L.2 Cell Potential and Reaction Gibbs Free Energy

- **Cell potential and emf**
  → *Cell potential, E_{cell}*: a measure of the ability of a cell reaction to move electrons through the external circuit
  → *Electromotive force* (emf): cell potential when a cell is operated reversibly
  → For our purposes, the cell potential E_{cell} will be taken to mean emf unless otherwise noted.
  → The amount of useful work a cell can perform depends on the cell potential and the amount of the limiting reactant present.

- **Units and definitions**

  → Current: *ampere*, A, the SI base unit of electric current

  → Charge: *coulomb*, C $\equiv$ A·s, the SI derived unit of charge

  → Energy: *joule*, J $\equiv$ kg·m^2·s^{-2}, the SI derived unit of energy

  → Potential: *volt*, V $\equiv$ J·C^{-1}, the SI derived unit of the cell potential

  → A charge of one *coulomb* falling through a potential difference of one *volt* produces one *joule* of energy.

  → Cell potential: Measured using a voltmeter connected to the two electrodes of a cell (see **Figure 6L.4** in the text)

  → Convention: Write galvanic cells with anode on the left and cathode on the right. This arrangement gives a positive cell potential.

- **Examples of galvanic cell reactions with measured cell potentials**

  → $Mg(s) + Zn^{2+}(aq, 1 \text{ M}) \rightarrow Zn(s) + Mg^{2+}(aq, 1 \text{ M})$ $E_{cell}° = 1.60$ V

  → $2H^+(aq, 1 \text{ M}) + Zn(s) \rightarrow Zn^{2+}(aq, 1 \text{ M}) + H_2(g, 1 \text{ bar})$ $E_{cell}° = 0.76$ V

  Note: For the preceding cells, all reactants and products are in their standard states and the potentials measured are called standard potentials, $E_{cell}°$.

- **Relation between reaction free energy, ΔG, and cell potential, E**

  → Reaction free energy, ΔG: *maximum nonexpansion* work, w_e, obtainable from a reaction at constant T and P: $\Delta G = w_e$

  → When n electrons move through a potential difference, E, the work done is the total charge times the potential difference: $w_e = QV$.

  → The charge on one mole of electrons is N_A times the charge of a single electron, $-e = -1.602\,177 \times 10^{-19}$ C: $Q = -eN_A$.

  → The Faraday constant is the *magnitude* of the charge of *one mole of electrons*: $F = -eN_A = 9.648\,531 \times 10^4$ C·mol^{-1}.

  → The work done and reaction free energy are given by the product of the total charge, $-nF$, and the cell potential, E: $\Delta G = w_e = -nFE$.

  → In the "molar" form, $\Delta G_r = -n_r FE$.

  → In this expression, n is the stoichiometric coefficient (pure number) of the *electrons* in the oxidation and reduction half-reactions that are combined to determine the cell reaction.

  → *Maximum* nonexpansion work can be obtained only if the cell operates *reversibly*.

  → *Reversibility* is realized only when the applied potential equals the potential generated by the cell.

  → All working cells operate *irreversibly* and produce lower potentials than the emf.

  → In sum, the relationship between the free energy and the cell potential for *reversible* electrochemical cells is one of the most important concepts in this chapter. It is given by

$$\Delta G_r = -n_r FE$$

- **Standard cell potential, $E_{cell}°$**

  → The standard state of a substance (s, l, or g) is the pure substance at a pressure of 1 bar.

  → For solutes in solution, we take a molar concentration of 1 mol·L^{-1} as the standard-state activity.

  → The standard cell potential, $E_{cell}°$, is the cell potential measured when all reactants and products are in their standard states.

  → The standard reaction free energy can be obtained from the standard cell potential (emf) and vice versa:

$$\Delta G° = -nFE_{cell}°$$

6L.2 The Notation for Cells

- **Cell diagram**

  → A symbolic representation of the cell components

  → A single vertical line | represents a phase boundary.

  → A double vertical line ‖ represents a salt bridge.

  → Reactants and products are represented by chemical symbols.

  → Components in the same phase are separated by commas.

  → The anode components are written first, followed by the salt bridge, if present, and then the cathode components.

  → Phase symbols (s, l, g) are normally used.

  → Concentrations of solutes and pressures of gases may also be given.

- **Cell potential**

  → Measured with an electronic voltmeter (see **Figure 6L.5** in the text)

  → Voltmeter draws negligible current.

  → Measured voltage is positive (+) when the positive terminal of the meter is connected to the cathode.

- **Cell diagrams with concentrations of reactants and products**

  → Often, the concentrations of reactants and products are included in the cell diagram to give a more complete description of the cell contents.

  Example: $Mg(s) + Zn^{2+}(aq, 1\ M) \rightarrow Zn(s) + Mg^{2+}(aq, 0.5\ M)$ cell reaction

  $Mg(s)\,|\,Mg^{2+}(aq, 0.5\ M)\,\|\,Zn^{2+}(aq, 1\ M)\,|\,Zn(s)$ cell diagram

  Standard cell diagram showing anode, cathode, and cell potential (see **Topic 6L.3** for standard electrode potentials)

  $Mg(s)\,|\,Mg^{2+}(aq, 1\ M)\,\|\,Zn^{2+}(aq, 1\ M)\,|\,Zn(s)$

  anode (−)   cathode (+) $E° = 1.60$ V

Electrons flow spontaneously from the anode (Mg) to the cathode (Zn), and the cell potential $E°$ is positive: $\Delta G° = -nFE° < 0$ for spontaneity.

Topic 6M: APPLICATIONS OF STANDARD POTENTIALS

6M.1 The Definition of Standard Potential

- **Standard cell potential**
  - → *Difference* between the two standard electrodes of an electrochemical cell
  - → The standard cell potential, $E°$, is the difference between the standard potentials of the electrode on the right side of the diagram and the electrode on the left side of the diagram.

$E_{cell}° = E°$(electrode on right of cell diagram) $- E°$(electrode on left of cell diagram)

or

$$E_{cell}° = E_R° - E_L°$$

  - → If $E_{cell}° > 0$, the cell reaction is spontaneous ($K_{eq} > 1$) under standard conditions as written.
  - → If $E_{cell}° < 0$, the reverse cell reaction is spontaneous ($K_{eq} < 1$) under standard conditions as written.
  - → The standard potential of an electrode is sometimes called the *standard electrode potential* or the *standard reduction potential*.

- **Standard electrode potentials, $E°$**
  - → The potential of a single electrode cannot be measured.
  - → A relative scale is required.
  - → By convention, the standard potential of the hydrogen electrode is assigned a value of zero *at all temperatures*: $E°(\text{H}^+/\text{H}_2) \equiv 0$.
  - → For the standard hydrogen electrode (SHE), the half-reaction for reduction is

    $2\,\text{H}^+(\text{aq, 1 M}) + 2\,\text{e}^- \rightarrow \text{H}_2(\text{g, 1 bar})$, and the half-cell diagram is

    $\quad\quad \text{H}^+(\text{aq, 1 M}) \,|\, \text{H}_2(\text{g, 1 bar}) \,|\, \text{Pt(s)} \quad\quad E°(\text{H}^+/\text{H}_2) \equiv 0$

  - → If an electrode is found to be the *anode* in combination with the SHE, it is assigned a *negative potential*. If it is the *cathode*, its standard potential is *positive*.
  - → Values of standard electrode potentials measured at 25 °C are given in **Table 6M.1** and **Appendix 2B** in the text.

- **Standard electrode potentials and free energy**
  - → The relation between free energy and potential also applies to standard electrode potentials:

$$\Delta G_r^\circ = -n_r F E^\circ$$

  - → ΔG_f° values in Appendix 2A can be used to determine ΔG° and then an unknown standard potential using this equation.

- **Calculating the standard potential of a redox couple from those of two related couples**
  - → If two redox couples are added or subtracted to give a third redox couple:
      The standard potentials in general cannot be added to give the unknown potential.
      The reason is that potential is an *intensive* property.
      Free energies are *extensive* and can be added.
  - → So to calculate the unknown potential, we convert potential values to free energy values.

- **Summary of the process of combining half-reactions**
  - → Combining two half-reactions (A and B) to obtain a third half-reaction (C) is accomplished by addition of the associated free energy changes.

  - → The equation $\Delta G_r^\circ = -n_r F E^\circ$ applies to each half-reaction and leads to the following equation:

$$E_C^\circ = \frac{n_A E_A^\circ + n_B E_B^\circ}{n_C}$$

    where the values of n_r correspond to the number of electrons in the two half-reactions (n_A, n_B) and in the half-reaction n_C.

- **Standard electrode potentials of half-reactions**
  - → Consider the half-reaction $M^{2+}(aq, 1\ M) + 2\ e^- \rightarrow M(s)$ $E = E^\circ(M^{2+}/M)$.
  - → Metals with *negative* standard potentials [$E = E^\circ(M^{2+}/M) < 0$] have a thermodynamic tendency to reduce H_3O^+ (aq) to H_2(g) under standard conditions (1 M acid).
  - → Metals with *positive* standard potentials [$E = E^\circ(M^{2+}/M) > 0$] cannot reduce hydrogen ions under standard conditions; in this instance, hydrogen gas is the stronger reducing agent, and will itself be oxidized.
  - → In general
      the more *negative* the standard electrode potential, the greater the tendency of a metal to *reduce* H^+.
      the more *positive* the standard electrode potential, the greater the tendency of a metal ion to *oxidize* H_2.

6M.2 The Electrochemical Series

  - → Standard half-reactions arranged in order of *decreasing* standard electrode potential and alphabetically (see both **Table 6M.1** and **Appendix 2B** in the text)
  - → Table of relative strengths of oxidizing and reducing agents (see the table that follows)

→ Going *up the table,* the oxidizing strength of *reactants* increases.

Example: $F_2(g)$ is an exceptionally *strong oxidizing agent* with a strong tendency to gain electrons and be reduced. $F_2(g)$ will oxidize any of the *product species* on the right side of the table in the reactions listed below it.

→ Going *down the table,* the reducing strength of *products* increases.

Example: Li(s) is an exceptionally *strong reducing agent* with a large tendency to lose electrons and be oxidized. Li(s) will reduce any of the *reactant species* on the left side of the table in the reactions listed above it.

→ *Standard* electrode potentials have values ranging from about +3 V to –3 V, a difference of about 6 V.

→ No single *standard* galvanic cell may have a potential larger than about 6 V because no known half-reactions can give a larger potential.

Reduction half-reaction		$E°(V)$
$F_2(g) + 2\,e^- \rightarrow 2\,F^-(aq)$	Reducing strength	+2.87
$Mn^{3+}(aq) + e^- \rightarrow Mn^{2+}(aq)$		+1.51
$I_2(s) + 2\,e^- \rightarrow 2\,I^-(aq)$		+0.54
$2\,H^+(aq) + 2\,e^- \rightarrow H_2(g)$		0
$Fe^{2+}(aq) + 2\,e^- \rightarrow Fe(s)$		−0.44
$Al^{3+}(aq) + 3\,e^- \rightarrow Al(s)$		−1.66
$Na^+(aq) + e^- \rightarrow Na(s)$		−2.71
$Li^+(aq) + e^- \rightarrow Li(s)$		−3.05

Oxidizing strength

- **Viewing half-reactions as conjugate pairs**

  → For a Brønsted–Lowry acid–base conjugate pair, the stronger the conjugate acid, the weaker the conjugate base.

  → For a given electrochemical half-reaction, the stronger the *conjugate* oxidizing agent, the weaker the *conjugate* reducing agent.

  Example: $F_2(g)$ is an extremely powerful oxidizing agent, whereas $F^-(aq)$ is an extremely weak reducing agent.

- **Uses of the electrochemical series**

  → Predict whether a reaction has $K > 1$ or $K < 1$ for species in their standard states.

  → Predict *relative* reducing and oxidizing strengths.

→ Predict which reactants in their standard states may react spontaneously in a redox reaction.

→ Calculate standard cell potentials and equilibrium constants.

Topic 6N: APPICATIONS OF STANDARD POTENTIALS

6N.1 Standard Potentials and Equilibrium Constants

- **Equilibrium constants**

  → Can be obtained from standard cell potentials

  → For redox, acid-base, dissolution–precipitation, and dilution (change in concentration) reactions (the reactions need not be redox reactions)

- **Quantitative aspects**

  → Combining $\Delta G_r° = -RT \ln K$ (from **Topic**) with $\Delta G_r° = -n_r F E_{cell}°$ yields

$$\ln K = \frac{n_r F E_{cell}°}{RT} = \frac{n_r E_{cell}°}{0.025\ 693\ \text{V}} \quad \text{at } T = 298.15\ \text{K}$$

  → In this equation we use the molar convention: n_r is dimensionless (a pure number). See "A Note on Good Practice" in **Topic 6L.2** in the text.

  → Standard electrode potentials may be used to calculate $E_{cell}°$ values.

  → Because $\Delta G_r°$ applies to *all* reactions, the equation may also be used for acid-base reactions, precipitation reactions, and dilution reactions.

  → For non-redox reactions, the overall cell reaction will not show the electron transfer that occurs in the half-reactions explicitly.

6N.2 The Nernst Equation

  → Describes the quantitative relationship between cell potential and the chemical composition of the cell

  → Is used to estimate the potential for the following:

     Cell potential from its chemical composition

     Half-cell potential *not* in its standard state

6N.3 Ion-Selective Electrodes

  → An electrode sensitive to the concentration of a particular ion

     Example: A metal wire in a solution containing the metal ion. The electrode potential for $E(Ag^+, Ag)$ or $E(Cu^{2+}, Cu)$ is sensitive to the concentration of Ag^+ or Cu^{2+}, respectively.

- **Measurement of pH**

  → An important application of the Nernst equation

→ Utilizes a galvanic cell containing an electrode sensitive to the concentration of the H^+ ion

→ A calomel electrode connected by a salt bridge to a hydrogen electrode:

Calomel electrode: $Hg_2Cl_2(s) + 2e^- \rightarrow 2Hg(l) + 2Cl^-(aq)$ $E° = +0.27$ V

Hydrogen electrode: $H_2(g) \rightarrow 2H^+(aq) + 2e$ $E° = 0.00$ V

Cell reaction: $Hg_2Cl_2(s) + H_2(g) \rightarrow 2H^+(aq) + 2Hg(l) + 2Cl^-(aq)$ $E° = +0.27$ V

→ Q in the Nernst equation for this cell reaction will depend only on the unknown concentration of H^+ if the Cl^- concentration in the calomel electrode is fixed at a value determined for a *saturated* solution of KCl and the pressure of H_2 gas is kept constant at 1 bar.

- **Glass electrode**
  → A thin-walled glass bulb containing an electrolyte with a potential proportional to pH
  → Used to replace the (messy) hydrogen electrode in modern pH meters
  → Modern pH meters use glass and calomel electrodes in a single unit (probe) for convenience. The probe contacts the test solution (unknown pH) through a small salt bridge.

- **pX meters**
  → Devices sensitive to other ions, such as Na^+ and CN^-
  → Used in industrial applications and in pollution control

6N.4 Corrosion

→ Unwanted oxidation of a metal

→ An electrochemical process that is destructive and costly

- **Retarding corrosion**
  → Coating the metal with paint (**painting**) or plastic. Nonuniform coverage or uneven bonding leads to deterioration of the coating and rust.
  → **Passivation**, the formation of a nonreactive surface layer (**Example:** Al_2O_3 on aluminum metal)
  → **Galvanization**, coating the metal with an unbroken layer of zinc. Zinc is preferentially oxidized, and the zinc oxide that forms provides a protective coating (passivation).
  → Use of a **sacrificial anode**, a metal more easily oxidized than the one to be protected, attached to it. Used for large structures such as bridges. A disadvantage is that sacrificial anodes must be replaced when completely oxidized (think of an underground pipeline).
  → **Cationic electrodeposition** coatings. A sacrificial metal (yttrium) is used as the primer for automobile and truck bodies to reduce corrosion.

- **Mechanism of corrosion**
  → Exposed metal surfaces can act as anodes or cathodes, points where oxidation or reduction can occur.
  → If the metal is wet (from dew, rain, or groundwater), a film of water containing dissolved ions may link the anode and cathode acting as a salt bridge.

→ The circuit is completed by electrons flowing through the metal, which acts as a wire in the external circuit of a galvanic cell.

- **Mechanism of rust formation on iron metal**

  → At the *anode,* iron is oxidized to iron(II) ions.

  (1) $Fe(s) \rightarrow Fe^{2+}(aq) + 2e^-$ $-E° = -(-0.44\text{ V}) = +0.44\text{ V}$

  → At the *cathode,* oxygen is reduced to water.

  (2) $O_2(g) + 4H^+(aq) + 4e^- \rightarrow 2H_2O(l)$ $E° = +1.23\text{ V}$

  → Iron(II) ions migrate to the cathode, where they may be oxidized by oxygen to iron(III).

  (3) $Fe^{2+}(aq) \rightarrow Fe^{3+}(aq) + e^-$ $-E° = -(+0.77\text{ V}) = -0.77\text{ V}$

  → The overall redox equation is obtained by adding reaction 1 four times, reaction 3 four times, and reaction 2 three times. The resulting cell reaction has $n = 12$.

  (4) $4Fe(s) + 3O_2(g) + 12H^+(aq) \rightarrow 4Fe^{3+}(aq) + 6H_2O(l)$ $E_{cell}° = +1.27\text{ V}$

  → Next, the iron(III) ions in reaction 4 precipitate as a hydrated iron(III) oxide or rust. The coefficient, x, of the waters of hydration in the oxide is not well defined.

  (5) $4Fe^{3+}(aq) + 2(3+x)H_2O(l) \rightarrow 2Fe_2O_3 \cdot xH_2O(s) + 12H^+(aq)$

  → Finally, the overall reaction for the formation of rust is obtained by adding reactions 4 and 5:

  (6) $4Fe(s) + 3O_2(g) + 2xH_2O(l) \rightarrow 2Fe_2O_3 \cdot xH_2O(s)$

  → From this mechanism, summarized in reaction 6, we expect corrosion to be accelerated by moisture, oxygen, and salts, which enhance the rate of ion transport.

Topic 6O: ELECTROLYSIS

6O.1 Electrolytic Cells

→ Drive a reaction in a nonspontaneous direction by using an electric current

→ Process conducted in an electrolytic cell

- **Electrolytic cell**

  → An electrochemical cell in which electrolysis occurs

  → Different in design from a galvanic cell

  → Both electrodes normally share the same compartment.

  → As in a galvanic cell, oxidation occurs at the anode and reduction occurs at the cathode.

  → Unlike in a galvanic cell, the anode has a *positive* charge and the cathode is *negative*.

  → In an electrolytic cell, electrons are forced to flow from the anode, where oxidation occurs, to the cathode, where reduction occurs.

- **Example of an electrolytic cell**

  → Sodium metal is produced by the electrolysis of molten rock salt (impure sodium chloride) and calcium chloride. Calcium chloride lowers the melting point of sodium chloride (recall freezing point depression), increasing the energy efficiency of the process.

  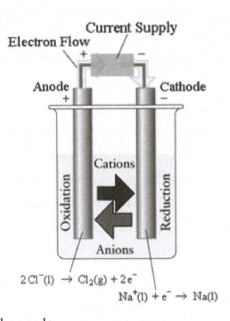

  → A schematic diagram of an electrolytic cell for the production of Na(l) and Cl_2(g) is shown nearby; also see text **Figure 6O.1** for a nonschematic diagram of this process.

  → The cell has one compartment that contains both the anode and cathode. The electrode material is often an inert metal such as platinum. Electrical current is supplied by an external source, such as a battery, to drive this nonspontaneous reaction. This current *forces* reduction to occur at the cathode and oxidation to occur at the anode.

The cathode is placed on the right side as in a galvanic cell. Sodium ions are reduced at the cathode and chloride ions are oxidized at the anode:

cathode: Na^+(l) + e^- → Na(l)

anode: $2Cl^-$(l) → Cl_2(g) + $2e^-$

Note that liquid sodium metal is formed in the electrolysis because the cell temperature of 600 °C is above the melting point of sodium (98 °C).

Notes: In the *molten* salt mixture, Na^+ is reduced in preference to Ca^{2+}. In *aqueous* salt solutions, care must be taken in predicting the products of electrolysis because of the possibility of oxidizing or reducing water instead of the anion or cation.

- **Overpotential and the products of electrolysis**

  → For electrolysis to occur, an external potential *at least as great as* that of the spontaneous cell reaction must be applied to an electrolytic cell.

  → The *actual* potential required is often *greater* than this minimum value. The necessary voltage required is called *overpotential*.

  → Overpotential depends on the structure of a solution near an electrode, which differs from that in the bulk solution. Near an electrode, electrons are transferred across an electrode–solution interface that depends on the solution and the condition of the electrode.

  → Because of overpotential, the observed electrolysis products may differ from those predicted using standard electrode potentials and the Nernst equation.

  → In a solution with more than one *reducible* species, the *reactant* with the *most positive* reduction potential will be reduced.

→ In a solution with more than one *oxidizable* species, the *product* with the *most negative* reduction potential will be oxidized.

6O.2 The Products of Electrolysis

- **Faraday's law of electrolysis**

  → The number of moles of product formed is *stoichiometrically equivalent* to the number of moles of electrons supplied.

  → The amount of product depends on the current, the time it is applied, and the number of moles of electrons in the half-reaction.

- **Quantitative aspects of Faraday's law**

  → Determine the stoichiometry of a half-reaction; for example, $M^{n+} + ne^- \rightarrow M(s)$.

  → The quantity of electricity, Q, passing through the electrolysis cell is determined by the electric current, I, and the time, t, of current flow. Recall that 1 ampere (A) = 1 coulomb (C) s^{-1}.

$$\boxed{\text{Charge supplied (C)} = \text{current (A)} \times \text{time (s)} \quad \text{or} \quad Q = It}$$

  → Because $Q = nF$, where n is the number of moles of electrons and F is the Faraday constant,

$$\boxed{\text{Moles of electrons} = \frac{\text{charge supplied (C)}}{F} = \frac{\text{current } (I) \times \text{time } (t)}{F} \quad \text{or} \quad n(e^-) = \frac{It}{F}}$$

  → The moles of product and the mass of product produced are determined from the number of moles of electrons passing through the cell, $n(e^-)$, the half-reaction, and the molar mass, M.

$$\text{Moles of product} = n(\text{p}) = \frac{It}{n(e^-)F}$$

$$\text{Mass of product} = m(\text{p}) = \left(\frac{It}{n(e^-)F} \right) M(\text{p})$$

where $M(\text{p})$ is the molar mass of the product p.

6O.3 Applications of Electrolysis

- **Uses of electrolysis**

  → Extracting metals from their salts. Examples include Al(s), Na (see **Figure 6O.4** in the text), and Mg.

  → Preparation of fluorine, chlorine, and sodium hydroxide

  → Refining (purifying) metals such as copper

  → Electroplating metals

- **Electroplating**

  → Electrolytic deposition of a thin metal film on an object

  → Object to be plated becomes the cathode of an electrolysis cell.

→ Cell electrolyte is a salt of the metal to be plated.

→ Cations are supplied either by the added salt or from oxidation of the anode, which is then made of the plating metal: for example, silver, gold, and chromium plating (see **Figure 6O.5** in the text)

INTERLUDE

Practical Cells

- **Practical galvanic cells**

  → Commonly called batteries

  → Should be inexpensive, portable, safe, and environmentally benign

  → Should have a high *specific energy* (reaction energy per kilogram) and stable current

  → See also **Interlude Table 1 Reactions in Commercial Batteries** in the text.

- **Primary cells**

  → cannot be recharged.

  → Some types of primary cells:

  Dry cells: (A, AA, C, D., batteries), used in flashlights, toys, remote control

  Cell diagram: $Zn(s) \mid ZnCl_2(aq), NH_4Cl(aq) \mid MnO(OH)(s) \mid MnO_2(s) \mid C(gr)$, 1.5 V

  Cell reaction: $Zn(s) + 2\,NH_4^+(aq) + 2\,MnO_2(s) \rightarrow Zn^{2+}(aq) + 2\,MnO(OH)(s) + 2\,NH_3(g)$

  Alkaline cells: similar to dry cells but with an alkaline electrolyte that increases battery life. Used in backup (emergency) power supplies and smoke detectors

  Cell diagram: $Zn(s) \mid ZnO(s) \mid OH^-(aq) \mid Mn(OH)_2(s) \mid MnO_2(s) \mid C(gr)$, 1.5 V

  Cell reaction: $Zn(s) + MnO_2(s) + H_2O(l) \rightarrow ZnO(s) + Mn(OH)_2(s)$

  Silver cells: have only solid reactants and products, high emf, long life; used in pacemakers, hearing aids, and cameras

  Cell diagram: $Zn(s) \mid ZnO(s) \mid KOH(aq) \mid Ag_2O(s) \mid Ag(s) \mid$ steel, 1.6 V

  Cell reaction: $Zn(s) + Ag_2O(s) \rightarrow ZnO(s) + 2\,Ag(s)$

- **Secondary cells**

  → Can be recharged

  → Must be charged before initial use

  → Some types of secondary cells:

  Lead–acid cell (automobile battery): has low specific energy but generates high current over short time to start vehicles

  Cell diagram: $Pb(s) \mid PbSO_4(s) \mid H^+(aq), HSO_4^-(aq) \mid PbO_2(s) \mid PbSO_4(s) \mid Pb(s)$, 2 V

  Cell reaction: $Pb(s) + PbO_2(s) + 2\,H_2SO_4(aq) \rightarrow 2\,PbSO_4(s) + 2\,H_2O(l)$

Lithium–ion cell: high energy density, high emf, can be recharged many times; used in laptop computers.

Sodium–sulfur cell: used to power electric vehicles, has all *liquid* reactants and products, a solid electrolyte, and relatively high voltage.

Cell diagram: $Na(l) \mid Na^+(\text{ceramic electrolyte}) \parallel S^{2-}(\text{ceramic electrolyte}) \mid S_8(l)$, 2.2 V

Cell reaction: $16\,Na(l) + S_8(l) \rightarrow 16\,Na^+(\text{electrolyte}) + 8\,S^{2-}(\text{electrolyte})$

Nickel–metal hydride (NiMH) cell: used in hybrid vehicles as a supplemental energy source; utilizes hydrogen storage in metal alloys, has a low mass, high energy density, high current load capability, and is nontoxic (environmentally friendly)

- **Fuel cells**
  - → Galvanic cells designed for continuous reaction
  - → Require a continuous supply of reactants
  - → **Alkaline–fuel cell:** used in the space shuttle

    Cell diagram: $Ni(s) \mid H_2(g) \mid KOH(aq) \mid O_2(g) \mid Ni(s)$, 1.23 V

    Cell reaction: $2\,H_2(g) + O_2(g) \rightarrow 2\,H_2O(l)$

Focus 7: KINETICS

Topic 7A: REACTION RATES

7A.1 Concentration and Reaction Rate

- **General definition of rate**

  → Rate is the change in a property divided by the time required for the change to occur.

- **Definition of rate in chemical kinetics**

  → Reaction rate is the change in molar concentration of a reactant or product divided by the time necessary for that change to occur.

- **Average versus instantaneous rates: an analogy to travel**

  → The *average* speed of an automobile trip is the length of the journey divided by the total time for the journey.

  → The *instantaneous* speed is obtained if the car is timed over a very short distance at some point in the journey. It is the number on the speedometer.

- **Average reaction rate**

  → For a general chemical reaction, $aA + bB \rightarrow cC + dD$, the average reaction rate is the *change* in molar concentration of a reactant or product divided by the time interval necessary for the change to occur.

  → The average reaction rate may be determined for *any* reactant or product.

  → Rates for different reactants and products may have different numerical values.

- **Unique average reaction rate**

  → For a general chemical reaction, $aA + bB \rightarrow cC + dD$, the *unique average reaction rate* is the average reaction rate of reactant or product divided by its stoichiometric coefficient used as a *pure number*.

  → Determined from *any* reactant or product

  → The same (has the same numerical value) for any reactant or product in a given reaction

  → For a general chemical reaction, the unique average reaction rate is defined as

  $$\text{Unique average reaction rate} = -\frac{1}{a}\frac{\Delta[A]}{\Delta t} = -\frac{1}{b}\frac{\Delta[B]}{\Delta t} = \frac{1}{c}\frac{\Delta[C]}{\Delta t} = \frac{1}{d}\frac{\Delta[D]}{\Delta t}$$

  → The minus signs for the terms involving reactants are required because the concentrations of reactants decrease as the reaction time increases.

Note: Two critical underlying assumptions for generating a unique reaction rate are that the overall reaction time be slow with respect to the buildup and decay of any intermediate and that the stoichiometry of the reaction be maintained throughout.

- **Reaction rates**

  → The time required to measure the concentration changes of reactants and products varies considerably according to the reaction.

  → Spectroscopic techniques are often used to monitor concentration, particularly for fast reactions. An important example is the *stopped-flow technique* shown in **Figure 7A.3** in the text. The fastest reactions occur on a time scale of femtoseconds (10^{-15} s). See **Box 7A.1** in the text.

7A.2 The Instantaneous Rate of Reaction

- **Reaction rates**

  → Most reactions slow down as they proceed and reactants are depleted.

  → If equilibrium is reached, the forward and reverse reaction rates are equal.

  → The reaction rate at a specific time is called the *instantaneous reaction rate*.

- **Instantaneous reaction rate**

  → For a product, the *slope* of a tangent line of a graph of concentration vs. time. The slope is always positive because product concentrations increase with time.

  → For a reactant, the *negative of the slope* of a tangent line on a graph of concentration vs. time. The slope is always negative because reactant concentrations decrease with time.

- **Mathematical form of the instantaneous rate**

  → The slope of the concentration–time curve is the derivative of the concentration with respect to time.

  → For the instantaneous disappearance of a reactant, R,

  $$\text{Reaction rate} = -\frac{d[R]}{dt} \qquad \text{(Note the minus sign.)}$$

  → For the instantaneous appearance of a product, P,

  $$\text{Reaction rate} = \frac{d[P]}{dt} \qquad \text{(Note the absence of a minus sign.)}$$

  → For the general reaction, $a\text{A} + b\text{B} \rightarrow c\text{C} + d\text{D}$,

  $$\boxed{\text{Unique instantaneous reaction rate} = -\frac{1}{a}\frac{d[A]}{dt} = -\frac{1}{b}\frac{d[B]}{dt} = \frac{1}{c}\frac{d[C]}{dt} = \frac{1}{d}\frac{d[D]}{dt}}$$

  → In the remainder of the chapter, the term "reaction rate" is used specifically to mean the *unique* instantaneous reaction rate.

  → The two conditions relating to *unique average* rates described in **Topic 7A.1** also apply here.

7A.3 Rate Laws and Reaction Order

- **Initial reaction rate**

  → Instantaneous rate at $t = 0$

  → In what is called the method of initial rates, analysis is simpler at $t = 0$ because products that may affect reaction rates are not present.

  → The method of initial rates is often used to determine the *rate law* of a reaction.

- **Rate law of a reaction**

  → Expression for the instantaneous reaction rate in terms of the concentrations of species that affect the rate, such as reactants or products

  → Always determined by experiment and usually difficult to predict

  → May have relatively simple or complex mathematical form

  → The exponents in a rate expression are *not* necessarily the stoichiometric coefficients of a reaction; therefore, we use *m* and *n* in place of *a* and *b* in rate equations.

- **Simple rate laws**

  → Form of a simple rate law: $\boxed{\text{Rate} = k[\text{A}]^{m}}$

  → k = rate constant, which is the reaction rate when all the concentrations appearing in the rate law are 1 M (recall that the symbol M stands for the units of mol·L^{-1}).

  → For a given reaction, k is *independent* of concentration(s) but is *dependent* on temperature, $k(T)$.

  → A refers to the reactant, and m, the power of [A], is called the order in reactant A and is determined experimentally.

  → Order is *not related to the reaction stoichiometry*. Exceptions include a special class of reactions, which are called *elementary* (see **Topic 7 C.1**).

  → Typical orders for reactions with one reactant are 0, 1, and 2. Fractional values are also common.

Zero-order reaction	$(m = 0)$:	Rate $= k$
First-order reaction	$(m = 1)$:	Rate $= k[\text{A}]$
Second-order reaction	$(m = 2)$:	Rate $= k[\text{A}]^2$

  → Remember that k has different values and units for different reactions.

- **More complex rate laws**

  → Many reactions have experimental rate laws containing concentrations of more than one species.

  → Form of a complex rate law: $\boxed{\text{Rate} = k[\text{A}]^{m}[\text{B}]^{n}...}$

  The overall order is the sum of the powers $m + n + ...$

  → The powers may also be negative values, and forms of even greater complexity with no overall order are found (see nearby table for an example and also **Topic 7E.3**).

→ More complex rate laws are treated in **Topic 7C.2** in the text and this study guide.

→ The following table lists experimental rate laws for a variety of reactions:

Reaction	Rate law	Order	Units of k
$2 NH_3(g) \rightarrow N_2(g) + 3 H_2(g)$	Rate $= k$	Zero	$mol \cdot L^{-1} \cdot s^{-1}$
$2 N_2O_5(g) \rightarrow 4 NO_2(g) + O_2(g)$	Rate $= k [N_2O_5]$	First	s^{-1}
$2 NO_2(g) \rightarrow NO(g) + NO_3(g)$	Rate $= k [NO_2]^2$	Second	$L \cdot mol^{-1} \cdot s^{-1}$
$2 NO(g) + O_2(g) \rightarrow 2 NO_2(g)$	Rate $= k [NO]^2 [O_2]$	First in O_2 Second in NO Third overall	$L^2 \cdot mol^{-2} \cdot s^{-1}$
$I_3^-(aq) + 2 N_3^-(aq)$ $\rightarrow 3 I^-(aq) + 3 N_2(g)$ in the presence of $CS_2(aq)$	Rate $= k [CS_2][N_3^-]$	First in CS_2 First in N_3^- Second overall	$L \cdot mol^{-1} \cdot s^{-1}$
$2 O_3(g) \rightarrow 3 O_2(g)$	Rate $= k \dfrac{[O_3]^2}{[O_2]}$ $= k [O_3]^2 [O_2]^{-1}$	Second in O_3 Minus one in O_2 First overall	s^{-1}

→ In general, the reaction order *does not follow* from the stoichiometry of the chemical equation.

→ In the case of the *catalytic* decomposition of ammonia on a hot platinum wire (first entry in the table), the reaction order is zero initially because the reaction occurs on the surface of the wire, and the surface coverage is independent of concentration. The rate of a zero-order reaction is independent of concentration until the reactant is nearly exhausted or until equilibrium is reached.

→ Reaction rates may depend on species that do not appear in the overall chemical equation.

→ The reaction of triodide ion with nitride ion, for example, is accelerated by the presence of carbon disulfide, which is neither a reactant nor a product. The concentration of carbon disulfide, a *catalyst*, appears in the rate law. Catalysis is discussed in **Topics 7E.1** to **7E.3** in the text and in this study guide.

→ In the last example in the table, the order with respect to the product O_2 is -1.

→ Negative orders often arise from reverse reactions in which products reform reactants, thereby slowing the overall reaction rate. Oxygen is present at the beginning of this reaction as commonly studied because ozone is made by an electrical discharge in oxygen gas. It is possible to study the kinetics of the decomposition of pure ozone. In the absence of oxygen, the initial rate law is quite different from the one listed in the table.

• **Procedure for determining the reaction orders and rate law**

→ The reaction order for a given species is determined by varying its initial concentration and measuring the initial reaction rate, keeping the concentrations of all other species constant.

→ The procedure is repeated for the other species until all the reaction orders are obtained.

→ Once the form of the rate law is known, a value of the rate constant is calculated for each set of initial concentrations. The reported rate constant is typically the average value for all the measurements. See **Table 7A.1** in the text for examples of rate laws and rate constants.

- **Pseudo-order reactions**

  → If a species in the rate - law equation is present in great excess, its concentration is effectively constant as the reaction proceeds.

  → The constant concentration may be incorporated into the rate constant, and the apparent order of the reaction changes.

  → Reactions in which the solvent appears in the rate law often show pseudo-order behavior because the solvent is usually present in large excess.

  → A technique called the *isolation method* is often used to determine the order in a particular reactant by keeping the concentration of other reactants in excess. The method is repeated for all reactants and the complete rate law is obtained.

  → A hidden danger is that occasionally the rate law is *not* the same in the extreme ranges of the concentrations of reactants. Different reaction mechanisms that yield different rate laws may occur under these extreme conditions.

Topic 7B: INTEGRATED RATE LAWS

7B.1 First-Order Integrated Rate Laws

- **Integration of a first-order rate law**

  → First-order rate law:
  $$-\frac{d[A]}{dt} = k[A] \qquad (1)$$

  → Integration with limits:
  $$-\int_{[A]_0}^{[A]_t} \frac{d[A]}{[A]} = \int_0^t k\,dt \qquad (2)$$

  → Result in logarithmic form:
  $$\ln\left(\frac{[A]_t}{[A]_0}\right) = -kt \qquad (3)$$

  → Result in exponential form:
  $$[A]_t = [A]_0 e^{-kt} \qquad (4)$$

  → If a plot of $\ln[A]_t$ vs. t gives a straight line with a negative slope, the rate law is first-order in [A].

  → The rearranged logarithmic expression $\ln[A]_t = \ln[A]_0 - kt$ shows the straight-line behavior, $y = \text{intercept} + (\text{slope} \times x)$, of a plot of $\ln[A]_t$ vs. t.

  → Plot of $\ln[A]_t$ versus t: slope $= -k$ *and* intercept $= \ln[A]_0$

- **Integrated first-order rate law**

  → [A] decays exponentially with time (**equation 4**).

  → Used to confirm that a reaction is first order and to measure its rate constant

  → Known k and $[A]_0$ values can be used to predict the value of [A] at any time t.

7B.2 Half-Lives for First-Order Reactions

- **Expression**

  → The half-life, $t_{1/2}$, of a substance is the time required for its concentration to fall to one-half its initial value.

  → At $t = t_{1/2}$, $[A]_{t_{1/2}} = \frac{1}{2}[A]_0$. The ratio of concentrations is $\dfrac{[A]_0}{[A]_{t_{1/2}}} = \dfrac{[A]_0}{\frac{1}{2}[A]_0} = 2$.

  → Solve for $t_{1/2}$ by rearranging the logarithmic form of the integrated rate law, **equation 3**, to give

  $t_{1/2} = \dfrac{1}{k} \ln\left(\dfrac{[A]_0}{[A]_{t_{1/2}}}\right)$. Then substitute $\dfrac{[A]_0}{[A]_{t_{1/2}}} = 2$.

  → Half-life expression:
  $$\boxed{t_{1/2} = \frac{\ln 2}{k}} \qquad (5)$$

- **Properties**

  → Independent of the initial concentration

  → Characteristic of the reaction

  → Inversely proportional to the rate constant, k

  → The logarithmic form of the rate law may be modified as follows:

  $$\boxed{\ln\left(\frac{[A]_0}{[A]_t}\right) = kt = (\ln 2)\left(\frac{t}{t_{1/2}}\right)} \qquad (6)$$

7B.3 Second-Order Integrated Rate Laws

- **Integration of a second-order rate law**

  → Second-order rate law:
  $$\boxed{-\frac{d[A]}{dt} = k[A]^2} \qquad (7)$$

  → Integration with limits:
  $$-\int_{[A]_0}^{[A]_t} \frac{d[A]}{[A]^2} = \int_0^t k\,dt \qquad (8)$$

  → Integrated form:
  $$\boxed{\frac{1}{[A]_t} - \frac{1}{[A]_0} = kt} \qquad (9)$$

→ Alternative form:

$$[A]_t = \frac{[A]_0}{1+[A]_0 kt} \qquad (10)$$

→ If a plot of $1/[A]_t$ vs. t gives a straight line, the rate law is second-order in [A].

→ The rearranged expression, $1/[A]_t = 1/[A]_0 + kt$, reveals the linear behavior of a plot of $1/[A]_t$ vs. t.

→ Plot of $1/[A]_t$ vs. t: slope $= + k$ *and*   intercept $= 1/[A]_0$

- **Integrated second-order rate law**

    → For the same initial rates, [A] decays more slowly with time for a second-order process than for a first-order one (see **Figure 7B.5** in the text).

    → Used to confirm that a reaction is second order and to determine its rate constant

    → If k and $[A]_0$ are known, the value of [A] at any time t can be predicted.

- **Summary of rate laws** → See **Table 7B.1** in the text.

- **Pseudo reaction order** → Treated in **Topic 7A.3** in this study guide

Topic 7C: REACTION MECHANISMS

7C.1 Elementary Reactions

- **Reactions**

    → A net chemical reaction is assumed to occur at the molecular level in a series of separable steps called *elementary reactions*.

    → A *reaction mechanism* is a proposed group of elementary reactions that accounts for the reaction's overall stoichiometry and experimental rate law.

- **Elementary reaction**

    → A single-step reaction, written *without* state symbols, that describes the behavior of *individual* atoms and molecules taking part in a chemical reaction

    → The *molecularity* of an elementary reaction specifies the number of *reactant* molecules or, more generally, particles or species that take part in it.

    → *Unimolecular:*   Only one reactant species is written (molecularity = 1).

    C_3H_6 (cyclopropane) → $CH_2=CH-CH_3$ (propene)

    Assumption: *A cyclopropane molecule decomposes spontaneously to form a propene molecule.*

→ *Bimolecular*:   Two reactant species combine to form products (molecularity = 2).

$$H_2 + Br \rightarrow HBr + H$$

Assumption: *A hydrogen molecule and a bromine atom collide to form a hydrogen bromide molecule and a hydrogen atom.*

→ *Termolecular*:   Three reactant species combine to form products (molecularity = 3).

$$I + I + H_2 \rightarrow HI + HI$$

Assumption: *Two iodine atoms and a hydrogen molecule collide simultaneously to form two hydrogen iodide molecules.*

- **Reaction mechanism**

  → *Sequence* of elementary reactions which, when added together, give the net chemical reaction and reproduce its rate law

  → *Reaction intermediate*, *not* a reactant or a product, is a species that appears in one or more of the elementary reactions in a proposed mechanism. It *usually* has a small concentration and does not appear in the overall rate law.

  → The plausibility of a reaction mechanism may be tested, but it cannot be proved.

  → The presence and behavior of a reaction intermediate are sometimes testable features of a proposed reaction mechanism.

  → Reverse elementary reactions can also be proposed in a reaction mechanism. They provide a mechanism for a reaction to reach equilibrium.

  Example: C_3H_6 (cyclopropane) $\rightarrow$ $CH_2=CH-CH_3$ (forward elementary reaction)

              $CH_2=CH-CH_3$ $\rightarrow$ C_3H_6 (cyclopropane) (reverse elementary reaction)

              At equilibrium, the forward and reverse reaction rates are equal.

7C.2 The Rate Laws of Elementary Reactions

- **Forms of rate laws for elementary reactions**

  → Order follows from the stoichiometry of the *reactants* in an *elementary* reaction. Note that products do *not* appear in the rate law of an *elementary* reaction. See **Table 7C.1** in the text.

  → *Unimolecular* elementary reactions are first-order.

      Reaction: A $\rightarrow$ products   Rate $= k\,[A]$

  → *Bimolecular* elementary reactions are second-order overall. If there are two different species present, the reaction is first-order in each one.

      Reaction: A + A $\rightarrow$ products   Rate $= k\,[A]^2$

      Reaction: A + B $\rightarrow$ products   Rate $= k\,[A][B]$

  → *Termolecular* elementary reactions are third-order overall. If there are two different species, the reaction is second-order in one of them and first-order in the other. If there are three different species, the reaction is first-order in each one.

Reaction: A + A + A → products   Rate = $k\,[A]^3$

Reaction: A + A + B → products   Rate = $k\,[A]^2[B]$

Reaction: A + B + C → products   Rate = $k\,[A][B][C]$

→ The molecularity of an elementary reaction that is written in the reverse direction also follows from its stoichiometry. The *products* have become *reactants*.

Forward reaction: A + B → P + Q + R   Forward rate = $k\,[A][B]$

Reverse reaction: P + Q + R → A + B   Reverse rate = $k'\,[P][Q][R]$

Note: The prime symbol, ′, is used here and in the following sections to denote a rate constant written for a *reverse* elementary reaction. The same symbol is used in **Topic 7D.1** to denote a rate constant for the *same* reaction (elementary or otherwise) measured at a *different* temperature. ***Do not confuse these two very different meanings of the prime symbol.***

- **Rate laws from reaction mechanisms**

  → The time evolution of each reactant, intermediate, or product may be determined by integrating the rate laws for the elementary reactions in a proposed reaction mechanism.

  → Recall that a successful mechanism must account for the stoichiometry of the overall reaction as well as the observed rate law.

  → Mathematical solution of kinetic equations often requires numerical methods for solving simultaneous differential equations. In a few cases, exact analytical solutions exist.

  → For multistep mechanisms, approximate methods are often used to determine a rate law consistent with a proposed reaction mechanism. A mechanism consistent with the experimentally determined rate law does not prove that the mechanism is correct. Further experiments, if possible, may be required.

7C.3 Combining Elementary Rate Laws

- **Approximate methods for determining rate laws**

  → Characteristics of reactions commonly encountered allow for the determination of rate laws with the use of three types of approximations.

  → The ***rate-determining step*** approximation is made to determine a rate law for a mechanism in which one step occurs at a rate *substantially* slower than any others.

  → The slow step is a bottleneck, and the overall reaction rate cannot be faster than the slow step.

  → The rate law for the rate-determining step is written first. If a reaction intermediate appears as a reactant in this step, its concentration term must be eliminated from the rate law.

  → The final rate law has concentration terms for reactants and products only.

  → The ***pre-equilibrium condition*** describes situations in which reaction intermediates are formed and removed in steps prior to the rate-determining one.

  → If the formation and removal of the intermediate in prior steps is rapid, an equilibrium concentration is established. If the intermediate appears in the rate-determining step, its relatively slow reaction in that step does not change its equilibrium concentration. The pre-equilibrium condition is sometimes called a *fast equilibrium*.

→ The **steady-state approximation** is a more general method for solving reaction mechanisms. The net rate of formation of any intermediate in the reaction mechanism is set equal to 0.

→ An intermediate is assumed to attain its steady-state concentration instantaneously, decaying slowly as reactants are consumed. An expression is obtained for the steady-state concentration of each intermediate in terms of the rate constants of elementary reactions and the concentrations of reactants and products.

→ The rate law for an elementary step that leads directly to product formation is usually chosen. The concentrations of all intermediates are removed from the chosen rate law, and a final rate law for the formation of product that reflects the concentrations of reactants and products is obtained.

→ Often, a predicted rate law does not quite agree with the experimental one. Reaction conditions that modify the form of the rate law predicted by the mechanism may lead to final agreement. (For an example, the Michaelis–Menten mechanism, see **Topic 7E.3** in the text and also in this study guide.) If complete exploration of conditions fails to reproduce the experimental result, the proposed reaction mechanism is rejected.

7C.4 Rates and Equilibrium

• **Elementary reaction**

→ At *equilibrium,* the forward and reverse reaction rates for an *elementary* reaction are equal.

For a given elementary reaction at equilibrium, $A + B \rightleftharpoons P + Q$.
Forward rate $= k\,[A][B] =$ reverse rate $= k'\,[P][Q]$

→ The *equilibrium constant* for an *elementary* reaction is equal to the ratio of the rate constants for the forward and reverse reactions.

In the example, $K = \dfrac{[P][Q]}{[A][B]} = \dfrac{(\text{reverse rate})/k'}{(\text{forward rate})/k} = \dfrac{k}{k'}$, and we conclude that

$$\boxed{K = \frac{k}{k'}}$$ for any elementary reaction.

• **Equilibrium constants from reaction mechanisms**

→ Consider an overall reaction with a multistep mechanism in which the rate constants for the elementary reactions are $k_1, k_2, k_3, \ldots$ in the forward direction and $k_1', k_2', k_3', \ldots$ in the reverse direction.

→ If each elementary step is in equilibrium, the overall reaction is obtained by summing the elementary reactions, and the equilibrium constant for the overall reaction is obtained by multiplying the equilibrium constants for the individual steps.

→ For the reaction at equilibrium, the procedure gives

$$K = \frac{k_1}{k_1'} \times \frac{k_2}{k_2'} \times \frac{k_3}{k_3'} \times \dots$$

→ Large equilibrium constants ($K \gg 1$) are expected if forward rate constants are much larger than reverse rate constants.

→ Note that that we have gained an understanding of when to expect large or small equilibrium constants from kinetic data alone, with no knowledge of the Gibbs free energy necessary!

7C.5 Chain Reactions

- **Chain reaction**

  → Series of linked elementary reactions that *propagate* in chain cycles

  → A reaction intermediate is produced in an *initiation* step.

  → Reaction intermediates are the *chain carriers,* which are often radicals.

  → In one propagation cycle, a reaction intermediate typically reacts to form a different intermediate, which in turn reacts to regenerate the original one.

  → In each cycle, some product is usually formed.

  → The cycle is eventually broken by a *termination* step that consumes the intermediate.

  → In *chain branching*, a chain carrier reacts to form two or more carriers in a single step.

  → Chain branching often leads to chemical explosions.

  → Branched chain reaction: $H^\bullet + O_2 \rightarrow HO^\bullet + {}^\bullet O^\bullet$ (two chain carriers produced from one)

- **Example in text of a chain reaction**

  → Reaction: $H_2(g) + Br_2(g) \rightarrow 2HBr(g)$

  → Initiation: $Br_2(g) \rightarrow Br^\bullet + Br^\bullet$ ($Br^\bullet$ is a chain carrier.)

  → Propagation: $Br^\bullet + H_2 \rightarrow HBr + H^\bullet$ ($H^\bullet$ is also a chain carrier.)

  $H^\bullet + Br_2 \rightarrow HBr + Br^\bullet$

  → Termination: $Br^\bullet + Br^\bullet \rightarrow Br_2$; $Br^\bullet + H^\bullet \rightarrow HBr$; $H^\bullet + H^\bullet \rightarrow H_2$

Topic 7D: MODELS OF REACTIONS

7D.1 The Effect of Temperature

- **Qualitative aspects**

  → A change in temperature results in a change of the rate constant of a reaction, and therefore its rate.

  → For *most* reactions, *increasing* the temperature results in a *larger value* of the rate constant.

  → For *many* reactions in organic chemistry, reaction rates *approximately* double for a 10 °C increase in temperature.

- **Quantitative aspects**

  → The temperature dependence of the rate constant of many reactions is given by the Arrhenius equation, shown here in two forms:

  Exponential form: $$k = A\mathrm{e}^{-\frac{E_a}{RT}}$$

  Logarithmic form: $$\ln k = \ln A - \frac{E_a}{RT}$$

  R is the universal gas constant ($8.314\,46$ J·K^{-1}·mol^{-1}).
  T is the absolute temperature (K).

  → The two constants, A and E_a, called the *Arrhenius parameters,* are *nearly* independent of temperature.

  → The parameter A is the *pre-exponential factor.* The parameter E_a is the *activation energy.* The two parameters depend on the reaction being studied.

  → The Arrhenius equation applies to all types of reactions, gas phase or in solution.

- **Obtaining Arrhenius parameters**

  → Plot $\ln k$ vs. $1/T$.

  → If the plot displays straight-line behavior, the *slope* of the line is $-E_a/R$ and the *intercept* is $\ln A$. Note that the *intercept* at ($1/T = 0$) implies $T = \infty$.

  → The parameter A *always* has a positive value. The *larger* the value of A, the *larger* is the value of k.

  → The parameter E_a *almost always* has a positive value. The *larger* the value of E_a, the *smaller* the value of k. If $E_a > 0$, the value of k *increases* as temperature *increases.*

  → The rate constant increases more quickly with increasing temperature if E_a is large.

→ If E_a and k are known at a given temperature, T, the logarithmic form of the Arrhenius equation can be used to calculate the rate constant, k', at another temperature, T'. Or alternatively, a value of E_a can be estimated from values of k' at T' and k at T.

$$\ln\frac{k'}{k} = \frac{E_a}{R}\left(\frac{1}{T} - \frac{1}{T'}\right)$$

- **Units of the Arrhenius parameters**

  → A has the same units as the rate constant of the reaction.

  → E_a has units of energy, and values are usually reported in kJ·mol^{-1}.

  Note: An equivalent SI unit for the liter is dm^3. Rate constants and Arrhenius pre-exponential factors incorporating volume units are sometimes reported using dm^3 or cm^3 instead of liters. Those incorporating amount units are sometimes reported in units of molecules instead of moles. For example, a second-order rate constant of 1.2×10^{11} L·mol^{-1}·s^{-1} is equivalent to 1.2×10^{11} dm^3·mol^{-1}·s^{-1} *or* 2.0×10^{-10} cm^3·molecule^{-1}·s^{-1}.

- **Selected Arrhenius parameters** → See **Table 7D.1**.

- **Temperature dependence of the equilibrium constant of an elementary reaction**

  → Follows from the temperature dependence of the ratio of the rate constant for the forward reaction to that of the reverse reaction, with both expressed in Arrhenius form

- **Reactions that are *exothermic***

  → The activation energy in the forward direction is smaller than in the reverse one. (See the two figures in this study guide in **Topic 7D.3** or **Fig. 7D.6** in the text.)

  → The rate constant with the larger activation energy increases more rapidly as temperature increases than the one with smaller energy.

  → As temperature is *rises*, k' is predicted to increase faster than k, and the equilibrium constant *decreases*.

  → *Reactants* are favored.

- **Reactions that are *endothermic***

  → The activation energy is larger in the forward direction than in the reverse one.

  → The rate constant with the larger activation energy increases more rapidly as temperature increases than the one with smaller energy.

  → As temperature is *rises*, k is predicted to increase faster than k', and the equilibrium constant *increases*.

  → *Products* are favored.

7D.2 Collision Theory

  → Model for how reactions occur at the molecular level

  → Applies to gases

  → Accounts for rate constants and the exponential form of the Arrhenius equation

→ Reveals the significance of the Arrhenius parameters A and E_a

- **Assumptions of collision theory**
  → Molecules must collide in order to react.
  → Only collisions with kinetic energy in excess of some minimum value, E_{min}, lead to reaction.
  → Only molecules with the correct orientation with respect to each other may react.

- **Total rate of collisions in a gas mixture**
  → Determined quantitatively given the kinetic model of a gas
  → Rate of collision = total number of collisions per unit volume per second

$$\text{Rate of collision} = \sigma \bar{v}_{rel} N_A^{\,2}[A][B]$$

  $\sigma =$ collision cross-section (area a molecule presents as a target during collision)
  Larger molecules are more likely to collide with other molecules than are smaller ones.

  $\bar{v}_{rel} =$ mean speed at which molecules approach each other in a gas

  → For a gas mixture of A and B with molar masses M_A and M_B, respectively,

$$\bar{v}_{rel} = \sqrt{\frac{8RT}{\pi\mu}} \qquad \mu = \frac{M_A M_B}{M_A + M_B}$$

  → Molecules collide more often at high temperature than at low temperature.
  → $N_A =$ Avogadro constant
  → [A] and [B] are the molar concentrations of A and B, respectively.

- **Total rate of reaction in a gas mixture**
  → Rate of reaction = number of collisions that lead to reaction per unit volume per second

$$\text{Rate of reaction} = P\sigma \bar{v}_{rel} N_A^{\,2}[A][B] \times e^{-E_{min}/RT}$$

  → $E_{min} =$ minimum kinetic energy necessary for a collision to lead to reaction. Collisions with energy less than the minimum do not lead to reaction.
  → $e^{-E_{min}/RT} =$ fraction of collisions with at least the energy E_{min}. Derived from the *Boltzmann distribution* (See **Figure 7D.5** in the text.)
  → $P =$ steric factor < 1. Reactive collisions often require preferred orientations of the colliding species.

- **Reaction rate constant**
  → Rate constant $= k = \dfrac{\text{rate of reaction}}{[A][B]}$

$$k = P\sigma \bar{v}_{rel} N_A^{\,2} e^{-E_{min}/RT}$$

→ Comparison with the Arrhenius equation in exponential form, $k = Ae^{-\frac{E_a}{RT}}$:

$E_a \approx E_{min}$ and $A \approx P\sigma \bar{v}_{rel} N_A^2$

→ Because the speed is proportional to $T^{1/2}$, the model predicts that the pre-exponential factor *and* activation energy both have a *slight* temperature dependence.

- **Summary**

→ According to collision theory, reaction occurs only if reactant molecules collide with a kinetic energy equal to or greater than the Arrhenius activation energy.

→ The reaction rate constant increases with the size and speed of the colliding species (molecules).

→ The steric factor P accounts for collisions that lead to no reaction because the orientation of the molecules during such collisions is unfavorable. See **Table 7D.2** in the text for some values of P.

7D.3 Transition - State Theory

→ Also often called *activated complex theory*

→ Applies to gas phase and solution reactions

→ Reacting molecules collide and distort.

→ During an encounter, the kinetic energy, E_K, of reactants decreases, the potential energy increases, the original chemical bonds lengthen, and new bonds begin to form.

→ Products may form if E_K for the collision $\geq E_a$, the activation energy, and if the reactants have the proper orientation.

→ Reactants reform if $E_K < E_a$, regardless of orientation.

→ If $E_K \geq E_a$, the molecules form either reactants or products. The lowest kinetic energy that may produce products is equal to E_a.

→ *Activated complex*: Combination of two reacting molecules at the transition point between reactants and products

Example: Reaction of the acid HA with water is shown schematically:

$$AH + H_2O \longrightarrow A^- \cdots H^+ \cdots O\overset{H}{\underset{H}{\diagdown}} \longrightarrow A^- + H_3O^+$$

Reactants   Activated Complex   Products

- **Reaction profile**

  → Shows how energy changes as colliding reactants form a *transition state* or an *activated complex* and then form products along the *minimum energy pathway*.

  → An analogous process is hiking through a mountain range and choosing the pathway of lowest elevation through it.

  → Progress of reaction refers to the relevant spatial coordinates of the molecules that follow bond breaking as reactants approach the energy maximum and bond formation as products form.

  → Reaction profiles for endothermic and exothermic reactions are shown in the following schematic drawings.

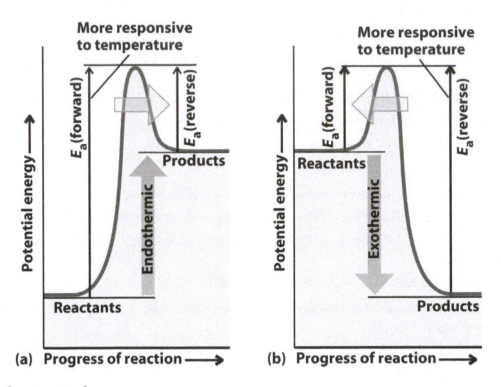

- **Potential energy surface**

  → Three-dimensional plot in which potential energy is plotted on the z-axis and the other two axes correspond to interatomic distances. See **Figure 7D.10** in the text for a potential energy surface of the $Br + H_2 \rightarrow HBr + H$ reaction.

  → Similar to a mountain pass, the *saddle point* corresponds to an activated complex with the *minimum* activation energy, E_a (forward), required for reactants to form products or products to form reactants, E_a (reverse). As shown in the figure, E_a (forward) and E_a (reverse) differ significantly in magnitude.

Topic 7E: CATALYSIS

7E.1 How Catalysts Work

- **Nature of a catalyst**
  - → Accelerates a reaction rate without being consumed
  - → Does not appear in the overall reaction
  - → Lowers the activation energy of a reaction by changing the pathway (mechanism)
  - → Accelerates *both* the forward and reverse reaction rates
  - → Does *not* change the final equilibrium composition of the reaction mixture
  - → May appear in the rate equation but usually does not

- **Homogeneous catalyst**
  - → Present in the same phase as the *reactants*

- **Heterogeneous catalyst**
  - → Present in a different phase from the *reactants*

7E.2 Industrial Catalysts

- **Automobile exhaust**
  - → Catalytic converters of automobiles use catalysts to bring about the complete and rapid combustion of unburned fuel.
  - → Mixture of gases leaving an engine includes carbon monoxide, unburned hydrocarbons, and the nitrogen oxides collectively referred to as NO_x, in addition to the expected carbon dioxide and water.
  - → Air pollution is decreased if the carbon compounds are oxidized to carbon dioxide and the NO_x reduced, by another catalyst, to nitrogen, which is harmless.
  - → An existing challenge is to find a catalyst or a combination of catalysts that will accelerate both the oxidation and the reduction reactions and be active when the car is first started and the engine is cool.

- **Microporous catalysts**
  - → Heterogeneous catalysts are used in catalytic converters and for many other specialized applications, because of their very large surface areas and reaction specificity.
  - → **Zeolites** are microporous aluminosilicates with three-dimensional structures characterized by hexagonal channels connected by tunnels (see **Figure 7E.5** in the text).
  - → The enclosed active sites in **zeolites** gives them a distinct advantage over other heterogeneous catalysts, because an intermediate can be held in place inside the channels until the products are formed.
  - → Channels allow products to grow only to a particular size.

- **Catalyst ZSM-5**
  - → Used to convert methanol to gasoline
  - → Pores of the **zeolite** are just large enough to allow product hydrocarbons consisting of about eight carbon atoms to escape, so the chains do not grow too long and equilibrium is never reached.

- **Poisoning a catalyst**

  → Catalysts can be poisoned, or inactivated.

  → For a heterogeneous catalyst, irreversible adsorption of a substance on the catalyst's surface may prevent reactants from reaching the surface and reacting on it.

  → For a homogeneous catalyst, a reactive site on a molecule may be similarly poisoned by the binding of a substance directly to the site or nearby either in a permanent or reversible way. The properties of the site are significantly altered to prevent catalytic action.

  → Some heavy metals, especially lead, are very potent poisons for heterogeneous catalysts, which is why lead-free gasoline must be used in engines fitted with catalytic converters.

  → The elimination of lead has the further benefit of decreasing the amount of poisonous lead in the environment.

7E.3 Living Catalysts: Enzymes

- **Enzyme, E**

  → A biological catalyst

  → Typically, a large molecule (protein) with crevices, pockets, and ridges on its surface (see **Figure 7E.6** in the text)

  → Catalyzes the reaction of a reactant known as the *substrate*, S, which produces a *product*, P:

$$E + S \rightarrow E + P$$

  → Is capable of increasing the rate of specific reactions enormously

- **Induced-fit mechanism**

  → Enzymes are highly specific catalysts that function by an *induced-fit mechanism* in which the substrate approaches a correctly - shaped pocket of the enzyme, known as the *active site.*

  → On binding, the enzyme distorts somewhat to accommodate the substrate, allowing the reaction to occur.

  → The shape of the pocket determines which substrate can react.

  → The induced-fit mechanism is a more realistic version of the lock-and-key mechanism, which assumes that the substrate and enzyme fit together like a lock and key.

- **Michaelis–Menten mechanism of enzyme reaction**

  → A two-step mechanism that accounts for the observed dependence of the reaction rate of many enzyme-substrate reactions

  → For an enzyme, E, and a substrate, S, the overall reaction and the Michaelis–Menten mechanism are

$$\text{Overall reaction:} \quad E + S \rightarrow E + P$$

$$\text{Mechanism:} \quad \text{Step 1} \quad E + S \rightleftharpoons ES$$
$$\text{Step 2} \quad ES \rightarrow E + P$$

→ In step 1, the enzyme and substrate form an enzyme–substrate complex, ES, that can dissociate to reform substrate (reverse of step 1) or decay to form product P (step 2). The rate of formation of product is given by step 2.

Rate of formation of $P = k_2[ES]$

→ The steady-state approximation may be applied to the intermediate, ES, and its concentration is given by

$$[ES] = \frac{k_1[E][S]}{k_1' + k_2}$$

→ Because the free enzyme concentration may be substantially diminished during the reaction, it is customary to express the rate in terms of the total enzyme concentration, $[E]_0 = [E] + [ES]$. Then $[E]$ is replaced by $[E]_0 - [ES]$ in the steady-state result, leading to

$$[ES] = \frac{k_1[E]_0[S]}{k_1' + k_2 + k_1[S]}$$

Rate of formation of $P = \dfrac{k_1 k_2 [E]_0[S]}{k_1' + k_2 + k_1[S]} = \dfrac{k_2[E]_0[S]}{K_M + [S]}$

→ $K_M = \dfrac{k_1' + k_2}{k_1}$ is known as the Michaelis constant.

→ At low concentration of substrate, $[S] \ll K_M$, and the rate law is given by

Rate of formation of $P = \dfrac{k_2[E]_0[S]}{K_M}$

→ The rate varies *linearly* with the concentration of substrate in this region.
   Physically, in this region, most enzyme sites lack substrate molecules, so doubling the substrate concentration doubles the number of occupied sites and the rate as well.

→ At high concentration of substrate, $[S] \gg K_M$ and the rate law becomes

Rate of formation of $P = \dfrac{k_2[E]_0[S]}{[S]} = k_2[E]_0$

→ In this region, the rate is *independent* of the substrate concentration (see **Figure 7E.8** in the text).

→ *Physically, all the enzyme sites are occupied with substrate molecules, so adding substrate has no effect on the rate.*

- **Enzyme poisoning**
  → Not all enzyme poisoning is detrimental.

   Example: Aspirin (acetylsalicylic acid) reduces inflammation by reacting irreversibly with cyclooxegenases to stop their catalytic activity. Cyclooxegenases produce prostaglandins and thromboxanes, which lead to inflammation. These enzymes are overactive in patients with chronic arthritis.

- **Summary**
  - → *Enzymes* are large proteins that function as *biological catalysts*.
  - → *Substrate* molecules (reactants) bind to *active sites* on enzyme molecules.
  - → The model of enzyme *binding* and *action* is called the *induced-fit mechanism*.
  - → Substrate molecules undergo reaction at the active site to form *product* molecules.
  - → Product molecules are released and the active site is restored.

Focus 8: MAIN-GROUP ELEMENTS

Topic 8A: PERIODIC TRENDS

8A.1 Atomic Properties

- **Atomic properties are mainly responsible for the properties of an element:**
  - → Atomic radius; first ionization energy; electron affinity; electronegativity; polarizability

- **Atomic and ionic radii**
  - → typically *decrease* from left to right across a period and *increase* down a group.

- **First ionization energies**
  - → typically *increase* from left to right across a period and *decrease* down a group.
  - → Increase across a period: Caused by increasing attraction of nuclear charge to electrons in the same valence shell
  - → Trend down a group: Concentric shells of electrons are progressively *more distant* from the nucleus.

- **Electron affinities (E_{ea})**
  - → measure the energy *released* when an anion is formed. The greater the E_{ea}, the more stable the anion formed.
  - → *Highest* values of E_{ea} are found at the top right of the periodic table. Cl has a greater *electron affinity* value than F, but F has a greater *electronegativity* value.
  - → Elements with high E_{ea} values are present as *anions* in compounds with *metallic* elements and commonly have *negative* oxidation states in covalent compounds with other *nonmetallic* elements. **Examples:** NaCl, an ionic salt; sulfur hexafluoride SF_6, in which F has an oxidation state of -1.

- **Electronegativities (χ)**
  - → measure the tendency of an atom to attract bonding electrons when it is part of a compound.
  - → Typically *increase* from left to right across a period and *decrease* down a group
  - → A useful guide to the type of bond an element may form (ionic, covalent, etc.)
  - → A predictor of the *polarity* of chemical bonds between nonmetallic elements

- **Polarizabilities**
  - → A measure of the ease of distorting an atom's electron cloud
  - → Typically *decrease* across a period (left to right) and *increase* within a group (top to bottom)
  - → Greatest for the more massive atoms in a group
  - → Anions are more polarizable than parent atoms.
  - → Diagonal neighbors in the periodic table have similar polarizability values and a similar amount of covalent character in the bonds they form.

- **Polarizing powers**
  - → Associated with ions of small size and high charge, such as Al^{3+} and Be^{2+}
  - → Cations with high polarizing power tend to exhibit covalent character in bonds formed with highly polarizable anions.

    Examples: $AlCl_3$ and BeI_2

8A.2 Bonding Trends

- **Elements bonded to other elements**
  - → Period 2 elements: Valence is determined by the number of valence-shell electrons and the octet rule.
  - → Period 3 or higher ($Z \geq 14$): The valence of an element may be increased through octet expansion allowed by access to empty d-orbitals.

    Examples: PCl_5 and SF_6 (see **Topic 2C.2** in the text)
  - → *Inert-pair effect:* Elements at the bottom of the p-block may display an oxidation number two less than the group number suggests.

    Example: Lead has an oxidation number of +2 in PbO and +4 in PbO_2. (see **Topic 1F.6** in the text)

- **Pure elements**
  - → With *low* ionization energies tend to have *metallic* bonds, for example, Li(s) and Mg(s)
  - → With *high* ionization energies are typically *molecular* and form *covalent* bonds

    Examples: $N_2(g)$, $O_2(g)$, and $F_2(g)$
  - → In or near the diagonal band of metalloids (intermediate ionization energies and three, four, or five valence electrons), elements tend to form network structures in which each atom is bonded to three or more other atoms.

    Example: The *network* structure of *black phosphorus*, in which each P atom is bonded to three others

8A.3 Trends Exhibited by Hydrides and Oxides

Chemical Properties: Hydrides

- **Binary hydrides**
  - → Binary compounds with hydrogen (hydrides)
  - → Formed by all main-group elements except the noble gases, possibly In and Tl, and several d-block elements
  - → Properties show periodic behavior.
  - → Formulas are directly related to the group number.

    Examples: SiH_4, NH_3, H_2S, and HCl in Groups 14, 15, 16, and 17, respectively

- **Saline (or salt-like) hydrides**
  - → Formed by all members of the s-block except Be

→ Ionic salts of strongly electropositive metals, with hydrogen present as the hydride ion H^-

→ White, high-melting solids with crystal structures resembling those of the corresponding halides

→ Prepared by heating the metal in hydrogen
   Example: $2\,Na(s) + H_2(g) \rightarrow 2\,NaH(s)$

→ React readily with water (H^- is a strong base), forming a basic solution and releasing $H_2(g)$
   Example: $H^-(aq) + H_2O(l) \rightarrow H_2(g) + OH^-(aq)$

- **Metallic hydrides**

   → Black, powdery electricity-conducting solids formed by certain d-block elements

   → Release hydrogen when heated or when treated with acids. A typical reaction in acid is:

$$CuH(s) + H_3O^+(aq) \rightarrow Cu^+(aq) + H_2(g) + H_2O(l)$$

- **Molecular hydrides**

   → Formed by nonmetals and consist of discrete molecules

   → Volatile; many are Brønsted acids or bases.

   → Gaseous compounds include the base $NH_3(g)$; the hydrogen halides $HX(g)$, X = F, Cl, Br, and I; and the lighter hydrocarbons, such as CH_4 and C_2H_4 (ethene).

   → Liquids include water and heavier hydrocarbons, such as benzene, C_6H_6, and octane, C_8H_{18}.

   → With *some exceptions* hydrides follow the pattern in the table below.

Block	s	d	p
Type	saline	metalllic	molecular

Chemical Properties: Oxides

- **Binary oxides**

   → Formed by all the main-group elements except the noble gases

   → Main-group binary oxides of *metals* are basic.
   Examples: MgO, Na_2O

   → Main-group binary oxides of *nonmetals* are acidic.
   Examples: NO_2, SO_3
   Exception: Al_2O_3 is *amphoteric*. Other *amphoteric oxides* include BeO, Ga_2O_3, SnO_2, and PbO_2.

   → Soluble ionic oxides: Formed from elements *on the left side* of the periodic table

   → Insoluble oxides with high melting points: Formed from elements *on the left side* of the p-block

   → Oxides with low melting points, often gaseous: Formed from elements *on the right side* of the p-block

- **Metallic elements and metalloids**

   → Ionic oxides: Formed from metallic elements with low ionization energies
   Example: Na_2O

   → Amphoteric oxides: Formed from metallic elements with intermediate ionization energies (Be, B, and Al) and from metalloids

   → Amphoteric oxides: Do not react with water but dissolve in acidic or basic solutions

- **Nonmetallic elements**

  → Many oxides of nonmetals are gaseous molecular compounds. Most can act as Lewis acids. They form acidic solutions in water and are called *acid anhydrides*.

    Examples: The acids HNO_3 and H_2SO_4 are derived from the oxides N_2O_5 and SO_3, respectively.

  → Oxides that do *not* react with water are called *formal anhydrides* of acids. The formal anhydride is obtained by removing the elements of water (H, H, and O) from the molecular formula of the acid. For example, the Lewis structure of the formal anhydride of acetic acid, CH_3COOH, is CH_2CO (ketene).

  → With some exceptions, hydrides follow the pattern in the table.

  → Some general properties of oxides are also summarized in the table.

Block	s	d	p
Type	ionic	ionic	covalent
Acid/base character	basic	basic to amphoteric	amphoteric to acidic

Topic 8B: HYDROGEN

8B.1 The Element

- **Hydrogen**

  → One valence electron but few similarities to the alkali metals

  → A nonmetal that resembles the halogens but has some different, unique properties

  → Does not fit clearly into any group and is therefore not assigned to one

- **Commercial production of hydrogen**

  → Formed (1) as a by-product of the *refining* of petroleum and (2) by the *electrolysis* of water

  → In the refining process, a Ni-catalyzed *re-forming reaction* produces products called *synthesis gas:*

    $$CH_4(g) + H_2O(g) \rightarrow CO(g) + 3\,H_2(g)$$

    In a second step, a *shift reaction* employing an Fe/Cu catalyst yields hydrogen gas:

    $$CO(g) + H_2O(g) \rightarrow CO_2(g) + H_2(g)$$

  → *Electrolysis of water* utilizes a salt solution to improve conductivity: $2\,H_2O(l) \rightarrow 2\,H_2(g) + O_2(g)$

- **Laboratory production and uses of hydrogen**

  → Produced by reaction of an active metal such as Zn with a strong acid such as HCl:

    $$Zn(s) + 2\,H_3O^+(aq) \rightarrow Zn^{2+}(aq) + H_2(g) + 2H_2O(l)$$

  → Produced by reaction of a hydride with water:

    $$CaH_2(s) + 2\,H_2O(l) \rightarrow Ca^{2+}(aq) + 2\,OH^-(aq) + 2\,H_2(g)$$

- **Liquid hydrogen**

  → Has a very low density and the highest specific density of any known fuel

8B.2 Compounds of Hydrogen

Reactant	Reaction with hydrogen
Group 1 metals (M)	$2M(s) + H_2(g) \rightarrow 2MH(s)$
Group 2 metals (M, not Be or Mg)	$M(s) + H_2(g) \rightarrow MH_2(s)$
some d-block metals (M)	$2M(s) + x\,H_2(g) \rightarrow 2MH_x(s)$
oxygen	$O_2(g) + 2H_2(g) \rightarrow 2H_2O(l)$
nitrogen	$N_2(g) + 3H_2(g) \rightarrow 2NH_3(g)$
halogen (X_2)	$X_2(g, l, s) + H_2(g) \rightarrow 2HX(g)$

- **Unusual properties of hydrogen**
  - → It can form a cation, H^+, or an anion, hydride, H^-.
  - → It can form covalent bonds with many other elements because of its intermediate electronegativity value of $\chi = 2.20$.

- **The hydride ion, H^-**
  - → Large, with a radius (154 pm) between those of F^- (133 pm) and Cl^- (181 pm) ions
  - → Highly polarizable, which adds covalent character in its bonds to cations
  - → The two electrons in H^- are weakly attached to the single proton, and an electron is easily lost, which makes H^- a good reducing agent.
  - → Saline hydrides are very strong reducing agents. The reduction half-reaction and standard reduction potential for the reduction of hydrogen are $H_2(g) + 2e^- \rightarrow 2H^-(aq)$ and $E° = -2.25$ V, respectively.
  - → The alkali metals are also strong reducing agents with reduction potentials similar to those of the hydride value.
  - → Hydride ions in saline hydrides reduce water upon contact: $H^-(aq) + H_2O(l) \rightarrow H_2(g) + OH^-(aq)$.

- **Hydrogen bonds**
  - → Relatively strong intermolecular bonds
  - → Formed by hydrides of N, O, and F
  - → Can be understood in terms of the coulombic attraction between the partial positive charge on a hydrogen atom and the partial negative charge of another atom
  - → The H atom is covalently bonded to a very electronegative N, O, or F atom, giving the H atom a partial positive charge.
  - → The partial negative charge is often the result of the lone-pair electrons on a different N, O, or F atom.
  - → A hydrogen bond is represented as three dots between atoms *not* covalently bonded to each other: $[O–H\cdots:O]$. It is typically about 5% as strong as a covalent bond.

Topic 8C: GROUP 1: THE ALKALI METALS

8C.1 The Group 1 Elements

Reactant	Reaction with alkali metal (M)
hydrogen	$2M(s) + H_2(g) \rightarrow 2MH(s)$
oxygen	$4Li(s) + O_2(g) \rightarrow 2Li_2O(s)$
	$2Na(s) + O_2(g) \rightarrow Na_2O_2(s)$
	$2M(s) + O_2(g) \rightarrow 2MO_2(s)$, M = K, Rb, Cs
nitrogen	$6Li(s) + N_2(g) \rightarrow 2Li_3N(s)$
halogen (X_2)	$2M(s) + X_2(g, l, s) \rightarrow 2MX(s)$
water	$2M(s) + 2H_2O(l) \rightarrow 2MOH(aq) + H_2(g)$

- **Alkali metals**
  - → Have the valence electron configuration ns^1, where n is the period number

- **Preparation**
  - → Pure alkali metals are obtained by electrolysis of the molten salts.

    Exception: K is prepared by reaction of molten KCl and Na vapor at 750 °C:

    $$KCl(l) + Na(g) \rightarrow NaCl(s) + K(g)$$

    This reaction is driven to form products by the condensation (removal) of K(g) because the equilibrium constant is unfavorable. Recall Le Chatelier's principle (see **Topic 5J** in the text).

- **Properties of alkali metals**
  - → Dominated by the ease with which their single valence electron is lost
  - → Soft silver-gray metals with low melting points, boiling points, and densities
  - → Melting and boiling points decrease down the group (see **Table 8C.1** in the text). Cs (melting point 28 °C) is very reactive and is transported in sealed ampoules. Fr is quite radioactive, and little is known about its properties.
  - → All alkali metals are highly reactive.

- **Applications**
  - → Li metal is used in the rechargeable lithium-ion battery, which holds a charge for a long time.
  - → Li is also used in thermonuclear weapons.

- **Other properties of alkali metals**
  - → Strong reducing agents (low first-ionization energies lead to their ease of oxidation)

    Example: Reduction of Ti(IV): $TiCl_4 + 4Na(l) \rightarrow 4NaCl(s) + Ti(s)$
  - → Reduce water to form *basic* solutions, releasing hydrogen gas

    Example: $2Na(s) + 2H_2O(l) \rightarrow 2NaOH(aq) + H_2(g)$
  - → Dissolve in $NH_3(l)$ to yield solvated electrons, which are used to reduce organic compounds
  - → React directly with almost all nonmetals, noble gases excluded
  - → However, only one alkali metal, Li, reacts with nitrogen to form the nitride (Li_3N).

- **Formation of compounds containing oxygen**
  - → Li forms mainly the oxide Li_2O.
  - → Na forms a pale yellow peroxide, Na_2O_2.
  - → K, Rb, and Cs form mainly a superoxide, for example, KO_2.

8C.2 Compounds of Lithium, Sodium, and Potassium

- **Lithium**
  - → Differs significantly from the other Group 1 elements, a common occurrence for elements at the head of a group
  - → For Group I, the differences originate in part from the small size of Li^+.
  - → Cations have strong polarizing power and a tendency toward covalency in their bonding.
  - → Has a diagonal relationship with Mg and is found in the minerals of Mg
  - → Is found in compounds with oxidation number +1; its compounds are used in ceramics, lubricants, and medicine.

    Examples: Lithium carbonate is an effective drug for bipolar (manic-depressive) disorder. Lithium soaps, which have higher melting points than sodium and potassium soaps, are used as thickeners in lubricating greases for high-temperature applications.

- **Sodium compounds**
  - → Important in part because they are *plentiful*, *inexpensive*, and water *soluble*
  - → NaCl is mined as *rock salt* and is used in the electrolytic production of $Cl_2(g)$ and NaOH.
  - → NaOH is an inexpensive starting material for the production of other sodium salts.
  - → $Na_2CO_3 \cdot 10\,H_2O$ was once used as *washing soda* to precipitate Ca^{2+} and Mg^{2+} ions.
  - → $NaHCO_3$ is also known as *sodium bicarbonate*, *bicarbonate of soda*, or *baking soda*.
  - → Weak acids (lactic acid (milk), citric acid (lemons), acetic acid (vinegar)) react with the hydrogen carbonate anion to release carbon dioxide gas:

    $$HCO_3^-(aq) + HA(aq) \rightarrow A^-(aq) + H_2O(l) + CO_2(g).$$

  - → Double-acting *baking powder* contains a solid weak acid as well as the hydrogen carbonate ion.

- **Potassium compounds**
  - → Usually more expensive than the corresponding sodium compounds
  - → Use may be justified because potassium compounds are generally less hygroscopic than sodium compounds.
  - → The K^+ cation is larger than the Na^+ ion and is less strongly hydrated by H_2O molecules.
  - → KNO_3 releases $O_2(g)$ when heated and is used to help matches ignite.
  - → KNO_3 is also the oxidizing agent in black gunpowder (75% KNO_3, 15% charcoal, and 10% sulfur).

→ KCl is used directly in some fertilizers as a source of potassium, but KNO_3 is required for crops that cannot tolerate high chloride ion concentrations.

→ Mineral sources of potassium are *sylvite*, KCl, and *carnallite*, $KCl \cdot MgCl_2 \cdot 6H_2O$.

Topic 8D: GROUP 2: THE ALKALINE EARTH METALS

8D.1 The Group 2 Elements

Reactant	Reaction with Group 2 (M)
hydrogen	$M(s) + H_2(g) \rightarrow MH_2(s)$, not Be or Mg
oxygen	$2M(s) + O_2(g) \rightarrow 2MO(s)$
nitrogen	$3M(s) + N_2(g) \rightarrow M_3N_2(s)$
halogen (X_2)	$M(s) + X_2(g, l, s) \rightarrow MX_2(s)$
water	$M(s) + 2H_2O(l) \rightarrow M(OH)_2(aq) + H_2(g)$, not Be

- **Alkaline earth metals**
    → Have the valence electron configuration ns^2, where n is the period number
    → Except for Be, Group 2 elements exhibit metal characteristics such as forming basic oxides.
    → Ca, Sr, and Ba are called alkaline earths because their oxides are basic (alkaline)
    → The name *alkaline earth* has been extended to all the Group 2 elements.

- **Preparation**
    → Pure metals: obtained by electrolysis of the molten salts (all) or by chemical reduction (except Be)
    → Ca(s), Sr(s), and Ba(s) may be obtained by reduction with Al(s) in a variation of the thermite reaction.
      Example: $3CaO(s) + 2Al(s) \rightarrow Al_2O_3(s) + 3Ca(s)$.
    → Be occurs in nature as *beryl*, $3BeO \cdot Al_2O_3 \cdot 6SiO_2$.
    → Mg is found in seawater as $Mg^{2+}(aq)$ and in the mineral dolomite, $CaCO_3 \cdot MgCO_3$.

- **Properties**
    → Silver-gray metals with much higher melting points, boiling points, and densities than those of the preceding alkali metals in the same period
    → Melting points decrease down the group with the exception of Mg, which has the lowest value. The trend in boiling point is irregular (see **Table 8D.1** in the text).
    → All Group 2 elements except Be reduce water.
      Example: $Sr(s) + 2H_2O(l) \rightarrow Sr^{2+}(aq) + 2OH^-(aq) + H_2(g)$
    → Be does not react with water. Mg reacts with hot water, and Ca reacts with cold water.
    → Be and Mg do not dissolve in HNO_3 because they develop a protective oxide film.
    → Mg burns vigorously in air because it reacts with $O_2(g)$, $N_2(g)$, and $CO_2(g)$.

→ Reactivity of the Group 2 metals with water and oxygen increases down the group, just as with the Group 1 metals.

- **Properties and applications of beryllium**
  - → Has a low density, making it useful for missile and satellite fabrication
  - → Used as a window for x-ray tubes
  - → Added in small amounts to Cu to increase its rigidity
  - → Be/Cu alloys are used in nonsparking tools (oil refineries and grain elevators) and for nonmagnetic parts that resist deformation and corrosion (electronics industry).

- **Properties and applications of magnesium**
  - → A silver-white metal protected from air oxidation by a white oxide film, making it appear dull gray
  - → Light and very soft, but its alloys have great strength and are used in applications where lightness and toughness are desired (airplanes)
  - → Limitation in widespread usage is related to its expense, low melting temperature, and difficulty in machining.

- **Properties and applications of calcium, strontium, and barium**
  - → Group 2 metals can be detected in burning compounds by the colors they give to flames. Ca burns orange-red, Sr crimson, and Ba yellow-green. Fireworks are made from their salts.

8D.2 Compounds of Beryllium, Magnesium, and Calcium

- **Beryllium**
  - → Differs significantly from the other Group 2 elements, as is typical of an element at the head of its group
  - → The difference originates in part from the small size of the Be^{2+} cation, which has strong polarizing power and a tendency toward covalency in its bonding.
  - → Has a diagonal relationship with Al. It is the only member of the group that, like Al, reacts in aqueous NaOH. In strongly basic solution, it forms the *beryllate ion* $[Be(OH)_4]^{2-}$.
  - → Compounds are extremely toxic.
  - → The high polarizing power of Be^{2+} produces moderately covalent compounds, while its small size allows no more than four groups to be attached, accounting for the prominence of the tetrahedral BeX_4 unit.
  - → $BeH_2(g)$ and $BeCl_2(g)$ condense to form solids with chains of tetrahedral BeH_4 and $BeCl_4$ units, respectively.
  - → In $BeCl_2$, the Be atoms act as Lewis acids and accept electron pairs from the Cl atoms, forming a chain of tetrahedral $BeCl_4$ units in the solid.

- **Magnesium**
  - → More metallic properties than Be
  - → Compounds are primarily ionic with some covalent character.

→ MgO and Mg_3N_2 are formed when magnesium burns in air.

→ MgO dissolves very slowly and only slightly in water. It can withstand high temperatures (*refractory*) because it melts at 2800 °C.

→ MgO conducts heat well but electricity only poorly; it is used as an insulator in electric heaters.

→ $Mg(OH)_2$ is a base. It is not very soluble in water but forms instead a white colloidal suspension, which is used as a stomach antacid called *milk of magnesia*. A side effect of stomach acid neutralization is the formation of $MgCl_2$, which acts as a purgative.

→ $MgSO_4$ (*Epsom salts*) is also a common purgative.

→ Chlorophyll is the most important compound of Mg. Magnesium ions in chlorophyll enable photosynthesis to occur.

→ Magnesium also plays an important role in energy generation in living cells. It is involved in the contraction of muscles.

- **Calcium**

→ Ca is more metallic in character than Mg, but its compounds share some similar properties.

→ $CaCO_3$, the most common compound of calcium, occurs naturally as chalk and limestone. Marble is a very dense form with colored impurities, most commonly Fe cations.

→ The most common forms of pure calcium carbonate are *calcite* and *aragonite*. All carbonates are the fossilized remains of marine life.

→ CaO is formed by heating $CaCO_3$: $CaCO_3(s) \rightarrow CaO(s) + CO_2(g)$.
  CaO is called *quicklime* because it reacts rapidly and exothermically with water:

$$CaO(s) + H_2O(l) \rightarrow Ca^{2+}(aq) + 2OH^-(aq)$$

→ $Ca(OH)_2$, the product of the equation, is known as *slaked lime* because in this form the thirst for water has been quenched, or slaked.

→ An aqueous solution of $Ca(OH)_2$, which is only slightly soluble in water, is called *lime water*. It is used as a test for carbon dioxide:

$$Ca(OH)_2(aq) + CO_2(g) \rightarrow CaCO_3(s) + H_2O(l).$$

- **Carbide, sulfate, and phosphate**

→ CaO may be converted into calcium carbide: $CaO(s) + 3C(s) \rightarrow CaC_2(s) + CO(g)$.

→ Calcium sulfate dihydrate (*gypsum*), $CaSO_4 \cdot 2H_2O$, is used in construction materials.

→ Ca is found in the rigid structural components of living organisms, either as $CaCO_3$ (shellfish shells) or as $Ca_3(PO_4)_2$ (bone).

→ Tooth enamel is a *hydroxyapatite,* $Ca_5(PO_4)_3OH$, which is subject to attack by acids produced when bacteria act on food residues. A more resistant coating is formed when the OH^- ions are replaced by F^- ions. The resulting mineral is called *fluoroapatite:*

$$Ca_5(PO_4)_3OH(s) + F^-(aq) \rightarrow Ca_5(PO_4)_3F(s) + OH^-(aq)$$

Tooth enamel is strengthened by the addition of fluorides (fluoridation) to drinking water and by the use of fluoridated toothpaste containing tin(II) fluoride, SnF_2, or sodium monofluorophosphate (MFP), Na_2FPO_3.

Topic 8E: GROUP 13: THE BORON FAMILY

8E.1 The Group 13 Elements

→ Have the valence electron configuration $ns^2\, np^1$

→ Common (maximum) oxidation number is +3, but heavier elements in the group, such as indium and thallium, also exhibit oxidation number +1 (inert pair effect; see **Topic 1F.6** in the text).

- **Preparation of boron**

  → Mined as the hydrates *borax* and *kernite*, $Na_2B_4O_7 \cdot xH_2O$, with $x = 10$ and 4, respectively

  → The ore is converted to the oxide B_2O_3 and extracted in its amorphous elemental form by reaction of the oxide with $Mg(s)$:

  $$B_2O_3(s) + 3\,Mg(s) \rightarrow 2\,B(s) + 3\,MgO(s).$$

  → A purer form of B is obtained by reduction of volatile $BBr_3(g)$ or $BCl_3(g)$ with $H_2(g)$.

- **Preparation of aluminum**

  → Mined as *bauxite*, $Al_2O_3 \cdot xH_2O$, where x ranges to a maximum of 3

  → The ore, which contains considerable iron oxide impurity, is processed to obtain alumina, Al_2O_3.

  → Al is then extracted by an electrolytic process (*Hall process*). Al_2O_3 is mixed with *cryolite*, Na_3AlF_6, to lower the melting point from 2050 °C (alumina) to 950 °C (mixture). The half-cell and cell reactions are:

Cathode:	$Al^{3+}(melt) + 3\,e^- \rightarrow Al(l)$
Anode:	$2\,O^{2-}(melt) + C(s,gr) \rightarrow CO_2(g) + 4\,e^-$
Cell:	$4\,Al^{3+}(melt) + 6\,O^{2-}(melt) + 3\,C(s,gr) \rightarrow 4\,Al(l) + 3\,CO_2(g)$

- **Properties of boron**

  → First element of the p-block

  → Boron forms perhaps the most extraordinary structures of all the elements.

  → Element is attacked only by the strongest oxidizing agents.

  → A nonmetal in most of its chemical properties

  → Has acidic oxides and forms a fascinating array of binary molecular hydrides

→ It exists in several allotropic forms in which B atoms attempt to share eight electrons, despite the small size of B and its ability to contribute only three electrons. In this sense, B is regarded as *electron deficient*.

→ One common form of B is a gray-black, nonmetallic, high-melting solid.

→ Another common form is a dark brown powder with a structure (icosahedral with 20 faces) that has clusters of 12 atoms. The bonds create a very hard three-dimensional structure.

- **Properties of aluminum**

  → Aluminum is the most abundant metallic element in the Earth's crust.

  → Low density, but a strong metal with excellent electrical conductivity

  → Easily oxidized, but its surface is passivated by a protective oxide film

  → If the thickness of the oxide layer is increased electrolytically, *anodized aluminum* results.

  → Al is amphoteric, reacting with both *nonoxidizing* acids and hot aqueous bases. The reactions are:

$$2\,Al(s) + 6\,H^+(aq) \rightarrow 2\,Al^{3+}(aq) + 3\,H_2(g)$$

$$2\,Al(s) + 2\,OH^-(aq) + 6\,H_2O(l) \rightarrow 2\,Al(OH)_4^-(aq) + 3\,H_2(g)$$

- **Applications**

  → Boron fibers are incorporated into plastics, forming resilient material stiffer than steel and lighter than Al(s).

  → Boron carbide is similar in hardness to diamond, and boron nitride is similar in structure and mechanical properties to graphite, but unlike graphite, boron nitride does not conduct electricity.

  → Al has widespread use in construction and aerospace industries. Because it is a soft metal, its strength is improved by alloy formation with Cu and Si.

  → Because of its high electrical conductivity, Al is used in overhead power lines. Its high negative electrode potential has led to its use in fuel cells.

- **Gallium**

  → A by-product of the production of Al

  → Common doping agent for semiconductors

  → GaAs, gallium arsenide, is used in light-emitting diodes (LEDs).

- **Thallium**

  → Poisonous heavy metal sometimes used as a rat poison

8E.2 Group 13 Oxides, Halides, and Nitrides

- **Boron oxide**

  → Boron, like most nonmetals, has *acidic* oxides. Hydration of B_2O_3 yields boric acid, H_3BO_3, or $B(OH)_3$, in which B has an incomplete octet and can therefore act as a Lewis acid:

$$(OH)_3B + :OH_2 \rightarrow (OH)_3B-OH_2$$

The complex formed is a weak *monoprotic* acid ($pK_a = 9.14$).

→ H_3BO_3 is a mild antiseptic and pesticide. It is also used as a fire retardant in home insulation and clothing.

→ The main use of H_3BO_3 is as a source of boron oxide, the anhydride of boric acid:

$$2\,H_3BO_3(s) \xrightarrow{\text{heat}} B_2O_3(s) + 3\,H_2O(l)$$

B_2O_3, which melts at 450 °C and dissolves many metal oxides, is used as a flux for soldering or welding. It is also used in the manufacture of fiberglass and borosilicate glass, such as Pyrex.

- **Aluminum oxide**

    → Al_2O_3, known as *alumina*, exists in several crystal forms, each of which is used in science and in commerce.

    → α-Alumina is a very hard substance, *corundum*, used as an abrasive known as *emery*. It is a rock-forming mineral and a naturally transparent material. Impurities cause colors to appear. Used as gems in ruby (red) and sapphire (green).

    → γ-Alumina is absorbent and is used as the stationary phase in chromatography. It is produced by heating $Al(OH)_3$ and is moderately reactive and amphoteric:

$$Al_2O_3(s) + 6\,H_3O^+(aq) + 3\,H_2O(l) \rightarrow 2\,[Al(H_2O)_6{}^{3+}](aq)$$

$$Al_2O_3(s) + 2\,OH^-(aq) + 3\,H_2O(l) \rightarrow 2\,Al(OH)_4{}^-(aq)$$

The strong polarizing effect of the small, highly charged Al^{3+} ion on the water molecules surrounding it gives the $[Al(H_2O)_6{}^{3+}]$ ion acidic properties.

    → Aluminum sulfate is prepared by reaction of Al_2O_3 with sulfuric acid:

$$Al_2O_3(s) + 3\,H_2SO_4(aq) \rightarrow Al_2(SO_4)_3(aq) + 3\,H_2O(l)$$

$Al_2(SO_4)_3$ (*papermaker's alum*) is used in the preparation of paper.

    → Sodium aluminate, $NaAl(OH)_4$, is used along with $Al_2(SO_4)_3$ in water purification.

- **Halides**

    → Boron halides are made either by direct reaction of the elements at high temperature or from the oxide.

    Examples: $B_2O_3(s) + 3\,CaF_2(s) + 3\,H_2SO_4(l) \xrightarrow{\text{heat}} 2\,BF_3(g) + 3\,CaSO_4(s) + 3\,H_2O(l)$

    $B_2O_3(s) + 3\,C(s) + 3\,Cl_2(g) \xrightarrow{\text{heat}} 2\,BCl_3(g) + 3\,CO(g)$

    → All boron halides are trigonal-planar, covalently bonded, and electron-deficient, with an incomplete octet on B, as a consequence of which they are strong Lewis acids.

    → BCl_3 and $AlCl_3$ are widely used as catalysts.

    → $AlCl_3$ is formed as an ionic compound by direct reaction of the elements or of alumina with chlorine in the presence of carbon:

$$2\,Al(s) + 3\,Cl_2(g) \rightarrow 2\,AlCl_3(s)$$

$$Al_2O_3(s) + 3\,Cl_2(g) + 3\,C(s, gr) \rightarrow 2\,AlCl_3(s) + 3\,CO(g)$$

Ionic $AlCl_3$ melts at 192 °C to form a molecular liquid, Al_2Cl_6.
This dimer of $AlCl_3$, formed by donation of an unshared electron
pair on Cl to a vacant orbital on an adjacent Al atom, contains
$AlCl_4$ tetrahedral units and two bridging Cl atoms.

→ Aluminum halides react with water in a highly exothermic reaction.

→ Lithium aluminum hydride, a white solid prepared from $AlCl_3$, is an important reducing agent in
organic chemistry:

$$4\,LiH + AlCl_3 \rightarrow LiAlH_4 + 3\,LiCl$$

- **Nitrides**

  → When boron is heated to white heat in ammonia, it forms boron nitride:

  $$2\,B(s) + 2\,NH_3(g) \rightarrow 2\,BN(s) + 3\,H_2(g)$$

  The structure of the solid is similar that of graphite, but the planes of hexagons of carbon atoms are
  replaced by hexagons of alternating B and N atoms (see **Figure 8E.3** in the text).

  → Boron nitride is a fluffy, slippery powder. It can be pressed into a solid rod that is easy
  to machine, and boron nitride, unlike graphite, is a good insulator. Under high pressure,
  boron nitride is converted to a diamond like crystalline material called Borazon. Boron
  nitride nanotubes similar to those formed by carbon have been found to have
  semiconducting properties.

  → Boron trifluoride reacts with ammonia to produce the Lewis acid–base complex F_3BNH_3, which is
  a solid at room temperature.

  $$F_3B(g) + :NH_3(g) \rightarrow F_3B-NH_3(s)$$

  The solid is stable up to about 125 °C, at which point it decomposes to form boron nitride and
  ammonium tetrafluoroborate:

  $$4\,F_3BNH_3(s) \rightarrow BN(s) + 3\,NH_4BF_4(s)$$

8E.3 Boranes, Borohydrides, and Borides

- **Boranes**

  → Compounds of boron and hydrogen, such as B_2H_6 and $B_{10}H_{14}$

  → Electron-deficient compounds with three-center, two-electron B–H–B bonds

  → Valid Lewis structures cannot be written.

  → Molecular orbital theory provides a framework for
  understanding the delocalized behavior of an electron pair
  associated with three atoms, as shown for B_2H_6.

  → Such bonds are called 2-electron, 3-center bonds.

  → Diborane, B_2H_6, is highly reactive. When heated, it decomposes to hydrogen and boron:

  $$B_2H_6(g) \rightarrow 2\,B(s) + 3\,H_2(g)$$

  It reacts with water, reducing hydrogen and forming boric acid:

  $$B_2H_6(g) + 6\,H_2O(l) \rightarrow 2\,B(OH)_3(aq) + 6\,H_2(g)$$

- **Borohydrides**
  - → Anionic versions of boranes. The most important borohydride is BH_4^-, as in sodium borohydride, $NaBH_4$, which is produced by the reaction of NaH with BCl_3:

$$4\,NaH + BCl_3 \rightarrow NaBH_4 + 3\,NaCl$$

  The borohydride anion BH_4^- is an effective and important reducing agent.
  - → Diborane is produced by the reaction of sodium borohydride with boron trifluoride:

$$3\,BH_4^- + 4\,BF_3 \rightarrow 3\,BF_4^- + 4\,B_2H_6$$

  - → Diborane is spontaneously flammable in air.

- **Borides**
  - → Many borides of the metals and nonmetals are known. Formulas of borides are usually unrelated to the position of the two elements in the periodic table. Examples given in the text are AlB_2, CaB_6, $B_{13}C_2$, $B_{12}S_2$, Ti_3B_4, TiB, and TiB_2.
  - → Boron atoms in borides commonly form extended structures such as zigzag chains, branched chains, or networks of hexagonal rings.

Topic 8F: GROUP 14: THE CARBON FAMILY

8F.1 The Group 14 Elements

- → Have the valence electron configuration $ns^2\,np^2$
- → The lighter elements have all four valence electrons available for bonding.
- → Heavier elements (Sn, Pb) display the inert-pair effect and exhibit oxidation numbers 2 and 4.
- → For Pb, the most common oxidation number is 2.
- → Metallic character increases from C to Pb.
- → Carbon is central to life and natural intelligence.
- → Silicon and germanium are central to electronic technology and artificial intelligence.

- **General properties**
  - → Carbon forms covalent compounds with nonmetals and ionic compounds with metals.
  - → The oxides of C and Si are acidic.
  - → Ge, a metalloid, exhibits metallic or nonmetallic properties, depending on the compound.
  - → Sn and Pb are classified as metals, but tin has some intermediate properties.

→ Sn is amphoteric and reacts with hot, concentrated hydrochloric acid and hot alkali:

$$Sn(s) + 2 H_3O^+(aq) \rightarrow Sn^{2+}(aq) + H_2(g) + 2 H_2O(l)$$

$$Sn(s) + 2 OH^-(aq) + 2 H_2O(l) \rightarrow Sn(OH)_4{}^{2-}(aq) + H_2(g)$$

→ Carbon, at the top of the group, differs greatly in its properties from the other members.

→ Carbon tends to form multiple bonds, whereas silicon does not.

→ CO_2 is a molecular gas, while $SiO_2(s)$ (the mineral silica) contains networks of –O–Si–O– groups.

→ Si compounds can act as Lewis acids, whereas C compounds usually cannot. Si may expand its octet, whereas C does not.

→ Allotropes of solid carbon include graphite and diamond. There are other forms of carbon, such as fullerenes, graphene, and nanotubes.

- **Graphite**

  → The thermodynamically stable allotrope of carbon (Soot contains small crystals of graphite.)

  → Produced pure commercially by heating C rods in an electric furnace for several days

  → Contains sp^2-hybridized C atoms

  → Consists structurally of large sheets of fused benzenelike hexagonal units stacked one upon the other. A π-bonding network of delocalized electrons accounts for the high electrical conductivity of graphite.

  → When certain impurities are present, graphitic sheets can slip past one another, and graphite becomes an excellent dry lubricant.

  → Only soluble in a few liquid metals

  → Soot and carbon black have commercial applications in rubber and inks. Activated charcoal is an important, versatile purifier. Unwanted compounds are adsorbed onto its microcrystalline surface.

- **Diamond**

  → The hardest substance known and an excellent conductor of heat

  → An excellent abrasive, and the heat generated by friction is rapidly conducted away

  → Synthetic diamonds are produced from graphite at high pressures (> 80 kbar) and temperatures (> 1500 °C) and by thermal decomposition of methane.

  → Soluble in liquid metals, such as Cr and Fe, but less soluble than graphite. This solubility difference is utilized in synthesizing diamond at high pressure and high temperature.

  → Carbon in diamond is sp^3-hybridized, and each C atom in a diamond crystal is bonded directly to four other C atoms and indirectly interconnected to all of the other C atoms in the crystal through C–C single σ-bonds.

  → The σ-bonding network of localized electrons accounts for the electrical insulating properties of diamond.

- **Fullerenes**

  → Buckminsterfullerene, C_{60}, is a soccer-ball–like molecule first identified in 1985. Numerous fullerenes, containing 44 to 84 carbon atoms, were discovered quickly thereafter.

→ Fullerenes are molecular electrical insulators and are soluble in organic solvents such as benzene.

→ Have relatively low ionization energies *and* large electron affinities, so they readily lose or gain electrons to form ions.

→ Contain an even number of carbon atoms, differ by a C_2 unit, and have pentagonal and hexagonal ring structures. Solid samples of fullerenes are called *fullerites*.

→ C_{60} has the atomic arrangement of a *truncated icosahedron* (soccer ball), with 32 faces, 12 pentagons, and 20 hexagons. The interior of the molecule is large, and other atoms may be inserted to produce compounds with unusual properties. The crystal structure of C_{60} is face-centered cubic.

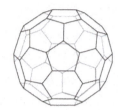

→ The anion C_{60}^{3-} is very stable, and K_3C_{60} is a superconductor below 18 K.

- **Graphene**
  → A single sheet of carbon atoms in a graphitelike hexagonal honeycomb array
  → Of interest because of its electrical, optical, and mechanical properties
  → Extremely thin but still highly opaque
  → Thermal conductivity higher than that of diamond
  → One of the strongest materials known
  → Potential use in nanoscience for miniaturized electronic circuits, antibacterial action, and solar cells

- **Carbon nanotubes** (see also **Topic 3J.7** in the text)
  → Concentric tubes with graphenelike walls rolled into cylinders.
  → Strong, conducting fibers with large surface areas

- **Coke**
  → Impure form of carbon
  → Obtained from destructive distillation of coal
  → Inexpensive, used in steel production

- **Carbon black**
  → Produced by heating gaseous hydrocarbons to approximately 1000 °C in the absence of air
  → Used to reinforce rubber and for printing ink

- **Activated charcoal**
  → Granules of microcrystalline carbon
  → Produced by heating waste organic matter in the absence of air
  → Very high surface area
  → Used to removed organic impurities from liquids and gases, as with gas masks, aquarium filters, and drinking water purification

- **Silicon**
  → Second most abundant element in the Earth's crust
  → Occurs naturally as silicon dioxide $SiO_2(s)$ and as silicates, which contain SiO_3^{2-} units

→ Forms of SiO_2 include quartz, quartzite, and sand.

→ Pure Si, which is widely used in semiconductors, is obtained from quartzite in a three-step process. Silicon produced in the first step is crude and requires further purification in the next two steps:

$$SiO_2(s) + 2\,C(s) \xrightarrow{\text{heat}} Si(s) + 2\,CO_2(g)$$

$$Si(s) + 2\,Cl_2(g) \rightarrow SiCl_4(l)$$

$$SiCl_4(l) + 2\,H_2(g) \rightarrow Si(s) + 4\,HCl(g)$$

→ To be used in semiconductors, silicon must be purified further by zone refining (see **Figure 8F.3** in the text) to obtain ultrapure Si.

→ Amorphous Si, used in photovoltaic devices, is prepared from the decomposition of silane, SiH_4, in an electric discharge.

- **Germanium**

    → Mendeleev predicted the existence of Ge, "eka-silicon," before its discovery.

    → Occurs as an impurity in Zn ores

    → Used mainly in the semiconductor industry to make very fast integrated circuits

    → Obtained easily from its ore

- **Tin**

    → Known since antiquity

    → Reduced from its ore, *cassiterite* (SnO_2), through reaction with C at 1200 °C:

$$SnO_2(s) + C(s) \rightarrow Sn(l) + CO_2(g)$$

    → Used to plate other materials and for the production of alloys

    → Expensive, not very strong, but resists corrosion

- **Lead**

    → A toxic heavy metal

    → The principal ore is *galena*, PbS, which is roasted in air to yield the oxide. It in turn is reduced with coke to produce the metal:

$$2\,PbS(s) + 3\,O_2(g) \rightarrow 2\,PbO(s) + 2\,SO_2(g)$$

$$PbO(s) + C(s) \rightarrow Pb(s) + CO(g)$$

    → Dense, malleable, and chemically inert (passivated)

    → Used as an electrode material in rechargeable car batteries, to transport concentrated sulfuric acid, and as a shield for radiation

8F.2 Oxides of Carbon and Silicon

→ Hugely important in industry

→ Oxides of carbon are all gases.

→ Oxides of silicon are all solids.

- **Carbon dioxide, CO$_2$**
  - → A nonmetal oxide and the acid anhydride of carbonic acid, H$_2$CO$_3$ (The equilibrium between CO$_2$ and H$_2$CO$_3$ favors CO$_2$.)
  - → Used to make carbonated beverages with high partial pressures of CO$_2$
  - → CO$_2$(aq) is an equilibrium mixture of CO$_2$, H$_2$CO$_3$, HCO$_3^-$, and a very small amount of CO$_3^{2-}$.
  - → Solid CO$_2$ (*dry ice*) sublimes to the gas phase, making it useful as a refrigerant and cold pack.

- **Carbon monoxide, CO**
  - → A highly toxic colorless, odorless, flammable gas that is nearly insoluble in water.
  - → Has an extremely large bond enthalpy (1074 kJ·mol^{-1}).
  - → Produced when an organic compound or C itself is burned in a limited amount of oxygen
  - → Commercially produced as *synthesis gas*:
  $$CH_4(g) + H_2O(g) \rightarrow CO(g) + 3H_2(g)$$
  - → CO, the *formal* anhydride of formic acid, is produced in the laboratory by dehydration of formic acid with hot, concentrated sulfuric acid:
  $$HCOOH(l) \rightarrow CO(g) + H_2O(l)$$
  - → CO is a moderately good Lewis base. One example is the reaction of CO with a d-block metal or ion:
  $$Ni(s) + 4CO(g) \rightarrow Ni(CO)_4(l)$$
  - → CO can also act as a Lewis acid, accepting electrons into its vacant antibonding π-orbitals.
  - → Complex formation is responsible for the toxicity of CO, which attaches more strongly than O$_2$ to iron in hemoglobin, thereby blocking the acceptance of O$_2$ by red blood cells.
  - → CO can act as a reducing agent, yielding CO$_2$.

- **Silica or silicon dioxide, SiO$_2$**
  - → Hard, rigid water-insoluble network solid
  - → Derives its strength from its covalently bonded network structure
  - → Occurs naturally in pure form only as quartz and sand
  - → Flint is silica with carbon impurities that make it black.

- **Silicates, compounds containing SiO$_4^{4-}$ units** (see **Topic 3I.2** in the text)
  - → Orthosilicic acid (H$_4$SiO$_4$) and metasilicic acid (H$_2$SiO$_3$) are weak acids.
  - → When sodium orthosilicate is acidified in solution, *silica gel* is produced. It is used as a drying agent, a support for catalysts, packing for chromatography columns, and so on. The reaction is:
  $$4H_3O^+(aq) + SiO_4^-(aq) + xH_2O(l) \rightarrow SiO_2(s) \cdot xH_2O(gel) + 6H_2O(l)$$

- **Aluminosilicates and cements** (see **Topic 3I.4** in the text)

8F.3 Other Important Group 14 Compounds

- **Carbides**

  → Can be saline (*saltlike*), covalent, or interstitial (three types)

  → *Saline carbides* are most commonly formed from Group 1 and 2 metals, Al, and a few other metals. s-Block metals form saline carbides when their oxides are heated with C.

  → Anions present in carbides are normally *methanide*, C^{4-}, or *acetylide*, C_2^{2-}, both very strong Brønsted bases. Methanides react with water to produce a basic solution and methane gas:

  $$Al_4C_3(s) + 12\,H_2O(l) \rightarrow 4\,Al(OH)_3(s) + 3\,CH_4(g)$$

  Acetylides release acetylene, $HC\equiv CH$, upon reaction with water:

  $$CaC_2(s) + 2\,H_2O(l) \rightarrow Ca(OH)_2(s) + C_2H_2(g)$$

  → *Covalent carbides* are formed by metals with electronegativity values similar to that of carbon, such as silicon and boron. Silicon carbide, SiC, known as *carborundum*, is an extremely hard, covalent carbide and is an excellent abrasive (see **Figure 8F.6** in the text). It is produced by the reaction of silicon dioxide with carbon at 2000 °C.

  $$SiO_2(s) + 3\,C(s) \rightarrow SiC(s) + 2\,CO(g)$$

  → *Interstitial carbides* are compounds formed by the direct reaction of a *d*-block metal with carbon at temperatures above 2000 °C. Carbon atoms lie in holes in close-packed arrays of metal atoms. Bonds between the carbon atoms and metal atoms stabilize the metal lattice and help to hold the metal atoms together. Examples are WC, Cr_3C, and Fe_3C, an important component of steel.

- **Tetrachlorides**

  → All Group 14 elements form *liquid* molecular tetrachlorides, of which $PbCl_4$ is the least stable. Carbon tetrachloride, a carcinogenic liquid, is formed by the reaction of chlorine and methane:

  $$CH_4(g) + 4\,Cl_2(g) \rightarrow CCl_4(l) + 4\,HCl(g)$$

  → Si reacts directly with chlorine to form silicon tetrachloride. Unlike CCl_4, $SiCl_4$ reacts with water as a Lewis acid, accepting a lone pair of electrons from H_2O:

  $$SiCl_4(l) + 2\,H_2O(l) \rightarrow SiO_2(s) + 4\,HCl(aq)$$

- **Cyanides**

  → Cyanides are strong Lewis bases that form a range of complexes with d-block metal ions. They are extremely poisonous, combining with cytochromes involved with the transfer of electrons and the supply of energy in cells.

  → The cyanide ion, CN^-, is the conjugate base of the acid HCN, which is made by heating methane, ammonia, and air in the presence of a Pt catalyst at 1100 °C:

  $$2\,CH_4(g) + 2\,NH_3(g) + 3\,O_2(g) \rightarrow 2\,HCN(g) + 6\,H_2O(g)$$

- **Methane and silane**

  → C readily bonds with itself to form chains, rings, and a multitude of compounds with hydrogen-named hydrocarbons (see **Focus 11** in the text).

→ Si forms a much smaller number of compounds with hydrogen; the simplest is the analogue of methane, *silane*, SiH_4, formed by the reaction of the reducing agent $LiAlH_4$ with silicon halides in ether:

$$SiCl_4 + LiAlH_4 \rightarrow SiH_4 + LiCl + AlCl_3$$

→ SiH_4 is more reactive than CH_4, igniting spontaneously in air. Higher silanes are unstable, decomposing on standing.

Topic 8G: GROUP 15: THE NITROGEN FAMILY

8G.1 The Group 15 Elements

→ Atoms of Group 15 elements have the valence electron configuration $ns^2\, np^3$.

• **Nitrogen**

→ *Nitrogen* differs greatly from other group members owing to its high electronegativity ($\chi = 3.04$), small size (radius of 75 pm), ability to form multiple bonds, and lack of available d-orbitals.

→ Found with oxidation numbers from –3 to +5.

→ *Elemental nitrogen*, N_2, is found in air (78.1% by volume, 76% by mass) and is prepared by fractional distillation of liquid air.

→ The strong triple bond in N_2 ($\Delta H_B = 944$ kJ·mol^{-1}) makes it very stable.

→ *Nitrogen fixation* is the process by which N_2 is converted into compounds usable by plants.

→ The *Haber synthesis* of ammonia is the major industrial method for fixing nitrogen, but it requires high temperatures and pressures and is therefore expensive.

→ Lightning produces some oxides of nitrogen, which are washed into soil by rain.

→ Bacteria found in the root nodules of legumes also fix N_2. An important area of current research is the search for catalysts that can mimic bacteria and fix nitrogen at ordinary temperatures.

→ The chemical and physical properties of nitrogen are different from those of its congeners because of the small size and unavailability of d-orbitals in nitrogen.

→ Nitrogen is highly electronegative ($\chi = 3.04$), forms strong hydrogen bonds, and forms multiple bonds with other Period 2 elements (e.g., cyanide, $C\equiv N^-$).

• **Phosphorus**

→ *Phosphorus* differs from nitrogen in its lower electronegativity ($\chi = 2.19$), larger size (radius of 110 pm), and the presence of available d-orbitals. Phosphorus may form as many as six bonds, whereas nitrogen can form a maximum of four.

→ *Phosphorus* is prepared from the *apatites*, mineral forms of *calcium phosphate*, $Ca_3(PO_4)_2$. The rocks are heated in an electric furnace with carbon and sand. The phosphorus vapor formed condenses as *white phosphorus*, a soft white poisonous molecular solid consisting of tetrahedral P_4 molecules. It bursts into flame upon exposure to air and is normally stored under water.

→ *Red phosphorus* is less reactive than white phosphorus and is used in match heads because it can be ignited by friction. It is a network solid consisting, most likely, of linked P_4 tetrahedral units.

→ *Red phosphorus* is the most thermodynamically stable form of phosphorus, yet *white phosphorus* crystallizes from the vapor or from the liquid phase. It is the form chosen to have an enthalpy and free energy of formation equal to 0. For *red phosphorus*, $\Delta H_f° = -17.6$ kJ·mol^{-1} and $\Delta G_f° = -12.1$ kJ·mol^{-1} at 25 °C.

→ *Black phosphorus* is a third form of the element; it has a network structure in which each P atom is bonded to three others at approximately right angles.

- **Arsenic and antimony**

   → *Arsenic* and *antimony* are metalloids. They are elements known since ancient times because they are easily reduced from their ores.

   → As elements, *arsenic* and *antimony* are used in lead alloys in the electrodes of storage batteries and in the semiconductor industry. *Gallium arsenide* is used as a light-detecting material with near-infrared response and in lasers, including those used for CD players.

- **Bismuth**

   → *Bismuth* is a metal with large, weakly bonded atoms. Bismuth has a low melting point, and it is used in alloys that serve as fire detectors in sprinkler systems.

   → *Bismuth* is also used to make low-temperature castings. Like ice, solid bismuth is less dense than the liquid. Bismuth ($Z = 83$) is the last element in the periodic table with a stable isotope, ^{209}Bi.

8G.2 Compounds with Hydrogen and the Halogens

- **Ammonia, NH$_3$, and phosphine, PH$_3$**

   → *Ammonia* is prepared in large amounts by the Haber process.

   → Ammonia is a pungent toxic gas that condenses to a colorless liquid at −33 °C. It acts as a solvent for a wide range of substances.

   → Autoprotolysis occurs to a much smaller extent in liquid ammonia (am) than in pure water.
   $$2\,NH_3(am) \rightleftharpoons NH_4^+(am) + NH_2^-(am) \qquad K_{am} = 1 \times 10^{-33} \text{ at } -35 \text{ °C}$$
   Very strong bases that are protonated by water survive in liquid ammonia.

   → *Ammonia* is a weak Brønsted base in water; it is also a fairly strong Lewis base, particularly toward d-block elements. One reaction is:
   $$Cu^{2+}(aq) + 4\,NH_3(aq) \rightarrow Cu(NH_3)_4^{2+}(aq)$$
   In this reaction, unshared electron pairs on nitrogen interact with the Lewis acid Cu^{2+} to form four Cu–NH$_3$ bonds.

   → *Ammonium salts* decompose when heated; if the salt contains a nonoxidizing anion, NH$_3$ is produced. The reaction is:
   $$(NH_4)_2SO_4(s) \xrightarrow{\text{heat}} 2\,NH_3(g) + H_2SO_4(l)$$

Decomposing ammonium carbonate has a pungent odor and is used as smelling salts to revive people who have fainted.

→ *Ammonium nitrate* is used as an explosive and as a fertilizer. It is one component of dynamite.

→ *Phosphine* is a poisonous gas that smells faintly of garlic and bursts into flame in air if it is impure. It is much less stable than ammonia. Because it does not form hydrogen bonds, it is not very soluble in water and is a very weak base ($pK_b = 27.4$). Its aqueous solution is nearly neutral.

- **Hydrazine, N_2H_4**

  → *Hydrazine* is an oily, colorless liquid. It is prepared by gentle oxidation of ammonia with alkaline hypochlorite solution:

  $$2\,NH_3(aq) + ClO^-(aq) \rightarrow N_2H_4(aq) + Cl^-(aq) + H_2O(l)$$

  Its physical properties are similar to those of water; its melting point is 1.5 °C, and its boiling point is 113 °C. It is dangerously explosive and is stored in aqueous solution.

  → A mixture of *methylhydrazine* and liquid N_2O_4 is used as a rocket fuel because these two liquids ignite on contact and produce a large volume of gas.

  $$4\,CH_3NHNH_2(l) + 5\,N_2O_4(l) \rightarrow 9\,N_2(g) + 12\,H_2O(g) + 4\,CO_2(g)$$

- **Nitrides, N^{3-}, and phosphides, P^{3-}**

  → *Nitrides* are solids that contain the nitride ion. Nitrides are stable only with small cations such as lithium, magnesium, or aluminum. Nitrides dissolve in water to form ammonia and the corresponding hydroxide:

  $$Zn_3N_2(s) + 6\,H_2O(l) \rightarrow 2\,NH_3(g) + 3\,Zn(OH)_2(aq)$$

  → *Boron nitride*, BN, is an important ceramic.

  → *Phosphides* are solids that contain the phosphide ion. Because PH_3 is a very weak parent acid of the strong Brønsted base P^{3-}, water is a sufficiently strong proton donor to react with phosphides to form phosphine:

  $$2\,P^{3-}(s) + 6\,H_2O(l) \rightarrow 2\,PH_3(g) + 6\,OH^-(aq)$$

- **Azides, N_3^-**

  → The *azide ion* is a highly reactive anion. Sodium azide is prepared by the reaction of dinitrogen oxide, N_2O, with molten sodium amide, $NaNH_2$.

  → *Sodium azide* is used in automobile air bags. When detonated, it decomposes into elemental sodium and nitrogen.

  → Some *azides*, such as AgN_3, $Cu(N_3)_2$, and $Pb(N_3)_2$, are shock sensitive. The azide ion is a weak base, and its conjugate acid, hydrazoic acid, HN_3, is a weak acid similar in strength to acetic acid.

- **Halides**

  → *Nitrogen* has an oxidation number of +3 in the *nitrogen halides*.

  → Nitrogen trifluoride, NF_3, is the most stable and does not react with water. Nitrogen trichloride, however, reacts with water to form ammonia and hypochlorous acid, HClO. Nitrogen triiodide is so unstable that it decomposes explosively upon contact with light.

  → *Phosphorus* forms chlorides with an oxidation number of +3 in PCl_3 and +5 in PCl_5.

→ *Phosphorus trichloride*, a liquid, is formed by direct chlorination of phosphorus. It is a major intermediate in the production of pesticides, oil additives, and flame retardants.

→ *Phosphorus pentachloride*, a solid, is made by allowing the trichloride to react with additional chlorine. It exists in the solid as tetrahedral PCl_4^+ cations and octahedral PCl_6^- anions, but it vaporizes to a gas of trigonal bipyramidal PCl_5 molecules.

→ *Phosphorus pentabromide* is also molecular in the vapor and ionic as a solid; but in the solid, the anions are Br^- anions, presumably because of the difficulty in fitting six large Br atoms around a central P atom.

→ Both *phosphorus trichloride* and *phosphorus pentachloride* react with water in a *hydrolysis reaction*, in which new element-oxygen bonds are formed and there is no change in oxidation state. An *oxoacid* and *hydrogen chloride* gas are formed in the reaction.

$$PCl_3(l) + 3\,H_2O(l) \rightarrow H_3PO_3(s) + 3\,HCl(g)$$

$$PCl_5(s) + 4\,H_2O(l) \rightarrow H_3PO_4(l) + 5\,HCl(g)$$

8G.3 Nitrogen Oxides and Oxoacids

→ All nitrogen oxides are acidic oxides, and some are the anhydrides of nitrogen oxoacids.

Oxidation number	Oxide formula	Oxide name	Oxoacid formula	Oxoacid name
5	N_2O_5	dinitrogen pentoxide	HNO_3	nitric acid
4	NO_2*	nitrogen dioxide	--------	
	N_2O_4	dinitrogen tetroxide	--------	
3	N_2O_3	dinitrogen trioxide	HNO_2	nitrous acid
2	NO	nitrogen monoxide, nitric oxide	-------- ------——	
1	N_2O	dinitrogen monoxide, nitrous oxide	$H_2N_2O_2$	hyponitrous acid

* $2\,NO_2(g) \rightleftharpoons N_2O_4(g)$

• **Dinitrogen oxide, N_2O**

→ *Dinitrogen oxide* is commonly called *nitrous oxide* (laughing gas), and it is the oxide of nitrogen with the lowest oxidation number for nitrogen (+1). It is formed by gentle heating of ammonium nitrate:

$$NH_4NO_3(s) \rightarrow N_2O(g) + 2\,H_2O(g)$$

→ N_2O is fairly unreactive, but it is toxic if inhaled in large amounts.

→ N_2O is used as a foaming agent for whipped cream.

- **Nitrogen oxide, NO**

  → *Nitrogen oxide* is commonly called *nitric oxide*; the nitrogen atom has an oxidation number of +2. Nitrogen oxide, a colorless gas, is produced industrially by the catalytic oxidation of ammonia at 1000 °C:
  $$4\,NH_3(g) + 5\,O_2(g) \rightarrow 4\,NO(g) + 6\,H_2O(g)$$

  → The endothermic formation of NO from the oxidation of N_2 occurs readily at the high temperatures that exist in automobile engines and turbine engine exhausts:
  $$N_2(g) + O_2(g) \rightarrow 2\,NO(g)$$

  → NO is further oxidized to NO_2 upon exposure to air. In this form, it contributes to smog, acid rain, and the destruction of the stratospheric ozone layer.

  → Interestingly, in small amounts, NO acts as a neurotransmitter to dilate blood vessels.

- **Nitrogen dioxide, NO$_2$**

  → *Nitrogen dioxide* is a choking, poisonous brown gas that contributes to the color and odor of smog. The oxidation number of nitrogen in NO_2 is +4. NO_2, like NO, is paramagnetic. It disproportionates in water to form nitric acid and nitrogen oxide:
  $$3\,NO_2(g) + H_2O(l) \rightarrow 2\,HNO_3(aq) + NO(g)$$

  → *Nitrogen dioxide* is prepared in the laboratory by heating lead(II) nitrate:
  $$2\,Pb(NO_3)_2(s) \rightarrow 4\,NO_2(g) + 2\,PbO(s) + O_2(g)$$
  And it exists in equilibrium with its colorless dimer, dinitrogen tetroxide:
  $$2\,NO_2(g) \rightleftharpoons N_2O_4(g)$$
  Only the dimer exists in the solid, so the brown gas condenses to a colorless solid.

- **Nitrous acid, HNO$_2$**

  → *Nitrous acid* can be produced in aqueous solution by mixing its anhydride, dinitrogen trioxide, with water. The oxidation number of nitrogen in HNO_2 is +3:
  $$N_2O_3(g) + H_2O(l) \rightarrow 2\,HNO_2(aq)$$

  → *Nitrous acid* has not been isolated as a pure compound, but it has many uses in aqueous solution. It is a weak acid with $pK_a = 3.4$. The conjugate base, nitrite ion, NO_2^-, forms ionic solids that are soluble in water and mildly toxic.

  → Despite their toxicity, nitrites are used in the curing of meat products because they retard bacterial growth.

- **Nitric acid, HNO$_3$**

  → *Nitric acid*, a widely used industrial and laboratory acid, is produced by the three-step *Ostwald process*:
  $$4\,NH_3(g) + 2\,O_2(g) \rightarrow 4\,NO(g) + 6\,H_2(g)$$
  $$2\,NO(g) + O_2(g) \rightarrow 2\,NO_2(g)$$
  $$3\,NO_2(g) + H_2O(l) \rightarrow 3\,HNO_3(aq) + NO(g)$$

  → The oxidation number of nitrogen in HNO_3 is +5.

  → Nitric acid has been isolated as a pure compound; it is a colorless liquid with a boiling point of 83 °C. It is usually used in aqueous solution. It is an excellent oxidizing agent as well as a strong acid. Its strength as an oxidizing agent is due in part to the strong triple bond in N_2, to which it is reduced.

8G.4 Phosphorus Oxides and Oxoacids

- **Phosphorus(III) oxide, P_4O_6, and phosphorous acid, H_3PO_3**
  - → *Phosphorus(III) oxide* is made by heating white phosphorus in a limited supply of air. P_4O_6 molecules consist of P_4 tetrahedral units, with each oxygen atom lying between two phosphorus corner atoms. Six oxygen atoms lie along the six edges of the tetrahedron, one per edge. The edge bonds are represented as –P–O–P–.
  - → P_4O_6 is the anhydride of *phosphorous acid*: $4\,H_3PO_3(aq) \rightarrow P_4O_6(s) + 6\,H_2O(l)$
    Phosphorous acid is a *diprotic acid*.

- **Phosphorus(V) oxide, P_4O_{10}, and phosphoric acid, H_3PO_4**
  - → *Phosphorus(V) oxide* is made by heating white phosphorus in an excess supply of air. P_4O_{10} molecules are similar to P_4O_6 but have an additional oxygen attached to each phosphorus at the apices of the tetrahedron.
  - → P_4O_{10} is the *anhydride* of *phosphoric acid*, H_3PO_4. It traps and reacts with water very efficiently and is widely used as a drying agent. Phosphoric acid is a *triprotic acid*. It is only mildly oxidizing, despite the fact that phosphorus is in a high oxidation state.
  - → Phosphoric acid is the parent acid of phosphate salts that contain the tetrahedral phosphate ion PO_4^{3-}. Phosphates are generally not very soluble, which makes them suitable structural material for bones and teeth.

- **Polyphosphates**
  - → *Polyphosphates* are compounds made up of linked PO_4^{3-} tetrahedral units. The simplest structure is the *pyrophosphate ion* $P_2O_7^{4-}$, in which two phosphate ions are linked by an oxygen atom through –O–P–O– bonds.
  - → More complicated structures form with longer chains or rings. The biochemically most important polyphosphate is *adenosine triphosphate*, ATP, which contains three phosphorus tetrahedral units linked by –O–P–O– bonds. Hydrolysis of ATP to *adenosine diphosphate*, ADP, by the rupture of an O–P bond releases energy that is used by cells to drive biochemical reactions within the cell:

    $$ATP + H_2O \rightarrow ADP + HPO_4^{2-} \qquad \Delta H = -41 \text{ kJ}$$

    ATP is often called the fuel of life because it provides the energy for many of the biochemical reactions occurring in cells.

Topic 8H: GROUP 16: THE OXYGEN FAMILY

8H.1 The Group 16 Elements

→   Members of Group 16 are called chalcogens.

  →   Valence shell configuration $ns^2 np^4$

  →   Anions favored; only two electrons are needed to complete the valence shell.

  →   Oxygen is the most abundant element in the Earth's crust.

- **Oxygen**

  →   *Elemental oxygen*, O_2, is found in air (21.0% by volume) and is prepared by fractional distillation of liquid air.

  →   Oxygen is a colorless, odorless, *paramagnetic* gas.

  →   A satisfactory Lewis structure consistent with its paramagnetism cannot be drawn for oxygen. The molecular orbital theory accounts for the paramagnetic properties (see **Topic 2G.2** of the text).

  →   *Elemental oxygen* exists in two allotropic forms, O_2 and O_3 (*ozone*). Inorganic chemists often use the term *dioxygen* to refer to molecular oxygen, O_2.

  →   Ozone is produced by an electric discharge in oxygen gas; its pungent odor is apparent near sparking electrical equipment and lightning strikes.

- **Sulfur**

  →   *Sulfur* is found in ores and as elemental sulfur.

  →   *Sulfur* is widely distributed as sulfide ores, which include *galena*, PbS; *cinnabar*, HgS; *iron pyrite*, FeS_2 (fool's gold); and *spahlerite*, ZnS.

  →   *Sulfur* is also found in elemental form as deposits, called *brimstone*, which are made by bacterial action on H_2S.

  →   *Elemental sulfur* is a tasteless, almost odorless insoluble nonmetallic yellow molecular solid of crownlike S_8 rings.

  →   The more stable allotrope, *rhombic sulfur*, forms beautiful yellow crystals, whereas the less stable allotrope, *monoclinic sulfur*, forms needlelike crystals. The two allotropes differ in the manner in which the S_8 rings are stacked together.

  →   For simplicity, the elemental form of sulfur is often represented as S(s) rather than $S_8(s)$.

  →   One method of recovering sulfur is the *Claus process*, in which some of the H_2S that occurs in oil and natural gas is first oxidized to sulfur dioxide:

$$2 H_2S(g) + 3 O_2(g) \rightarrow 2 SO_2(g) + 2 H_2O(l)$$

  →   Sulfur dioxide oxidizes the remaining hydrogen sulfide, and both are converted to elemental sulfur at 300 °C in the presence of an alumina catalyst:

$$2 H_2S(g) + SO_2(g) \rightarrow 3 S(s) + 2 H_2O(l)$$

  →   *Sulfur* is used primarily to produce *sulfuric acid* and to vulcanize rubber.

  →   Vulcanization increases the toughness of rubber by introducing cross-links between the polymer chains of natural rubber.

→ An important property of sulfur is its ability to form chains of atoms, *catenation*. The –S–S– links that connect different parts of the chains of amino acids in proteins are an important example. These *disulfide links* contribute to the shapes of proteins (see **Topic 11E.5** in the text).

- **Selenium and tellurium**

  → *Selenium* and *tellurium* are found in sulfide ores; they are also recovered from the anode sludge formed during the electrolytic refining of copper.

  → Both elements have several allotropic forms; the most stable one consists of long *zigzag chains* of atoms.

  → The allotropes look like a silver-white metal, but electrical conductivity is poor. Exposing selenium to light increases its electrical conductivity, so it is used in solar cells.

- **Polonium**

  → *Polonium* is a radioactive, low-melting metalloid. It was identified in uranium ores in 1898 by the Curies. Its most stable isotope, ^{209}Po, has a half-life of 103 y.

  → *Polonium* decays by emission of an alpha particle, $^{4}He^{2+}$. It is used in antistatic devices in textile mills to counteract the buildup of negative charges on fast-moving fabric.

8H.2 Compounds with Hydrogen

→ Hydrogen reacts with all the Group 16 elements.

- **Water, H_2O, and hydrogen sulfide, H_2S**

  → *Water* is a remarkable compound possessing a unique set of physical and chemical properties.

  → *Water* is purified in domestic water supplies in a multistep procedure that includes *aeration*, addition of *slaked lime*, removal of particles by *coagulation* and *flocculation*, addition of CO_2, *filtration*, reduction of pH, and addition of chlorine *disinfectant* to kill bacteria. High-purity water is obtained by *distillation*.

  → *Water* has a higher boiling point than expected on the basis of its molar mass because of the extensive hydrogen bonding between H_2O molecules.

  → Hydrogen bonding also causes a more open structure and therefore, a lower density for the solid than for the liquid. Because of its high polarity, water is an excellent solvent for ionic compounds.

  → Chemically, water is *amphiprotic*, and it can both donate and accept protons; so it is both a *Brønsted acid* and a *Brønsted base*.

  $$H_2O(l, acid) + NH_3(aq) \rightarrow NH_4^+(aq) + OH^-(aq)$$
  $$CH_3COOH(aq) + H_2O(l, base) \rightarrow H_3O^+(aq) + CH_3COO^-(aq)$$

  → *Water* can act as a *Lewis base*, donating its unshared pair of electrons on the oxygen atom:

  $$Fe^{3+}(aq) + 6H_2O(l) \rightarrow Fe(H_2O)_6^{3+}(aq)$$

  → *Water* can act as an *oxidizing agent* and a *reducing agent*.

  $$2Na(s) + 2H_2O(l, ox.\ agent) \rightarrow 2NaOH(aq) + H_2(g)$$
  $$2H_2O(l, red.\ agent) + 2F_2(g) \rightarrow 4HF(aq) + O_2(g)$$

→ *Hydrolysis reaction*: Water as a reactant forms a bond between oxygen and another element. Hydrolysis can occur *with* or *without* a change in oxidation number:

$$PCl_5(s) + 4H_2O(l) \rightarrow H_3PO_4(aq) + 5HCl(aq) \qquad (1)$$
$$Cl_2(g) + H_2O(l) \rightarrow HClO(aq) + HCl(aq) \qquad (2)$$

Reaction (1) has *no* change in oxidation number. Reaction (2) is a *disproportionation reaction* in which the oxidation number of chlorine changes from 0 in Cl_2 to +1 in HClO and −1 in HCl.

→ *Hydrogen sulfide* is an example of a Group 16 binary compound with hydrogen. It is a toxic gas with an offensive odor (rotten eggs). Egg proteins contain sulfur and release H_2S when they decompose. Iron(II) sulfide sometimes appears as a pale green discoloration at the boundary between the white and the yolk.

→ *Hydrogen sulfide* is prepared by the direct reaction of hydrogen and sulfur at 600 °C or by protonation of the sulfide ion, which is a Brønsted base:

$$FeS(s) + 2HCl(aq) \rightarrow FeCl_2(aq) + H_2S(g)$$

→ *Hydrogen sulfide* dissolves in water and is slowly oxidized by dissolved air to form a colloidal dispersion of particles of sulfur, S_8.

→ *Hydrogen sulfide*, a *weak diprotic acid*, is the parent acid of the hydrogen sulfides, HS^-, and the sulfides, S^{2-}.

→ Sulfides of s-block elements are moderately soluble in water, whereas the sulfides of the heavy p- and d-block metals are generally very insoluble.

• **Hydrogen peroxide, H_2O_2, and polysulfanes, $HS–S_n–SH$**

→ *Hydrogen peroxide* is a highly reactive, pale blue liquid that is appreciably denser ($1.44\ g \cdot mL^{-1}$) than water at 25 °C. Its melting point is -0.4 °C and its boiling point is 152 °C.

→ Chemically, *hydrogen peroxide* is more acidic than water ($pK_a = 11.75$). It is a strong oxidizing agent, but it can also function as a reducing agent.

$$2Fe^{2+}(aq) + H_2O_2(aq,\ ox.\ agent) + 2H^+(aq) \rightarrow 2Fe^{3+}(aq) + 2H_2O(l)$$
$$Cl_2(g) + H_2O_2(aq,\ red.\ agent) + 2OH^-(aq) \rightarrow 2Cl^-(aq) + 2H_2O(l) + O_2(g)$$

→ The O–O bond in *hydrogen peroxide* is very weak. The oxidation state of oxygen in hydrogen peroxide is −1.

→ A *polysulfane* is a *catenated* molecular compound of composition $HS–S_n–SH$, where n can take values from 0 through 6.

→ The sulfur analogue of hydrogen peroxide occurs with $n = 0$.

→ Two polysulfide ions obtained from polysulfanes are found to occur in the mineral *lapis lazuli* (see **Figure 8H.11** in the text); its color derives from impurities of S_2^- (blue) and S_3^- (hint of green).

→ The S_2^- ion is an analogue of the *superoxide* ion, O_2^-; and S_3^- is an analogue of the *ozonide* ion, O_3^-.

H.3 Sulfur Oxides and Oxoacids

- **Sulfur dioxide, SO_2, sulfurous acid, H_2SO_3, and sulfite, SO_3^{2-}**

  → *Sulfur dioxide,* a poisonous gas, is made by burning sulfur in air. Volcanic activity, combustion of fuels contaminated with sulfur, and oxidation of hydrogen sulfide are the major sources of atmospheric SO_2. The equation for oxidation of hydrogen sulfide in the atmosphere is:
  $$2\,H_2S(g) + 3\,O_2(g) \rightarrow 2\,SO_2(g) + 2\,H_2O(g)$$

  → *Sulfurous acid* is formed by the reaction of its acid anhydride, SO_2, and water:
  $$SO_2(g) + H_2O(l) \rightarrow H_2SO_3(aq)$$

  → *Sulfurous acid* is an equilibrium mixture of $(HO)SHO_2$ and $(HO)SO(OH)$. These molecules are also in equilibrium with SO_2 molecules, each of which is surrounded by a cage of water molecules.

  → When the solution is cooled, crystals with a composition of roughly $SO_2 \cdot 7\,H_2O$ form. This substance is an example of a *clathrate*, in which a molecule sits in a cage of other molecules. Methane, carbon dioxide, and the noble gases also form clathrates with water.

  → In *sulfite ions* and *sulfur dioxide*, the oxidation number of the sulfur atom is +4. Because this value is intermediate in sulfur's range of −2 to +6, these compounds can act as either oxidizing agents or reducing agents.

  → The most important reaction of sulfur dioxide is its slow oxidation to sulfur trioxide, SO_3, in which the oxidation number of the sulfur atom is +6.
  $$2\,SO_2(g) + O_2(g) \rightarrow 2\,SO_3(g)$$
  This reaction is catalyzed by metal cations in droplets of water or by certain surfaces. The atmosphere also has indirect pathways for the conversion of SO_2 to SO_3.

- **Sulfur trioxide, SO_3, sulfuric acid, H_2SO_4, and sulfate, SO_4^{2-}**

  → At normal temperatures, *sulfur trioxide* is a volatile liquid with a boiling point of 45 °C. Its shape is trigonal planar, with bond angles of 120°.

  → In the solid, and to some extent in the liquid, the molecules aggregate to form *trimers*, S_3O_9, as well as larger groupings.

  → *Sulfuric acid* is produced commercially by the *contact process*, in which sulfur is first burned in oxygen at 1000 °C. The resulting sulfur dioxide reacts with oxygen to form sulfur trioxide over a V_2O_5 catalyst at 500 °C.

  → Because sulfur trioxide forms a corrosive acid with water, it is absorbed in 98% concentrated sulfuric acid to give a dense, oily liquid called *oleum*, $H_2S_2O_7$:
  $$SO_3(g) + H_2SO_4(l) \rightarrow H_2S_2O_7(l)$$

  → *Sulfuric acid* is a colorless, corrosive, oily liquid that boils (decomposes) at 300 °C. It is a strong *Brønsted acid*, a powerful *dehydrating agent*, and a mild *oxidizing agent*.

→ In water, the first ionization of sulfuric acid is complete, but HSO_4^- is a *weak acid* with $pK_a = 1.92$. As a *dehydrating agent*, sulfuric acid is able to extract water from a variety of compounds, including sucrose, $C_{12}H_{22}O_{11}$, and formic acid, HCOOH.

$$C_{12}H_{22}O_{11}(s) \rightarrow 12\,C(s) + 11\,H_2O(l)$$

$$HCOOH(l) \rightarrow CO(g) + H_2O(l)$$

→ Because of the large exothermicity associated with dilution, mixing sulfuric acid with water can cause violent splashing, so for reasons of safety, sulfuric acid (which is denser than water) is always carefully added to water, not the water to acid.

→ The *sulfate ion* is a mild *oxidizing agent*.

$$4\,H^+(aq) + SO_4^{2-}(aq) + 2\,e^- \rightarrow SO_2(g) + 2\,H_2O(l) \quad E° = 0.17 \text{ V}$$

Topic 8I: GROUP 17: THE HALOGEN FAMILY

8I.1 The Group 17 Elements

→ The halogens have the valence configurations $ns^2\,np^5$. Except for fluorine, their oxidation numbers range from −1 to +7.

- **Fluorine**

  → *Fluorine* is a reactive, almost colorless gas of F_2 molecules.

  → The properties follow from its high electronegativity, small size, and lack of available d-orbitals for bonding.

  → It is the most electronegative element, with an oxidation number of −1 in all its compounds.

  → Elements combined with fluorine are often found with their highest oxidation numbers, such as +7 for I in IF_7.

  → The small size of the fluoride ion results in high lattice enthalpies for fluorides, a property that makes them less soluble than other halides.

  → *Fluorine* occurs widely in many minerals, including *fluorspar*, CaF_2; *cryolite*, Na_3AlF_6; and the *fluoroapatites*, $Ca_5F(PO_4)_3$.

  → *Elemental fluorine* is obtained by the electrolysis of an anhydrous molten KF–HF mixture at 75 °C, using a carbon electrode.

  → Fluorine is highly reactive and highly oxidizing; it is used in the production of SF_6 for electrical equipment and volatile UF_6 for processing nuclear fuel (see **Focus 10** in the text).

  → The strong bonds formed by *fluorine* make many of its compounds relatively inert.

  → Fluorine's ability to form *hydrogen bonds* results in relatively high melting points, boiling points, and enthalpies of vaporization for many of its compounds.

- **Chlorine**

  → One of the chemicals manufactured in the greatest amount worldwide.

→ *Chlorine* is a pale yellow-green reactive gas of Cl_2 molecules that condenses at -34 °C.

→ *Chlorine* reacts directly with all the elements except carbon, nitrogen, oxygen, and the noble gases.

→ Used in a large number of industrial processes, including the manufacture of plastics, solvents, and pesticides

→ Also used as a disinfectant to treat water supplies.

→ *Elemental chlorine* is obtained by the electrolysis of molten or aqueous NaCl. It is a strong oxidizing agent and oxidizes metals to high oxidation states:

$$2\,Fe(s) + 3\,Cl_2(g) \rightarrow 2\,FeCl_3(s)$$

- **Bromine**

  → *Bromine* is a corrosive reddish-brown liquid of Br_2 molecules. It has a penetrating odor.

  → Organic bromides are used in textiles as fire retardants and in pesticides. Inorganic bromides, particularly silver bromide, are used in photographic emulsions.

  → *Elemental bromine* is produced from brine wells by the oxidation of Br^- by Cl_2.

$$2\,Br^-(aq) + Cl_2(g) \rightarrow Br_2(l) + 2\,Cl^-(aq)$$

- **Iodine**

  → A lustrous blue-black solid of I_2 molecules that readily sublimes to form a purple vapor

  → Occurs as I^- in seawater and as an impurity in Chile saltpeter, KNO_3

  → The best source is the brine from oil wells; *elemental iodine* is produced from brine wells by the oxidation of I^- by Cl_2.

  → *Elemental iodine* is slightly soluble in water, but it dissolves well in iodide solutions because it reacts with aqueous I^- to form the triiodide ion, I_3^-.

  → *Iodine* is an essential trace element in human nutrition, and *iodides* are often added to table salt. This iodized salt prevents iodine deficiency, which leads to the enlargement of the thyroid gland in the neck (a condition called *goiter*).

- **Astatine (not in the text)**

  → *Astatine* is a radioactive element that occurs in uranium ores, but only to a tiny extent. Its most stable isotope, ^{210}At, has a half-life of 8.3 h. The isotopes formed in uranium ores have much shorter lifetimes. The properties of astatine are surmised from spectroscopic measurements.

  → It is made by bombarding bismuth with alpha particles in a cyclotron, which accelerates particles to high speed.

8I.2 Compounds of the Halogens

- **Interhalogens**

  → *Interhalogen* compounds consist of a heavier halogen atom at the center of the compound and an odd number of lighter ones on the periphery. An exception is I_2Cl_6, which contains two (–I–Cl–I–) bridges between the two central iodine atoms. The two bridging chlorine atoms act as Lewis bases (electron pair donors) in a manner similar to that of Al_2Cl_6.

→ *Interhalogens* are typically formed by reaction between stoichiometric amounts of the elements:

$$I_2(g) + 7\,F_2(g) \rightarrow 2\,IF_7(g); \quad Cl_2(g) + 3\,F_2(g) \rightarrow 2\,ClF_3(g)$$

→ As the central halogen becomes heavier, bond enthalpy decreases.

$$(X\text{–}F): \quad ClF_5\ (154\ kJ\cdot mol^{-1}) < BrF_5\ (187\ kJ\cdot mol^{-1}) < IF_5\ (269\ kJ\cdot mol^{-1})$$

→ The physical properties of the *interhalogens* lie between those of their parent halogens.

→ The fluorides of the heavier halogens are very reactive.

Example: $BrF_3(g)$ is so reactive that asbestos will burn in it.

→ **Table 8I.2** in the text gives a compilation of the known *interhalogens*.

- **Compounds with hydrogen**

  → *Hydrogen halides* are prepared by direct reaction of the elements or by the action of nonvolatile acids on *metal halides*:

  $$Cl_2(g) + H_2(g) \rightarrow 2\,HCl(g)$$

  $$CaF_2(s) + H_2SO_4(aq,\ conc.) \rightarrow Ca(HSO_4)_2(aq) + 2\,HF(g)$$

  → Because Br^- and I^- are oxidized by sulfuric acid, phosphoric acid is used in the preparation of HBr and HI.

  → *Hydrogen halides* are pungent colorless gases, but *hydrogen fluoride* is liquid below 20 °C. Its low volatility is a sign of extensive hydrogen bonding, and short zigzag chains up to about $(HF)_5$ persist in the vapor.

  → *Hydrogen halides* dissolve readily in water to produce acidic solutions. *Hydrogen fluoride* has the unusual property of attacking and dissolving silica glasses, and it is used for glass etching (see **Figure 8I.5** in the text).

  → *Halides* of metals tend to be ionic unless the metal has an oxidation number greater than +2.

  Examples: Sodium chloride and copper(II) chloride are ionic compounds with high melting points, whereas titanium(IV) chloride and iron(III) chloride sublime as molecules.

- **Halogen oxides and oxoacids**

  → *Hypohalous acids* HXO or HOX consist of the halogen atom as a *terminal* atom with an oxidation number of +1. They are prepared by the direct reaction of a halogen with water; the corresponding *hypohalite* (XO^-) salts are prepared by the reaction of a halogen with aqueous alkali solution:

  $$Cl_2(g) + H_2O(l) \rightarrow HOCl(aq) + HCl(aq)$$

  $$Cl_2(g) + NaOH(aq) \rightarrow NaOCl(aq) + HCl(aq)$$

  → *Hypochlorites* cause the oxidation of organic material in water by producing oxygen in aqueous solution, which readily oxidizes organic materials. The production of O_2 occurs in two steps; the net result is:

  $$2\,OCl^-(aq) \rightarrow 2\,Cl^-(aq) + O_2(g)$$

  → *Chlorates*, ClO_3^-, contain a *central* chlorine atom with an oxidation number of +5. They are prepared by the reaction of chlorine with hot aqueous alkali:

  $$3\,Cl_2(g) + 6\,OH^-(aq) \rightarrow ClO_3^-(aq) + 5\,Cl^-(aq) + 3\,H_2O(l)$$

→ *Chlorates* decompose at elevated temperatures; the final product is determined by the presence of a catalyst:

$$4\,KClO_3(l) \rightarrow 3\,KClO_4(s) + KCl(s) \quad \textit{no catalyst}$$

$$2\,KClO_3(l) \rightarrow 2\,KCl(s) + 3\,O_2(g) \quad \textit{MnO}_2 \textit{ catalyst}$$

→ *Chlorates* are good oxidizing agents and are also used to produce the important compound ClO_2. Sulfur dioxide is a convenient reducing agent for this reaction:

$$2\,NaClO_3(aq) + SO_2(g) + H_2SO_4(aq, \text{dilute}) \rightarrow 2\,NaHSO_4(aq) + 2\,ClO_2(g)$$

→ *Chlorine dioxide*, ClO_2, contains a *central* chlorine atom with an oxidation number of +4. It has an odd number of electrons and is a paramagnetic yellow gas that is used to bleach paper pulp. It is highly reactive and may explode violently under some conditions.

→ *Perchlorates*, ClO_4^-, contain a *central* chlorine atom with an oxidation number of +7. They are prepared by the electrolysis of aqueous chlorates. The half-reaction is:

$$ClO_3^-(aq) + H_2O(l) \rightarrow ClO_4^-(aq) + 2\,H^+(aq) + 2\,e^-$$

→ *Perchlorates* and *perchloric acid*, $HClO_4$, are powerful oxidizing agents; *perchloric acid* in contact with small amounts of organic materials may explode.

→ The *oxidizing strength* and *acidity* of oxoacids both increase as the oxidation number of the halogen increases.

→ For oxoacids with the same number of oxygen atoms, both oxidizing and acid strength increase as the electronegativity value of the halogen increases. The general rules for the acid strength of oxoacids are summarized in **Topic 6C** in the text.

Topic 8J: GROUP 18: THE NOBLE GASES

8J.1 The Group 18 Elements

→ The noble gases have closed-shell electron configurations, $ns^2\,np^6$.

→ The first compound of a noble gas was synthesized in 1962; until that time they were called inert gases.

- **Helium**

   → *Helium* (Greek word *helios*, sun) is the second most abundant element in the universe after hydrogen. Helium atoms are light and travel at velocities sufficient to escape from the Earth's atmosphere. The first evidence for helium was discovered in the solar spectrum taken during a solar eclipse in 1868. *Helium* is found as a component of gases trapped in rocks.

   → Alpha particles, $^4He^{2+}$, are released by nuclear decay of naturally occurring uranium and thorium ores. They are high-energy helium-4 nuclei that capture two electrons to become helium atoms. Low density and lack of flammability of helium make it suitable for lighter-than-air ships such as blimps. *Helium* is found in the atmosphere with a concentration of 5.24 ppm (parts per million by volume).

→ Has the lowest melting point of any substance, freezes only under pressure, and is the only substance with two liquid phases. Below 2 K, liquid helium-II exhibits superfluidity, the ability to flow without viscosity. It is used as a cryogen for studies at low temperature.

- **Neon**

  → *Neon* (Greek word *neos*, new) is widely used in advertising display signs because it emits an orange-red glow when an electric current passes through it.

  → Of all the noble gases, the discharge of neon is the most intense at ordinary voltages and currents. The neon atom is the source of laser light in the helium-neon laser.

  → Has over 40 times more refrigerating capacity per unit volume than liquid helium and more than three times that of liquid hydrogen. It is compact, inert, and less expensive than helium when used as a *cryogenic* coolant.

  → Found in the atmosphere with a concentration of 18.2 ppm.

- **Argon**

  → *Argon* (Greek word *argos*, inactive) is mainly used to fill electric light bulbs at a pressure of about 400 Pa, where it conducts heat away from the filament. It is also used to fill fluorescent tubes.

  → Used as an inert gas shield (to prevent oxidation) for arc welding and cutting and as a protective atmosphere for growing silicon and germanium crystals.

  → The most abundant noble gas, with an atmospheric concentration of about 0.93% (or about 1 part per hundred by volume).

- **Krypton**

  → *Krypton* (Greek word *kryptos*, hidden) gives an intense white light when an electric discharge is passed through it, and it is used in airport runway lighting.

  → Produced in nuclear fission, and its atmospheric abundance provides one measure of worldwide nuclear activity.

  → Found in the atmosphere with a concentration of about 1 ppm.

- **Xenon**

  → *Xenon* (Greek word *xenos*, stranger) is used in halogen lamps, for automobile headlights, and in high-speed photographic flash tubes.

  → Used in the nuclear energy area in bubble chambers, probes, and other applications where a high atomic mass is desirable. Currently, the anesthetic properties of xenon are being explored.

  → Found in the atmosphere with a concentration of about 0.087 ppm.

- **Radon** (not in the text)

  → *Radon* (name derived from the element *radium*) is a radioactive gas. It is formed by radioactive decay processes deep in the Earth.

  → Uranium-238 decays very slowly to radium-226 which decays further by alpha particle emission to radon-222, and ultimately to lead-206 (see **Figure 10A.16** in the text). The half-life of radon-222 is 3.825 days. Shorter-lived isotopes are formed from the decay of thorium-232 and uranium-235.

→ Every square mile of soil to a depth of 6 inches is estimated to contain about 1 g of radium, which releases radon in tiny amounts into the atmosphere. Accumulation of radon in buildings can be dangerous to human health.

→ Used in implant seeds for the therapeutic treatment of localized tumors.

→ Found in the atmosphere with a concentration estimated to be 1 in 10^{21} parts of air.

8J.2 Compounds of the Noble Gases

- **Noble gases**

  → Ionization energies of the noble gases are very high and decrease down the group.

- **Helium, neon, and argon**

  → The ionization energies of helium, neon, and argon are so high that they form no stable *neutral* compounds. However, He and Ne have been trapped inside buckminsterfullerene cages.

  → If electrons are removed from noble gas atoms, the resulting ions are radicals, and they may form stable molecular ions such as $NeAr^+$, ArH^+, and $HeNe^+$ in the gas phase. If an electron is added to them, these ions rapidly dissociate into neutral atoms.

- **Krypton**

  → *Krypton* forms a thermodynamically unstable neutral compound, KrF_2 (Recall that diamond is also thermodynamically unstable.) It is a volatile white solid that decomposes slowly at room temperature.

  → In 1988, a compound with a Kr–N bond was discovered, but it is stable only at temperatures below −50 °C.

- **Xenon**

  → *Xenon* is the noble gas element with the richest chemistry. It forms several compounds with fluorine and oxygen and even forms compounds with Xe–N and Xe–C bonds. Xenon is found in compounds with oxidation numbers of +2, +4, +6, and +8, and XeF_2 is thermodynamically stable.

  → Direct reaction of *fluorine* with *xenon* at high temperature yields compounds with oxidation numbers of +2 (XeF_2, thermodynamically stable), +4 (XeF_4), and +6 (XeF_6). In the case of XeF_6, high pressure is required. All three fluorides are crystalline solids. In the gas phase, all are molecular compounds. Solid *xenon hexafluoride* is an ionic compound with a complex structure of XeF_5^+ cations bridged by F^- anions.

  → *Xenon fluorides* are powerful *fluorinating agents*, reagents that attach fluorine to other substances. *Xenon trioxide* is synthesized by the hydrolysis of *xenon tetrafluoride*.

  $$6\,XeF_4(s) + 12\,H_2O(l) \rightarrow 2\,XeO_3(aq) + 4\,Xe(g) + 3\,O_2(g) + 24\,HF(aq)$$

  → The trioxide is the anhydride of *xenic acid*, H_2XeO_4. In aqueous basic solution, the acid forms the *hydrogen xenate ion* $HXeO_4^-$, which further disproportionates to the *perxenate ion*, XeO_6^{4-}, in which xenon attains its highest oxidation number, +8.

  $$2\,HXeO_4^-(aq) + 2\,OH^-(aq) \rightarrow XeO_6^{4-}(aq) + Xe(g) + O_2(g) + 2\,H_2O(l)$$

- **Radon** (not in the text)

  → *Radon* chemistry is difficult to study because all of its isotopes are radioactive. The longest-lived isotope, radon-222, has a half-life of only 3.82 d (d = day).

  → A *radon fluoride* of unknown composition (likely RnF_2) is found to form readily, but it decomposes rapidly during attempts to vaporize it for further study.

Focus 9: THE d-BLOCK

Topic 9A: PERIODIC TRENDS OF THE d-BLOCK ELEMENTS

9A.1 Trends in Physical Properties

- **d-Block elements**

  → Electron configurations: $ns^2(n-1)d^m$, m = 1 – 10 (with 10 exceptions).

  → For most d-block elements, the $(n-1)$d-orbitals *fill after* ns^2, but the ns electrons are removed *before* $(n-1)$d-electrons when cations are formed. (There are also some exceptions to this rule for cations with a +1 charge.)

  → All are metals, most are good electrical conductors, particularly Ag, Cu, and Au.

  → Most are malleable, ductile, lustrous, and silver-white in color.
  Exceptions: Cu (red-brown) and Au (yellow)

  → Most have higher melting and boiling points than main group elements.
  A major exception is Hg, which is a liquid at room temperature.

- **Shapes of the d-orbitals**

  → Some properties of the d-block elements follow from the shapes of the d-orbitals.

  → d-Orbital *lobes* are relatively far apart from each other, so electrons in different d-orbitals on the same atom repel each other weakly.

  → Electron density in d-orbitals is low near the nucleus in part because of the high angular momentum of d-electrons. Because d-electrons are relatively far from the nucleus, they are not very effective in shielding other electrons from the positive charge of the nucleus.

- **Trends in atomic radii**

  Across a Period

  → *Nuclear charge* and the *number* of d-electrons both increase across a row from left to right.

  → The first five electrons added to a d-subshell are placed in different orbitals (Hund's rule).

  → Since repulsion between d-electrons in different orbitals is small, increasing *effective nuclear charge* is the dominant factor, so atoms tend to become smaller as Z increases in the first half of the d-block. **Exception:** Mn

  → Further across the block, radii begin to *increase* slightly with Z because *electron–electron repulsion* in doubly - filled orbitals outweighs the effect of increasing *effective nuclear charge*.

  → Attractions and repulsions are finely balanced, and the range of atomic d-metal radii is small.

  → d-Metals form many alloys because atoms of one metal can easily replace atoms of another metal.

  Down a group: *Periods 4 and 5*

  → d-Metals in *Period 5* are typically larger than those in *Period 4*.

  → The effective nuclear charge on the outer electrons is roughly the same, whereas the n quantum number increases down the group. The usual pattern is an increase in size for group members from top to bottom. This pattern is followed in *Periods 4* and *5* for the d-metals. **Exception:** Mn

Across Period 6: the Lanthanide Contraction

→ d-Metal radii in *Period 6* are *approximately the same* as those in *Period 4*. They are smaller than expected because of the *lanthanide contraction,* the decrease in radius along the first row of the f-block.

→ There are 14 f-block elements before the first d-block element in *Period 6,* Lu. At Lu, the atomic radius has fallen from 224 pm for Ba to 173 pm for Lu.

→ *Period 6* d-block elements are substantially denser than those in *Period 5* owing to the *lanthanide* contraction. The radii are about equal, but the masses are almost twice as large. Ir and Os are the two densest elements (approximately 22.6 g·cm^{-3}).

→ A second effect of the *contraction* is the *low reactivity* of Pt and Au, whose valence electrons are so tightly bound that they are not readily available for chemical reaction.

9A.2 Trends in Chemical Properties

• **Oxidation numbers (states)**

→ We will use *oxidation state* and *oxidation number* interchangeably.

→ Most d-block elements have more than one common oxidation number and one or more less common oxidation numbers (see **Figure 9A.7** in the text).

→ Oxidation numbers range from negative values to +8.

Example: In $[Fe(CO)_4]^{2-}$, Fe has an oxidation number of –2, whereas in OsO_4, Os has an oxidation number of +8.

→ Oxidation numbers show greatest variability for elements in the middle of the d-block.

Example: Mn, at the center of its row, has seven oxidation numbers, whereas Sc, at the beginning of the same row, has only one (in addition to zero).

→ *Period 5* and *Period 6* d-block elements tend to have higher oxidation numbers than those in *Period 4.*

Example: The highest oxidation number of Ni (Group 10, *Period 4*) is +4, whereas that of Pt (Group 10, *Period 6*) is +6.

• **Chemical properties and oxidation states**

→ Species containing d-block elements with *high oxidation numbers* tend to be good *oxidizing agents.*

Examples: MnO_4^- with Mn +7 and CrO_4^{2-} with Cr +6

→ Species containing d-block elements with *high oxidation numbers* tend to exhibit *covalent bonding.*

Example: Mn_2O_7 with Mn +7 is a covalent *liquid* at room temperature.

→ Species containing d-block elements with *high oxidation numbers* tend to have *oxides* that are *acid anhydrides.*
   Example: CrO_3 (with Cr +6) + H_2O → 2 H_2CrO_4 (acidic)

→ Species containing d-block elements with *low oxidation numbers* tend to be good *reducing agents.*
   Examples: CrO with Cr +2 and $FeCl_2$ with Fe +2

→ Species containing d-block elements with *low oxidation numbers* tend to exhibit *ionic bonding.*
   Example: Mn_3O_4, with Mn +2 and Mn +3, is an ionic solid.

→ Species containing d-block elements with *low oxidation numbers* tend to have *oxides* that are *basic anhydrides*.
   Example: CrO with Cr +2 is a *basic* oxide.

Topic 9B: SELECTED d-BLOCK ELEMENTS: A SURVEY

9B.1 Scandium through Nickel

- **Properties**
  → Chemical properties of the d-block elements are more diverse than their physical properties.

- **Properties of the metals** (see **Table 9B.1** in the text)
  → Densities increase from 2.99 $g \cdot cm^{-3}$ for Sc to 8.91 $g \cdot cm^{-3}$ for Ni.
  → Melting points increase from 1540 °C for Sc to 1920 °C for V, then show irregular behavior to 1455 °C for Ni.
     Exception: Mn has an anomalously low melting point, 1250 °C.
  → Boiling points increase from 2800 °C for Sc to 3400 °C for V, then show irregular behavior to 2150 °C for Ni.
  → Valence electron configurations follow the filling rule: 4s fills before 3d from Sc, [Ar] $3d^1 4s^2$, to Ni, [Ar] $3d^8 4s^2$, with the exception of Cr, [Ar] $3d^5 4s^1$ (two half-filled subshells).

- **Scandium, Sc, [Ar] $3d^1 4s^2$**
  → First isolated in 1937
  → A highly reactive metal that reacts with water as vigorously as Ca
  → Has only a few commercial uses and is thought *not* to be essential to life
  → Has one oxidation state, +3, in compounds
  → Small, highly charged Sc^{3+} is strongly hydrated in water, forming the $Sc(OH_2)_6^{3+}$ ion, which is a weak acid (about as strong as acetic acid).

- **Titanium, Ti, [Ar] $3d^2 4s^2$**
  → A light, strong metal passivated by an oxide coating, which masks its inherent reactivity
  → Commonly found with the +4 oxidation number in its ores: *rutile,* TiO_2, and *ilmenite,* $FeTiO_2$
  → Strong reducing agents are necessary to extract Ti from its ores

→ The metal is used in jet engines and dental appliances.

→ Titanium oxide, TiO_2, is used as a white pigment in paint and paper, also as a semiconductor in solar cells.

→ Forms a series of oxides called *titanates*

→ *Barium titanate*, $BaTiO_3$, is a *piezoelectric* material (develops an electrical signal when stressed).

- **Vanadium, V, [Ar] $3d^3\,4s^2$**

  → A soft, silver-gray metal produced by reducing its oxide or chloride

  → V metal is used in *ferroalloys* (Fe, V, C) to make tough steels for automotive springs

  → Has positive oxidation numbers ranging from +1 to +5

  → Common oxidation numbers are +4 and +5.

  → Many colored compounds containing vanadium species are used as ceramic glazes.

  → Orange-yellow *vanadium(V) oxide*, V_2O_5, is used as an oxidizing agent and catalyst in the contact process for the manufacture of sulfuric acid (see **Topic 8H.3** in the text).

- **Chromium, Cr, [Ar] $3d^5\,4s^1$**

  → A bright, corrosion-resistant metal

  → Used to make stainless steel and for chromium plating

  → Possible use in spintronics, for which information is stored as nuclear spin orientation rather than electron spin magnetization

  → Obtained from *chromite*, $FeCr_2O_4$, by reduction with C in an electric arc furnace at high temperature:

  $$FeCr_2O_4(s) + 4\,C(s) \rightarrow Fe(l) + 2\,Cr(l) + 4\,CO(g)$$

  → An exception to the normal filling order of electrons ($3d^5\,4s^1$, not $3d^4\,4s^2$)

- **Compounds of Cr**

  → Positive oxidation numbers range from +1 to +6.

  → Common oxidation numbers are +3 and +6.

  → *Chromium(IV) oxide*, CrO_2, is used as a *ferromagnetic* coating for "chrome" magnetic tapes.

  → *Sodium chromate*, $Na_2CrO_4(s)$, is used to prepare many other Cr compounds. In acid solution, CrO_4^{2-} forms the *dichromate ion* $Cr_2O_7^{2-}$; in both ions, the oxidation number of Cr is +6.

  $$2\,CrO_4^{2-}(aq) + 2\,H_3O^+(aq) \rightarrow Cr_2O_7^{2-}(aq) + 3\,H_2O(l)$$

  → Cr compounds are used as pigments, corrosion inhibitors, fungicides, and ceramic glazes.

- **Manganese, Mn, [Ar] $3d^5\,4s^2$**

  → A gray metal that corrodes easily and is rarely used alone

  → Important in alloys such as steel and bronze

  → Obtained from the ore *pyrolusite*, mostly MnO_2, at high temperature:

  $$3\,MnO_2(s) + 4\,Al(s) \rightarrow 3\,Mn(l) + 2\,Al_2O_3(s)$$

- **Compounds of Mn**
  - → Positive oxidation numbers range from +1 to +7.
  - → The stablest oxidation number is +2, but +4, +7, and, to some extent, +3 are also common.
  - → The most important compound of Mn, MnO_2, or *manganese dioxide,* is a brown-black solid used as a decolorizer in glass, to prepare other Mn compounds, and in dry cells.
  - → The *permanganate ion* MnO_4^-, with Mn in the +7 oxidation state, is an important oxidizing agent and a mild disinfectant.

- **Iron, Fe, [Ar] $3d^6 4s^2$**
  - → The most abundant element on Earth, the second most abundant element in the Earth's crust, and the most widely used d-metal
  - → Obtained from the principal ores *hematite,* Fe_2O_3, and *magnetite* Fe_3O_4
  - → Reactive metal that readily corrodes in moist air, forming rust
  - → Passivated by oxidizing acids such as HNO_3 but reacts with nonoxidizing acids
  - → Forms corrosion-resistant alloys and steels (see **Table 9B.2** in the text)
  - → A *ferromagnetic* metal, as are its oxides and many of its alloys

- **Pig iron and steel**
  - → Steels are very important industrial products and a measure of a country's economic power.
  - → *Pig iron*, a precursor to steel, contains 3% to 5% C, about 2% Si, and other impurities. It is produced by reduction of iron ores, such as Fe_2O_3, in a blast furnace through a series of redox and Lewis acid–base reactions.
  - → To make steel, *pig iron* is purified to *lower* the carbon content and to remove other impurities. Then, appropriate metals are *added* to form the desired steel.
  - → Steels typically contain 2% or less carbon; their hardness, tensile strength, and ductility depend on the carbon content. Higher carbon content produces harder steel.

- **Compounds of Fe**
  - → Positive oxidation numbers range from +2 to +6.
  - → Common oxidation numbers are +2 and +3.
  - → Salts of iron vary in color from pale yellow to dark green.
  - → Color in aqueous solution is dominated by $[FeOH(H_2O)_5]^{2+}$, the conjugate base of $Fe(H_2O)_6]^{3+}$.
  - → *Fe(II)* is readily oxidized to *Fe(III)*; the reaction is slow in acid solution and rapid in base.
  - → Iron forms compounds, such as $Fe(CO)_5$, in which the oxidation number of iron is zero.
  - → A healthy adult human has about 3 g of iron, mostly as hemoglobin.

- **Cobalt, Co, [Ar] $3d^7 4s^2$**
  - → A silver-gray metal used mainly in Fe alloys
  - → Permanent magnets such as those used in loudspeakers are made of *alnico steel,* an alloy of Fe, Ni, Co, and Al.
  - → Co steels are hard and are used for drill bits and surgical tools.

- **Compounds of Co**
  - → Positive oxidation numbers range from +1 to +4.
  - → Common oxidation numbers are +2 and +3.
  - → *Co(II) oxide*, CoO(s), is a deep blue salt used to color glass and ceramic glazes.

- **Nickel, Ni, [Ar] $3d^8 4s^2$**
  - → A hard, silver-white metal used mainly in stainless steel (see Section 7.3 in the text)
  - → Used as a catalyst for the *hydrogenation* of organic molecules
  - → Obtained as a pure metal by the *Mond process* (the first and third steps require heat):

$$NiO(s,ore) + H_2(g) \rightarrow Ni(s, impure) + H_2O(g)$$

$$Ni(s, impure) + 4CO(g) \rightarrow Ni(CO)_4(l, pure)$$

$$Ni(CO)_4(l, pure) \rightarrow Ni(s, pure) + 4CO(g)$$

- **Compounds of Ni**
  - → Positive oxidation numbers range from +1 to +4.
  - → Common oxidation numbers are +2 and +3; +2 is the most stable.
  - → The green color of aqueous solutions of nickel salts arises from $[Ni(H_2O)_6]^{2+}$ ions.
  - → In nickel–cadmium (nicad) batteries, *Ni(III)* is reduced to *Ni(II)*.

- **Carbonyl compounds**
  - → Many transition metals form compounds with CO.
  - → In *carbonyl compounds*, both the metal and carbonyl (CO) groups are regarded as *nearly* neutral species, and the metal is assigned an oxidation state of 0.
  - → When Fe is heated in CO, it reacts to form trigonal - bipyramidal *iron pentacarbonyl*, $Fe(CO)_5$, a yellow molecular liquid that melts at −20 °C, boils at 103 °C, and decomposes in visible light.
  - → Other examples of transition metal carbonyls are *nickel tetracarbonyl*, $Ni(CO)_4$, a colorless, toxic, flammable liquid that boils at 43 °C, and *chromium hexacarbonyl*, $Cr(CO)_6$, a colorless crystal that sublimes readily. $Ni(CO)_4$ is *tetrahedral*; and $Cr(CO)_6$ is *octahedral*.
  - → Some metal carbonyl compounds can be reduced to produce anions in which the metal has a *negative* oxidation number.

    Example: In the $[Fe(CO)_4]^{2-}$ anion, produced by reduction of $Fe(CO)_5$, Fe has an oxidation number of −2:

$$Fe(CO)_5 \xrightarrow[\text{tetrahydrofuran}]{\text{Na}} [Fe(CO)_4]^{2-} + CO$$

9B.2 Groups 11 and 12

- **Properties of the metals** (see **Table 9B.3** in the text)
  - → All the elements of Groups 11 and 12 are metals with completely filled d-subshells.
  - → The Group 11 metals, *copper, silver, and gold*, are called the *coinage metals* and have valence electron configurations of $(n − 1)d^{10} ns^1$.

→ The low reactivity of the coinage metals derives partly from the poor shielding abilities of the d-electrons, and hence the strong attraction of the nucleus on the outermost electrons.

→ The effect is enhanced in Period 6 by the lanthanide contraction, accounting for the inertness of gold.

→ The Group 12 metals, *zinc*, *cadmium*, and *mercury*, have valence electron configurations of $(n-1)d^{10}ns^2$. Zinc and, to a lesser extent, cadmium resemble beryllium or magnesium in their chemistry.

- **Copper, Cu, $[Ar]3d^{10}4s^1$**

  → Found as the metal and in sulfide ores such as *chalcopyrite*, $CuFeS_2$

  → Can be extracted from its ores by *pyrometallurgical* (high temperature) or *hydrometallurgical* (aqueous solution) methods:

  → *Pyrometallurgical example*: $2CuFeS_2(s) + 3O_2(g) \rightarrow 2CuS(s) + 2FeO(s) + 2SO_2(g)$
  The CuS is smelted to obtain blister copper: $CuS(s) + O_2(g) \rightarrow Cu(l) + SO_2(g)$.
  Both steps require high temperature.

  → *Hydrometallurgical example*: Sulfuric acid oxidizes Cu in ores to give soluble Cu^{2+} salts, which are then reduced: $Cu^{2+}(aq) + H_2(g) \rightarrow Cu(s) + 2H^+(aq)$ $\Delta G^o = -65$ kJ

  → Used as an electrical conductor in wires

  → Forms alloys such as brass and bronze, which are used in plumbing and casting

  → Corrodes in moist air, forming green *basic copper carbonate*:
  $$2Cu(s) + H_2O(l) + O_2(g) + CO_2(g) \rightarrow Cu_2(OH)_2CO_3(s)$$

  → Is oxidized by *oxidizing* acids such as HNO_3

- **Compounds of Cu**

  → In water common oxidation numbers are +1 and +2; +2 is more stable.

  → *Cu(I)* disproportionates in water to form metallic copper and copper(II).

  → In water *Cu(II)* forms a pale - blue hydrated ion, $[Cu(H_2O)_6]^{2+}$.

  → Cu is essential for animal metabolism. In mammals, Cu-containing enzymes are required for healthy nerve and connective tissue. In species such as octopi and lobsters, Cu (not Fe) is used to transport oxygen in the blood.

- **Silver, Ag, $[Kr]4d^{10}5s^1$**

  → Rarely found as the free metal; mostly obtained as a by-product of the refining of Cu and Pb

  → Reacts with sulfur to produce a black tarnish on silver dishes and cutlery

  → Like Cu, Ag is oxidized by *oxidizing* acids:
  $$3Ag(s) + 4H_3O^+(aq) + NO_3^-(aq) \rightarrow 3Ag^+(aq) + NO(g) + 6H_2O(l)$$

- **Compounds of Ag**

  → Oxidation numbers +1, +2, and +3 are known, but +1 is most stable.

  → Ag(I) does *not* disproportionate in water.

→ Except for $AgNO_3$ and AgF, Ag salts are insoluble in water.

→ Silver halides are used in photographic film.

- **Gold, Au, [Xe] $4f^{14} 5d^{10} 6s^1$**

    → So inert that it is usually found as the free metal

    → Highly malleable, easily made into thin foil (gold leaf)

    → *Cannot* be oxidized by nitric acid, but it dissolves in *aqua regia*, a mixture of sulfuric and hydrochloric acids:

    $$Au(s) + 6H^+(aq) + 3NO_3^-(aq) + 4Cl^-(aq) \rightarrow AuCl_4^-(aq) + 3NO_2(g) + 3H_2O(l)$$

    → A carat is a 1/24 part; 10- and 14-carat gold contain 10/24 and 14/24 parts of gold by mass, respectively.

    → Positive oxidation numbers are +1, +2, and +3, but +1 and +3 are most common.

- **Zinc, Zn, [Ar] $3d^{10} 4s^2$**

    → A silvery reactive metal found mainly in the sulfide ore, *sphalerite*, ZnS

    → Obtained from the ore by *froth flotation*, followed by *smelting* with coke, both at high *T*:

    $$2ZnS(s) + 3O_2(g) \rightarrow 2ZnO(s) + 2SO_2(g)$$
    $$ZnO(s) + C(s) \rightarrow Zn(l) + CO(g)$$

    → Used primarily for *galvanizing* Fe

    → Like Cu, Zn is protected by a hard film of *basic carbonate*, $Zn_2(OH)_2CO_3$.

- **Compounds of Zn**

    → Common positive oxidation number is +2.

    → An *amphoteric* metal, Zn dissolves in both acidic and basic solutions.

    Acid: $Zn(s) + 2H_3O^+(aq) \rightarrow Zn^{2+}(aq) + H_2(g) + 2H_2O(l)$

    Base: $Zn(s) + 2OH^-(aq) + 2H_2O(l) \rightarrow [Zn(OH)_4]^{2-} + H_2(g)$

    $Zn(OH)_4^{2-}$ is called the *zincate ion*. These reactions indictae that galvanized containers should not be used to transport either acids or alkalis.

    → Zn is an essential element for human health; it occurs in many enzymes.

    → Zn is toxic only in very large amounts.

- **Cadmium, Cd, [Kr] $4d^{10} 5^2$**

    → A silvery reactive metal

- **Compounds of Cd**

    → Common positive oxidation number is +2.

    → *Cadmiate ions* (analogous to *zincate ions*) are known, but Cd does not react with strong bases. Like Zn, Cd reacts with nonoxidizing acids.

    → Unlike Zn, Cd salts are *deadly poisons*.

    → Cd disrupts human metabolism by replacing other essential metals in the body, such as Zn and Ca, leading to soft bones and to kidney and lung disorders.

    → Cd and Zn are alike chemically in many ways, but both are quite different from Hg.

→ A d^N complex with the minimum number of unpaired electrons is called a *low-spin complex*. Low-spin complexes are expected for *strong-field ligands*.

→ The d^4 through d^7 octahedral complexes may be high- or low-spin.

- **Tetrahedral complexes**

→ Because Δ_T is small with respect to the energy required to pair an electron in the e-orbitals, t_2-orbitals are always accessible for tetrahedral complexes and *tetrahedral complexes* are *almost always* high-spin.

9D.3 The Colors of Complexes

- **Absorption of light and transmission or reflection**

→ Visible light contains all wavelengths from about 400 nm (blue) to 700 nm (red).

→ Transition metal complexes often *absorb* some visible light and *transmit* or *reflect* the rest.

→ *Transmitted* or *reflected* light (the *complementary color* of the light absorbed) is the light we see and the light that determines the color of an object.

→ The color wheel (see **Figure 9D.7** in the text) provides a simple way of relating color *absorbed* to color *seen* in some systems.

→ The color seen is the *opposite* of the color absorbed.

- **Electronic transitions in complexes**

→ Light absorption in complexes is often associated with d–d *transitions*. An electron is excited from a t_{2g}- to an e_g-orbital in an *octahedral complex* or from an e- to a t_2-orbital in a *tetrahedral complex*.

→ In *charge-transfer transitions* an electron is excited from a ligand-centered orbital to a metal-centered orbital or vice versa. These transitions are often quite intense.
   Example: The deep purple color of the permanganate ion, MnO_4^-

→ Color arises when a substance absorbs light from a portion of the visible spectrum and transmits or reflects the rest. Both d–d and charge-transfer transitions can occur in the visible region, and either or both processes can contribute to the color of a transition metal complex.

- **Color, d–d transitions, and the spectrochemical series**

→ *Weak-field* complexes have *small* Δ values and absorb *low-energy* (*red* or *infrared*) radiation.

→ Absent strong charge-transfer transitions, solutions of these complexes appear *green* (the complementary color of *red*) or *colorless* (if the absorption is in the *infrared*).

→ *Strong-field* complexes have *large* Δ values and absorb *high-energy* (*blue, violet,* or *ultraviolet*) radiation.

→ In the absence of complicating charge-transfer transitions, solutions of these complexes appear *orange* or *yellow* (the complementary colors of *blue* and *violet*, respectively) or *colorless* (if absorption is in the *ultraviolet*).

→ The ligand field splitting Δ can be determined from the electronic spectrum of a given complex, but, in most cases, it is not related simply to the color of its solution.

→ The absorption spectrum of a complex is characterized by its *molar absorption coefficient ε:*

$$A = \log \frac{I_0}{I}$$, where I_0 represents the incident intensity and I represents the transmitted intensity.

* The absorbance A is proportional to the path length L and molar concentration c

$$A \propto Lc \quad \text{or} \quad A = \varepsilon Lc$$

→ In terms of intensities

$$I = I_0 10^{-\varepsilon cL} \quad \text{(Replacing } A \text{ in the first equation and taking the antilogarithm, } 10^x\text{).}$$

→ This important equation, called *Beer's law*, is widely used in chemistry.

9D.4 Magnetic Properties of Complexes

- **Complexes may be diamagnetic or paramagnetic**

  → *Diamagnetic complexes* have no unpaired electrons and are repelled by a magnetic field.

  → *Paramagnetic complexes* have one or more unpaired electrons and are attracted by a magnetic field. The greater the number of unpaired electrons, the greater the attraction

- **Magnetic properties of complexes**

  → Determined from the *d*-electron configuration of the complex

  → Strong-field ligands produce weakly paramagnetic low-spin complexes.

  → Weak-field ligands produce strongly paramagnetic high-spin complexes.

9D.5 Ligand Field Theory

→ Molecular orbitals (MOs) are generated from the available atomic orbitals (AOs) in a complex.

→ Only one AO is used for each monodentate ligand (two for bidentate, etc.).

Examples: For a chloride ligand, a Cl 3p-orbital, directed toward the metal, is used. For an ammonia ligand, the sp³ lone-pair orbital is chosen. For bidentate ethylenediamine(en), two nitrogen sp³ lone-pair orbitals are chosen.

→ The treatment of π-*bonding* in complexes provides further insight into the *spectrochemical series*.

→ Ligand field theory explains why uncharged species such as CO are strong-field ligands, whereas negatively charged species such as Cl⁻ are weak-field ligands.

→ Based on electrostatic considerations, we would expect the *opposite* to be true.

- **Octahedral complexes**

  → For first-row transition elements, the 4s-, 4p-, and 3d-orbitals (nine total orbitals) of the metal ion are chosen because they have similar energies.

  → The nine metal and six ligand orbitals (one from each ligand) yield a total of 15 AOs, which overlap to form 15 MOs (see **Figure 9D.14** in the text).

  → The final result is that six MOs are bonding, six are antibonding, and three are nonbonding.

→ The 12 ligand electrons and the available d-valence electrons are placed in the MOs in accordance with the building-up principle to determine the ground-state electron configuration of the complex.

→ In transition metal ions, the valence s- and p-orbitals are typically vacant, whereas the d-orbitals are partially occupied.

→ The first 12 electrons form six metal-ligand sigma bonds. The d-electrons of the metal enter the t_{2g}- and e_g-orbitals in the same way they did in the crystal field theory. Ligand field theory, however, identifies the t_{2g}-orbitals as *nonbonding* and the e_g-orbitals as *antibonding*.

→ π-*bonding* in octahedral complexes arises from overlap of the metal t_{2g}-orbitals with p- or π-ligand orbitals to form *bonding* and *antibonding* MOs, π and π^*, respectively.

→ The details of this interaction are shown in **Figure 9D.15** in the text.

- **Weak-field ligand** (see **Figure 9D.15(a)** in the text)

  → The *antibonding* ligand π^*-orbital is too high in energy to take part in bonding.

  → The *bonding* ligand π-orbital combines with a metal t_{2g}-orbital, and the combination *raises* the energy of the MO that the metal d-electron occupies.

  → The electron from the metal enters an *antibonding* MO. The energy of the e_g-orbitals is unchanged, but the splitting between the t_{2g}- and e_g-orbitals is reduced.

  → The ligand field splitting, Δ_O, is therefore *decreased* in this case.

  → Some examples of weak-field ligands are Cl^- and Br^-, which have a valence shell p-orbital with two electrons in it, oriented perpendicular to the metal-ligand sigma bond.

- **Strong-field ligand (see Figure 9D.15(b) in the text)**

  → The bonding π-orbital of the ligand is too low in energy to take part in the bonding. The *empty* antibonding π^*-orbital of the ligand combines with a t_{2g}-orbital of the metal, and the combination *lowers* the energy of the MO that the metal d-electron occupies. The electron from the metal enters a *bonding* MO.

  → The energy of the e_g-orbitals is unchanged, and the splitting between the t_{2g}- and e_g-orbitals is increased.

  → The ligand field splitting, Δ_O, is *increased* in this case.

  → Some examples of strong-field ligands are CO and CN^-, each of which has an unoccupied valence shell π^*-orbital oriented perpendicular to the metal-ligand sigma bond.

Focus 10: NUCLEAR CHEMISTRY

Topic 10A: NUCLEAR DECAY

10A.1 The Evidence for Nuclear Decay

- **Radioactivity**
  - → Spontaneous emission of radiation or particles by a nucleus undergoing decay
  - → Includes the emission of α particles, β particles, and γ radiation
  - → These three types, the most common forms of radioactivity, were the first ones to be discovered.

- **Nuclear radiation** (see **Table 10A.1** in the text)

  α Particles
  - → Helium-4 nuclei, $_2^4\text{He}^{2+}$, traveling at speeds equal to about 10% of the speed of light, c
  - → Deflected by electric and magnetic fields
  - → When α particles interact with matter, their speed is reduced and they are neutralized.
  - → Denoted by $_2^4\alpha$ or α

  β Particles
  - → Rapidly moving *electrons,* e⁻, emitted by nuclei at speeds less than 90% of the speed of light
  - → Deflected by electric and magnetic fields, sometimes called *negatrons*
  - → When traveling between positively and negatively charged plates, α and β particles are deflected in opposite directions.
  - → Denoted by $_{-1}^0\text{e}$, β^-, or β (charge of −1, negligible mass compared with protons and neutrons)

  γ Radiation
  - → High-energy *photons* (electromagnetic radiation) traveling at the speed of light
  - → Uncharged and not deflected by electric and magnetic fields
  - → Denoted by γ or $_0^0\gamma$ (no mass and no charge)

  β^+ Particles
  - → Called *positrons*
  - → Have the mass of an electron but carry a positive charge
  - → When an electron and a positron, its *antiparticle,* meet, they are annihilated and are completely transformed into energy, mostly γ radiation.
  - → Denoted by $_{+1}^0\text{e}$ or β^+

  p Particles
  - → *Protons* are positively charged subatomic particles found in nuclei. When emitted in nuclear decay, protons typically travel at speeds equal to about 10% of the speed of light.
  - → Denoted by $_1^1\text{H}^+$, $_1^1\text{p}$, or p

n Particles

→ *Neutrons* are uncharged subatomic particles found in nuclei. The mass of a neutron is slightly larger than the mass of a proton (see **Table B.1** in the text). When emitted in nuclear decay, neutrons typically travel at speeds less than about 10% of the speed of light.

→ Denoted by $^1_0$n or n

• **Penetrating power of nuclear radiation**

→ Power to penetrate matter decreases with increased charge and mass of the particle.

→ Uncharged γ radiation and neutrons are more penetrating than charged positrons or alpha particles.

→ Singly charged β particles (electrons) are more penetrating than the more massive doubly charged α particles, which are virtually nonpenetrating.

• **Antiparticles**

→ Subatomic particles with equal mass and opposite charge

→ Annihilate each other upon encounter, producing energy

10A.2 Nuclear Reactions

• **Nucleons and nuclides**

→ A *nucleon* is a proton or a neutron in the nucleus of an atom.

→ A *nuclide* is an atom characterized by its atomic number, mass number, and nuclear energy state.

→ Nuclides are often denoted by ^A_ZE , where Z is the atomic number, A is the mass number (protons plus neutrons), and E is the symbol for the element. $A - Z$ equals the number of neutrons in the nuclide.

Examples: $^{12}_6$C and $^{235}_{92}$U

→ Nuclear isomers are atomic nuclei of a particular nuclide that have equal proton number and equal mass number, differ in energy content, and are long-lived.

Examples: The two states of $^{99}_{43}$Tc observed in decay schemes. The longest-lived theoretically unstable nuclear isomer is tantalum-180m, which has a half-life in excess of 1000 trillion years and has never been observed to decay to tantalum-180.

→ *A nuclide is a neutral atom and therefore is uncharged.* There are about 256 nuclides in nature that are so stable that they have never been observed to decay. They occur among the 82 elements with one or more stable nuclides.

→ *Radioactive nuclei* can change their structure spontaneously by *nuclear decay*, the partial breakup of a nucleus.

• **Nuclear reactions**

→ Any transformations that a nucleus undergoes

→ Differ from chemical reactions in three important ways:

1) Isotopes of a given element often have different nuclear properties but always have similar chemical properties.

Example: ^{12}C and ^{14}C are difficult to separate chemically because they have similar chemical properties, yet the two isotopes have different nuclear stability. Carbon-12 is a stable nuclide, whereas carbon-14 decays with a half-life of 5730 a (see **Topic 10B.2** in the text or in this guide). **Note:** a = year, d = day

2) Nuclear reactions often produce a different element.

Example: Titanium-44 captures one of its electrons to produce the nuclide scandium-44.

3) Nuclear reactions involve enormous energies relative to chemical reactions.

- **Balancing nuclear reactions**

  → Like chemical reactions, *nuclear reactions* must be balanced with respect to both *charge* and *mass*.

  → **Charge balance:** For the several nuclides $^A_Z E$ in a nuclear reaction, the sum of the atomic numbers Z of the reactants must equal the sum of the Z values of the products.

  → **Mass balance:** The sum of the mass numbers A of the reactants must equal the sum of the mass numbers A of the products.

  → The result of *most* nuclear decay reactions is *nuclear transmutation*, the conversion of one element to another.

  → For nuclear decay reactions:

    A reactant nuclide is called a *parent nuclide*.

    A product nuclide is called a *daughter nuclide*.

10A.3 The Pattern of Nuclear Stability

- **Characterization of nuclei** (see **Figure 10A.12** in the text)

  Even - even

  → Nuclei with an even number of protons and neutrons; consequently A and Z are even numbers.

  → 157 even - even nuclides are stable according to the text.

    Examples: helium-4, $^4_2 He$; carbon-12, $^{12}_6 C$; oxygen-16, $^{16}_8 O$; and neon-22, $^{12}_{10} Ne$

  Even - odd

  → Nuclei with an even number of protons and an odd number of neutrons; consequently Z is even and A is odd

  → 53 even - odd nuclides are stable according to the text.

    Examples: helium-3, $^3_2 He$; berylium-9, $^9_4 Be$; carbon-13, $^{13}_6 C$; and neon-21, $^{21}_{10} Ne$

  Odd - even

  → Nuclei with an odd number of protons and an even number of neutrons; consequently both Z and A are odd numbers

  → 50 odd - even nuclides are stable according to the text.

    Examples: hydrogen-1, $^1_1 H$; lithium-7, $^7_3 Li$; boron-11, $^{11}_5 B$; and nitrogen-15, $^{15}_7 N$

Odd - odd

→ Nuclei with an odd number of protons and neutrons; consequently Z is odd and A is even

→ *Only 4 odd - odd nuclides are stable.*

→ The stable odd-odd nuclides are hydrogen-2 (deuterium), ^2_1H ; lithium-6, ^6_3Li ; boron-10, $^{10}_5B$; and nitrogen-14, $^{14}_7N$.

Summary

→ Even - even nuclides are the stablest and most abundant nuclides.

→ Odd - odd nuclides are the least stable and least abundant nuclides.

• **Strong force**

→ An attractive force that holds nucleons together in the nucleus

→ Overcomes the coulomb repulsion of protons

→ Acts only over a very short distance, approximately the diameter of the nucleus

• **Magic numbers**

→ 2, 8, 20, 50, 82, 114, 126, and 184 for *either* protons or neutrons

→ Nuclei with *magic numbers* of protons and/or neutrons are most likely to be stable.

→ Analogous to the pattern of electronic stability in atoms associated with electrons in filled subshells

→ *Doubly magic* nuclides are very stable.

   Examples: helium-4, ^4_2He ; oxygen-16, $^{16}_8O$; calcium-40, $^{40}_{20}Ca$; and lead-208, $^{208}_{82}Pb$

• **Band of stability and sea of instability** (see **Figure 10A.13** in the text)

→ On a plot of nuclide mass A (*y*-axis) versus atomic number Z (*x*-axis)

→ Stable nuclides are found in a narrow band, the *band of stability,* that ends at $Z = 83$ (bismuth).

→ All nuclides with $Z > 83$ are unstable.

→ Unstable nuclides are found in the *sea of instability*, a region above and below the *band of stability*.

→ Nuclides *above* the band of stability are *neutron rich.*

→ Neutron-rich nuclei are likely to emit β particles (neutron → proton + β).

→ Nuclides *below* the band of stability are *proton rich.*

→ Proton-rich nuclei are likely to emit positrons, β$^+$ particles (proton → neutron + β$^+$), or to capture electrons (proton + β → neutron).

→ Heavier nuclides below the band of stability and those with $Z > 83$ may also decay by emitting α particles.

• **Models of nuclear structure** (not in the text)

→ Three common models of nuclear structure

→ Description and order of sophistication

   1) The *liquid drop model*, in which nucleons are considered to be packed together in the nucleus like molecules in a liquid

2) The *independent particle model*, in which nucleons are described by quantum numbers and assigned to shells. Nucleons in the outermost shells are most easily lost as a result of radioactive decay.

3) The *collective model*, in which nucleons are considered to occupy quantized energy levels and to interact with each other by the strong force and the electrostatic (coulomb) force

10A.4 Predicting the Type of Nuclear Decay

- **Massive nuclei with $Z > 83$**
  - → Lose protons to reduce their atomic number and generally lose neutrons as well
  - → Stepwise decay gives rise to a *radioactive series,* a characteristic sequence of nuclides.

    Examples of radioactive series

    1) The ^{238}U series (see **Figure 10A.16** in the text) is one in which α (alpha) and β (beta) particles are successively ejected. The final step is the formation of a stable isotope of lead (magic number 82), ^{206}Pb.

    2) The ^{235}U series proceeds similarly and ends at ^{207}Pb. This series is sometimes called the actinium series.

    3) The ^{232}Th series ends at ^{208}Pb.

- **Neutron-rich nuclides**
  - → Nuclides that have high neutron-to-proton (n/p) ratios with respect to the band of stability
  - → Tend to decay by reducing the number of neutrons
  - → Commonly undergo β decay, which decreases the n/p ratio

- **Proton-rich nuclides**
  - → Nuclides that have high proton-to-neutron (p/n) ratios with respect to the band of stability
  - → Tend to decay by reducing the number of protons
  - → Commonly undergo electron capture, positron emission, or proton emission (least likely), which decreases the p/n ratio

10A.5 Nucleosynthesis

- **Formation of the elements**
  - → Elements are formed by *nucleosynthesis.*
  - → Nucleosynthesis occurs when particles collide vigorously, as in stars or particle accelerators.
  - → *Transmutation* is the conversion of one element to another.
  - → The first artificial transmutation was observed in 1919 by Rutherford, who converted $^{14}_{7}N$ to $^{17}_{8}O$ by bombarding the parent nuclide with high-speed α particles:

$$^{14}_{7}N + ^{4}_{2}\alpha \rightarrow ^{17}_{8}O + ^{1}_{1}p$$

The reactants must collide with enough energy to overcome the repulsive coulomb force between positively charged nuclei.

→ Nucleosynthesis reactions are commonly written in shorthand as:
target (incoming species, ejected species) product
In the Rutherford example, the shorthand notation is $^{14}_{7}N\,(\alpha,p)\,^{17}_{8}O$.

→ Nucleosynthesis can be effected using a variety of projectile species, including γ radiation, protons, neutrons, deuterium nuclides, α particles, and heavier nuclides.

→ Because they are uncharged, neutrons can travel relatively slowly and still react with nuclides.

- **Transuranium elements**

  → Elements following uranium ($Z = 92$) in the periodic table

  → Elements from rutherfordium (Rf, $Z = 104$) to meitnerium (Mt, $Z = 109$) were given official names in 1997.

  → Darmstadtium (Ds, $Z = 110$) was given an official name in 2003 and roentgenium (Rg, $Z = 111$) in 2004.

- **Recently named elements**

  → Rg = Roentgenium ($Z = 111$) *official* (in honor of Wilhelm Röntgen (Roentgen))

  → Cn = Copernicium ($Z = 112$) *official* (in honor of Nicolaus Copernicus)

  → Fl = Flerovium ($Z = 114$) *proposed* (in honor of Georgiy Flerov (1913-1990))

  → Lv = Livermorium ($Z = 116$) *proposed* (in honor of Lawrence Livermore Lab)

- **Transmeitnerium elements**

  → Temporary names are based on shorthand numbers derived from Latin (see **Table 10A.2** in the text).

  Examples: $Z = 114$ (Uuq, ununquadium) and $Z = 116$ (Uuh, ununhexium)

Topic 10B: RADIOACTIVITY

10B.1 The Biological Effects of Radiation

- **Nuclear radiation**

  → Sometimes called ionizing radiation because it can eject electrons from atoms

- **Penetrating power**

  → α, β, and γ particles are forms of *ionizing radiation* with different *penetration power*.

  → Different amount of shielding is required to prevent harmful effects (see **Table 10B.1** in the text).

  → α Particles have the least penetrating power and are absorbed by paper and the outer layers of skin. They are extremely dangerous if inhaled or ingested.

  → β Particles are 100 times more penetrating than α particles and are absorbed by 1 cm of flesh or 3 mm of aluminum, for example.

→ γ Radiation is 100 times more penetrating than β particles. It can pass through buildings and bodies, leaving a trail of ionized or damaged molecules, and it is absorbed by lead bricks or thick concrete.

→ The order of *increasing* penetrating power is α < β < γ.

- **Radiation damage to tissue**

→ Depends on the *type* and *strength* of the radiation, the *length* of exposure, and the *extent* to which the radiation can reach sensitive tissue

→ For example, Pu^{4+} is an α emitter, if ingested, it replaces Fe^{3+} in the body, inhibiting the production of red blood cells.

- **Absorbed dose of radiation**

→ Energy deposited in a sample, such as a human body, when exposed to radiation

→ Units of the absorbed dose of radiation:

1 rad = amount of radiation that deposits 0.01 J of energy per kilogram of tissue

1 gray (Gy) = an energy deposit of 1 J·kg^{-1} (SI unit)

1 Gy = 100 rad

Note: The name *rad* stands for radiation absorbed dose.

- **Dose equivalent**

→ Dose modified to account for different destructive powers of radiation

→ Relative destructive power is called the *relative biological effectiveness, Q*. For β$^-$ and γ radiation, $Q \equiv 1$. For α radiation, $Q \approx 20$.

→ The natural unit of dose equivalent is the rem (*roentgen equivalent man*).

→ Definition: Dose equivalent in rem = Q × absorbed dose in rad

→ SI Units: Dose equivalent in sievert (Sv) = Q × absorbed dose in gray (Gy)

1 Sv = 100 rem

- **Common radiation dose and harmful effects**

→ Typical annual human dose equivalent from natural sources is 0.2 rem·y^{-1}, a number with a wide range depending upon habitat and lifestyle.

→ A typical chest radiogram gives a dose equivalent of about 7 millirem.

→ 30 rad (30 rem) of γ radiation may cause a reduction in white blood cell count.

→ 30 rad (600 rem) of α radiation causes death.

10B.2 Measuring the Rate of Nuclear Decay

- **Activity of a sample**

→ Number of nuclear disintegrations per second

→ One nuclear disintegration per second is called 1 becquerel, Bq (SI unit).

→ The curie, Ci, an older unit of activity, is equal to 3.7×10^{10} Bq, the radioactive output of 1 g of radium-226.

→ Because the curie represents a large value of decay, the activity of typical samples is given in millicuries (mCi) or microcuries (μCi).

Note: See **Table 10B.2** in the text for a list of the radiation units.

- **Nuclear decay**

→ The decay of a nucleus can be written as:

Parent nucleus → daughter nucleus + radiation

→ This form is the same as a *unimolecular elementary reaction,* with an unstable nucleus taking the place of an excited molecule.

→ The rate of nuclear decay depends only on the identity of the nucleus (isotope), not on its chemical form or temperature.

- **Law of radioactive decay**

→ The rate of nuclear decay is proportional to the number of radioactive nuclei N:

$$\boxed{\text{Activity} = \text{rate of decay} = k \times N}$$

where k is the *decay constant* (or rate constant for the reaction).

→ The unit of the *decay constant* is $(\text{time})^{-1}$, the same as for any *first-order* rate constant.

- **Integrated rate law for nuclear decay and half-life**

→ Exponential form: $\boxed{N = N_0 e^{-kt}}$ The number of radioactive nuclei N decays exponentially with time. Large values of k correspond to more rapid decay.

→ Logarithmic form: $\boxed{\ln\left(\dfrac{N}{N_0}\right) = -kt}$

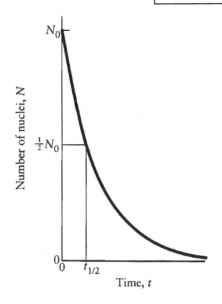

On the left is a plot of N versus t for nuclear decay. N_0 is the number of radioactive nuclei present at $t = 0$. N is the number of radioactive nuclei present at a later time t. The time required for the nuclei to decay to half their initial number is equal to the half-life, $t_{1/2}$. At this time,

$$N = \frac{1}{2}N_0$$

Substituting this value into the integrated rate law yields

$$\boxed{t_{1/2} = \frac{\ln 2}{k}}$$

Half-lives of radioactive nuclides span an enormous range, from picoseconds (^{215}Fr, 120 ps) to billions of years (^{238}U, 4.5×10^9 a or 4.5 Ga). See **Table 10B.3** in the text for a list of half-lives of common radioactive isotopes.

- **Isotopic dating**

→ Used to determine the ages of rocks and of archeological artifacts

→ Carried out by measuring the activity of a radioactive isotope in the sample

→ Useful isotopes for dating are ^{238}U, ^{40}K, $^{3}_{1}H$ (tritium), and ^{14}C.

- **Radiocarbon dating**
  - → Most important example of isotopic dating
  - → Uses the beta decay of ^{14}C, for which the half-life is 5730 a or 5.73 ka
- **Radioactive ^{14}C**
  - → Formed in the atmosphere by neutron bombardment of ^{14}N
    $$^{14}_{7}N + ^{1}_{0}n \rightarrow ^{14}_{6}C + ^{1}_{1}p$$
  - → Enters living organisms as $^{14}CO_2$ through photosynthesis and digestion
  - → Leaves living organisms by excretion and respiration
  - → Achieves a steady-state concentration in living organism with an activity of 15 disintegrations per minute per gram of *total* carbon
  - → Shows decreasing activity in dead organisms because they no longer ingest ^{14}C from the atmosphere
  - → The time of death of an organism can be estimated by measuring the activity of the sample and using the integrated rate law.

10B.3 Uses of Radioisotopes

- **Radioisotopes**
  - → Used to help cure disease, preserve food, and trace the mechanisms of chemical reactions
  - → Used to power spacecraft, locate sources of water, and determine the age of nonliving materials, including wood (^{14}C dating), rocks ($^{238}U/^{206}Pb$ ratio), and groundwater (tritium dating using the $^{1}H/^{3}H$ ratio)
- **Radioactive tracers**
  - → Radioactive isotopes used to track changes and locations

    Example: Phosphorus-32, a β emitter with $t_{1/2} = 14.28$ d, can be incorporated into phosphate-containing fertilizer to follow the mechanism of plant growth.

  - → In chemical reactions, tracers can help determine the mechanism of the reaction. Nonradioactive isotopes are also used for this purpose.

Topic 10C: NUCLEAR ENERGY

10C.1 Mass–Energy Conversion

- **Nuclear binding energy**
  - → Energy *released* when protons and neutrons join together to form a *nucleus:*
    $$(Z)^{1}_{1}p + (A - Z)\,^{1}_{0}n \rightarrow ^{A}_{Z}E^{Z+}$$

$\rightarrow$ Since *nuclides* are *neutral* atoms, the energy released is also given *approximately* by:

$$(Z)_1^1H + (A-Z)\,_0^1n \;\rightarrow\; _Z^A E$$

$\rightarrow$ The binding energy is then given by applying the Einstein mass-energy equation to each reactant and product species:

$$E_{bind} = |\Delta E| = |\textstyle\sum E(\text{products}) - \sum E(\text{reactants})|$$
$$= |\,E(\text{product}) - \textstyle\sum E(\text{reactants})\,| = |\,mc^2(\text{product}) - \sum mc^2(\text{reactants})\,|$$
$$= |\,m(\text{product}) - \textstyle\sum m(\text{reactants})\,| \times c^2$$

$$\boxed{E_{bind} = |\Delta m| \times c^2}$$

$\rightarrow$ *Binding energy* is a measure of the stability of the nucleus.

$\rightarrow$ The *binding energy per nucleon* is defined as E_{bind}/A and is a better measure of the stability of the nucleus.

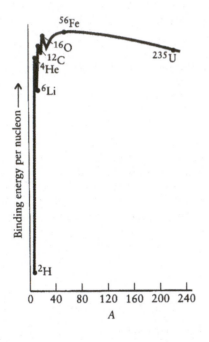

A plot of the binding energy per nucleon versus atomic mass number is shown on the left (see also **Figure 10C.1** in the text). Nucleons are bound together most strongly in the elements near iron and nickel, which accounts for the high abundance of iron and nickel in meteorites and on planets similar to the Earth. The figure shows that light nuclei become more stable when they fuse together to form heavier ones. Heavy nuclei become more stable when they undergo fission and split into lighter ones. Also, note the increased stability of the nuclei with magic numbers, particularly ^{16}O, which is *doubly magic*.

- **Units used in nuclear calculations**

    $\rightarrow$ *Mass* is usually reported in *atomic mass units*, u.

    $\rightarrow$ One atomic mass unit is defined as *exactly* one-twelfth the mass of one atom (nuclide) of ^{12}C:

    $$\boxed{1\,u = 1.660\,54 \times 10^{-27}\ kg}$$

    $\rightarrow$ The *atomic mass unit*, u, is numerically equivalent to the atomic mass constant m_u, and this older notation is sometimes used.

    $\rightarrow$ *Binding energy* is usually reported in electronvolts, eV, or millions (mega) of electronvolts, MeV.

→ An electronvolt is the change in potential energy of an electron (charge of $1.602\,18 \times 10^{-19}$ C) when it is moved through a potential difference of 1 V:

$$1\text{ eV} = 1.602\,18 \times 10^{-19}\text{ J}$$

10C.2 The Extraction of Nuclear Energy

→ Occurs when a nucleus breaks into two or more smaller nuclei

→ Releases a large amount of energy

→ May be *spontaneous* or *induced*

- **Spontaneous nuclear fission**

  → Occurs when oscillations in heavy nuclei lead them to break into two smaller nuclei of similar mass

  → Yields a variety of products for a given nuclide (see **Figure 10C.2** in the text)

  Example: Consider the spontaneous fission of americium-244. Two of the daughter nuclides formed are iodine-134 and molybdenum-107. The reaction is:

  $$^{244}_{95}\text{Am} \;\rightarrow\; ^{134}_{53}\text{I} \;+\; ^{107}_{42}\text{Mo} \;+\; 3\,^{1}_{0}\text{n}$$

- **Induced nuclear fission**

  → Caused by bombarding a heavy nucleus with neutrons

  → Yields a variety of daughter nuclides for a given parent

  Example: Consider the induced fission of plutonium-239 by neutron bombardment. A number of products are formed. Two important reactions:

  $$^{239}_{94}\text{Pu} \;+\; ^{1}_{0}\text{n} \;\rightarrow\; ^{98}_{42}\text{Mo} \;+\; ^{138}_{52}\text{Te} \;+\; 4\,^{1}_{0}\text{n}$$

  $$^{239}_{94}\text{Pu} \;+\; ^{1}_{0}\text{n} \;\rightarrow\; ^{100}_{43}\text{Tc} \;+\; ^{135}_{51}\text{Sb} \;+\; 5\,^{1}_{0}\text{n}$$

- **Energy changes in fission reactions**

  → Energy released during fission is calculated by using Einstein's equation.

  → For a balanced nuclear reaction,

  $$\Delta m = \sum m(\text{products}) - \sum m(\text{reactants}) \qquad \text{and} \qquad \Delta E = \Delta m \times c^2$$

- **Fissionable and fissile nuclei**

  → *Fissionable nuclei* can undergo induced fission.

  → *Fissile nuclei* can undergo induced fission *with slow-moving neutrons.*

  Examples of fissile nuclei: $^{235}_{92}\text{U}$, $^{233}_{92}\text{U}$, and $^{239}_{94}\text{Pu}$, which are nuclear power plant fuels.

  The nuclide $^{238}_{92}\text{U}$ is fissionable by fast-moving neutrons but is *not* fissile.

- **Nuclear branched chain reactions**

  → Neutrons are *chain carriers* in a branched chain reaction (see **Topic 7C.5** in the text).

  → Chain branching occurs when an induced fission reaction produces two or more neutrons.

Example: $^{235}_{92}U + ^1_0n \rightarrow ^{97}_{40}Zr + ^{137}_{52}Te + 2\,^1_0n$, in which two neutrons are produced for each ^{235}U nuclide that reacts. The two neutrons can either escape into the surroundings or be captured by (react with) other ^{235}U nuclides.

→ A *critical mass* is the mass of fissionable material above which so few neutrons escape from the sample that the fission chain reaction is sustained.

→ The critical mass for pure plutonium of normal density is about 15 kg.

→ A sample of plutonium with mass greater than 15 kg is *supercritical*; the reaction is self-sustaining and may result in an explosion.

→ A *subcritical* sample is one that has less than the critical mass for its density.

• **Controlled and uncontrolled (explosive) nuclear reactions**

→ Explosive nuclear reactions occur when a *subcritical* amount of fissile material is made *supercritical* so rapidly that the chain reaction occurs uniformly throughout the material.

→ Controlled nuclear reactions occur in nuclear reactors.

→ Controlled reactions are not explosive because there is a subcritical amount of fuel.

→ The rate of reaction is controlled by *moderators*, such as graphite rods.

→ Moderators slow the emitted neutrons so that a greater fraction can induce fission.

→ The difference between a fission explosion and controlled fission is shown in the figure. Each symbol ⊗ represents a fission event. The neutrons from the fission are shown but products are not.

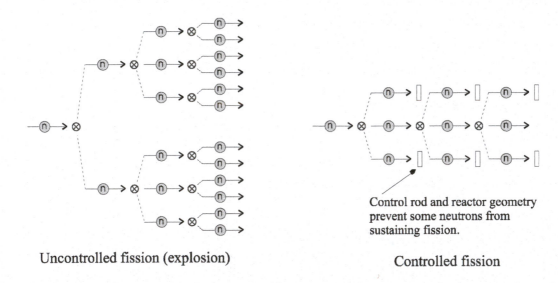

Uncontrolled fission (explosion)   Controlled fission

Note: The number of neutrons released in an event is variable, as is the identity of the products.

• **Breeder reactors**

→ Nuclear reactors used to synthesize fissile nuclides for fuel and weapon use.

→ Breeder reactors run very hot and fast because no moderator is present. As a result, they are more dangerous than nuclear reactors used for power generation.

10C.3 The Chemistry of Nuclear Power

- **Chemistry**
  - → The key to the safe use of nuclear power
  - → Used to prepare nuclear fuel
  - → Used to recover important fission products
  - → Used to dispose of nuclear waste safely

- **Uranium**
  - → The fuel of nuclear reactors
  - → Obtained primarily from the ore pitchblende, UO_2, by reducing the oxide to the metal and then *enriching*, or increasing, the fraction of the fissile nuclide, ^{235}U
  - → To be useful as a fuel, the percentage of ^{235}U must be increased from its natural abundance of 0.7% to about 3%.

- **Enrichment**
  - → Exploits the mass difference between ^{235}U and ^{238}U
  - → Separation is accomplished by repeated effusion of $^{235}UF_6$ and $^{238}UF_6$ vapor (see Graham's law of effusion, treated in **Topic 3D.1**).

$$\frac{\text{Rate of effusion of } ^{235}UF_6}{\text{Rate of effusion of } ^{238}UF_6} = \sqrt{\frac{M_{238}}{M_{235}}} = \sqrt{\frac{352.1}{349.0}} = 1.004$$

  Note: Because the ratio is close to 1, hundreds of effusion steps are necessary to get the desired separation.

- **Nuclear waste**
  - → Spent nuclear fuel called *nuclear waste* is still radioactive.
  - → Waste is a mixture of uranium and fission products.
  - → Nuclear waste must be stored safely for about 10 half-lives; it is generally buried underground.
  - → For underground burial, incorporation of the highly radioactive fission (HRF) products into a glass or ceramic material is better than placing it in metal storage drums.
  - → Storage drums can corrode, allowing the waste to seep into aquifers and/or contaminate large areas of soil.

→ When numbering atoms in the chain, the lowest numbers are given preferentially to (a) functional groups named by suffixes (see **Toolbox 11D.1** in the text), (b) double bonds, (c) triple bonds, and (d) groups named by prefixes.

     Example: 4-chloro-2-hexyne ($CH_3CH_2CH(Cl)C\equiv CCH_3$)

11A.2 Isomers

- **Structural isomers**

  → Molecules constructed from the same atoms connected differently

  → Have the *same* molecular formula but *different* structural formulas

       Example: Butane ($CH_3(CH_2)_2CH_3$) and methylpropane ($CH_3CH(CH_3)CH_3$) are structural isomers with the same molecular formula (C_4H_{10}). The molecules have a different *connectivity*.

- **Stereoisomers**

  → Molecules with the same connectivity but with some of their atoms arranged differently in space

  → *Geometrical isomers* are *stereoisomers* with different arrangements in space on either side of a double bond or above and below the ring of a cycloalkane.

  → *Optical isomers* are *stereoisomers* in which each isomer is the mirror image of the other *and* the images are not superimposable.

- **Geometrical isomers**

  → Compounds with the same molecular formula and with atoms bonded to the same neighbors but with a different arrangement of atoms in space

  → For molecules with a double bond, effective lack of rotation about the double bond makes this type of isomer possible.

  → *Geometrical isomers* of organic compounds are distinguished by italicized prefixes: *cis* (from the Latin word for *on this side*) and *trans* (from the Latin word for *across*).

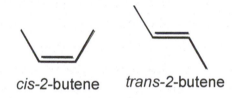

     cis-2-butene   *trans*-2-butene

       Example: Consider 2-butene, which exists as cis and trans geometrical isomers. In the trans isomer, the two methyl groups lie across the double bond from each other. In the cis isomer, they are on the same side of the double bond.

                    Rotation about the double bond does not normally occur because the π-bond is rigid and the two isomers are distinct compounds with different chemical and physical properties. If sufficient energy is added to either compound, a *cis–trans isomerization* reaction may occur in which the π-bond is partially broken and part of the sample is converted to the other isomer.

- **Optical isomers**

  → *Nonsuperimposable* mirror images

  → A *chiral* molecule has a mirror image that cannot be superimposed. Your hand is a chiral object; its mirror image is not superimposable.

  → An organic molecule is chiral if at least one of its C atoms has four *different* groups attached to it. Such a C atom is called a *stereogenic center*.

  → A pair of *enantiomers* consists of a chiral molecule and its mirror image.

  → Enantiomers have identical chemical properties (melting point, solubility, etc.), except when they react with other chiral species.

  → Enantiomers differ in only one physical property; chiral molecules display *optical activity,* the ability to rotate the plane of polarization of light.

  → Mixtures of enantiomers in equal proportions are *racemic mixtures,* which are not optically active.

  → A molecule that has a superimposable mirror image is said to be *achiral.*

- **Summary**

  → The three types of isomerism are displayed in **Figure 11A.1** in the text.

11A.3 Physical Properties of Alkanes and Alkenes

- **Physical properties**

  → *Alkanes* are best regarded as nonpolar molecules held together primarily by London forces, which increase with the number of electrons in a molecule.

  → Consequently, for unbranched *alkanes*, melting and boiling points increase with chain length.

  → Unbranched *alkanes* tend to have higher melting points, boiling points, and heats of vaporization than their branched structural isomers.

  → Molecules with unbranched chains can get closer together than molecules with branched chains. As a result, molecules with branched chains have *weaker* intermolecular forces than their isomers with unbranched chains.

  Example: Unbranched butane has a higher boiling point (−0.5 °C) than its branched-chain structural isomer, methylpropane (−11.6 °C).

- **Chemical properties**

  → *Alkanes* are not very reactive chemically. They were once called *paraffins*, derived from the Latin for little affinity.

  → *Alkanes* are unaffected by concentrated sulfuric acid, boiling nitric acid, strong oxidizing agents such as $KMnO_4$, and by boiling aqueous NaOH.

  → The C–C bond enthalpy (348 kJ·mol^{-1}) and the C–H bond enthalpy (412 kJ·mol^{-1}) are large, so there is little energy advantage in replacing them with most other bonds.

  → Notable exceptions are C=O (743 kJ·mol^{-1}), C–OH (360 kJ·mol^{-1}), and C–F (484 kJ·mol^{-1}).

- **Reactivity**

  → Alkanes undergo catalyzed reactions in the refining of petroleum (see **Topic 11B.1** and **Impact on Technology: Fuels** following **Focus 11**) as well as combustion and substitution reactions.

- **Combustion (oxidation) reactions**

  → *Alkanes* are used as fuels because their enthalpies of combustion are high.

  → The products of combustion are carbon dioxide and water. Strong C–H bonds are replaced by even stronger O–H bonds in H_2O, and the O=O bonds are replaced by two strong C=O bonds in CO_2.

  Example: Combustion of one mole of octane releases 5471 kJ of heat:
  $$C_8H_{18}(l) + 12.5\,O_2(g) \rightarrow 8\,CO_2(g) + 9\,H_2O(l) \qquad \Delta H_r^\circ = -5471 \text{ kJ·mol}^{-1}$$

- **Double bond**

  → The carbon–carbon double bond, C=C, consists of a σ-bond and a π-bond.

  → Each carbon atom is sp^2-hybridized, and one of the hybrid orbitals is used to form the σ-bond.

  → The unhybridized p-orbitals on each atom overlap with each other to form a π-bond.

  → All four atoms attached to the C=C group lie in the same plane and are fixed in that arrangement by resistance to twisting of the π-bond (see **Figure 11A.5** in the text).

  → Alkenes cannot roll up into a compact arrangement as alkanes can, so alkenes have lower melting points.

  → In C=C, the π-bond is weaker than the σ-bond. A consequence of this weakness is the reaction most common in alkenes: replacement of the π-bond by two new σ-bonds, as in an elimination reaction.

Topic 11B: REACTIONS OF ALIPHATIC HYDROCARBONS

11B.1 Alkane Substitution Reactions

- **Substitution type reactions**

  → *Alkanes* are used as raw materials for the synthesis of many reactive organic compounds.

  → Organic chemists introduce reactive groups into alkane molecules in a process called *functionalization*.

  → Functionalization of alkanes is achieved by a *substitution reaction*, in which an atom or group of atoms replaces an atom in the original molecule (hydrogen, in the case of alkanes).

  → Reaction of methane (CH_4) and chlorine (Cl_2) is an example of a substitution reaction. In the presence of ultraviolet light or temperatures above 300 °C, the gases react explosively:
  $$CH_4(g) + Cl_2(g) \xrightarrow{\text{light or heat}} CH_3Cl(g) + HCl(g)$$

  Chloromethane (CH_3Cl) is only one of four products; the others are dichloromethane (CH_2Cl_2), trichloromethane ($CHCl_3$), and tetrachloromethane (CCl_4), which is carcinogenic.

- **Radical chain mechanism**

  → Kinetic studies suggest that alkane substitution reactions such as this one proceed by a *radical chain* mechanism.

  → The *initiation step* is the dissociation of chlorine:
  $$Cl_2 \xrightarrow{\text{light or heat}} 2\,Cl\cdot$$

  → Chlorine atoms proceed to attack methane molecules and abstract a hydrogen atom:
  $$Cl\cdot + CH_4 \rightarrow HCl + \cdot CH_3$$
  Because one of the products is a radical, this reaction is a *propagation step*.

  → In a second propagation step, the methyl radical may react with a chlorine molecule:
  $$Cl_2 + \cdot CH_3 \rightarrow CH_3Cl + Cl\cdot$$

  The chlorine atom may take part in the other *propagation step* or attack a CH_3Cl molecule and eventually form CH_2Cl_2. $CHCl_3$ and CCl_4 can also be formed by a continuation of this process.

  → A termination step occurs when two radicals combine to form a nonradical product:
  $$Cl\cdot + \cdot CH_3 \rightarrow CH_3Cl$$

  The substitution reaction is not very clean and the product is usually a mixture of compounds.

  One may limit the production of the more highly substituted alkanes by using a large excess of the alkane.

11B.2 Synthesis of Alkenes and Alkynes

- **Formation of alkenes by elimination reactions**

  → In the petrochemical industry, alkanes are converted to more reactive alkenes by a catalytic process called *dehydrogenation*:
  $$CH_3CH_3(g) \xrightarrow{Cr_2O_3} CH_2{=}CH_2(g) + H_2(g)$$

  This is an example of an *elimination reaction*, one in which two groups on neighboring C atoms are removed from a molecule, leaving a multiple bond (see **Figure 11B.2** in the text).

  → In the laboratory, alkenes are produced by *dehydrohalogenation* of haloalkanes, the removal of a hydrogen atom and a halogen atom from neighboring carbon atoms:
  $$CH_3CH_2Br \xrightarrow{CH_3CH_2O^- \text{ in ethanol at } 70\,°C} CH_2{=}CH_2 + HBr$$

  This elimination reaction is carried out in hot ethanol containing sodium ethoxide, CH_3CH_2ONa.

  → *Dehydrohalogenation* occurs by attack of the ethoxide ion on a hydrogen atom of the methyl group. A H atom is removed as a proton and $CH_3CH_2O–H$ is formed. When the methyl C atom forms a second bond to its neighbor, the Br^- ion departs.

  → States of reactants and products in organic reactions are often not given because the reaction may take place on a catalyst surface or in a nonaqueous solvent, as in the reaction mentioned earlier.

11B.3 Electrophilic Addition

- **Addition reaction**

  → Characteristic reaction of alkenes, in which atoms supplied by the reactant form σ-bonds to the two C atoms joined by the π-bond, which is broken (see **Figure 11B.3** in the text).

  → Almost all addition reactions are exothermic.

  → A *hydrogenation* reaction is the addition of two hydrogen atoms at a double bond:
  $$CH_3CH=CHCH_3 + H_2 \rightarrow CH_3CH_2-CH_2CH_3$$

  → A *halogenation* reaction is the addition of two halogen atoms at a double bond:
  $$CH_3CH=CHCH_3 + Cl_2 \rightarrow CH_3CHCl-CHClCH_3$$

  → A *hydrohalogenation* reaction is the addition of a hydrogen atom and a halogen atom at a double bond:　　$CH_3CH=CHCH_3 + HCl \rightarrow CH_3CH_2-CHClCH_3$

- **Estimating the reaction enthalpy of an addition reaction**

  → Use bond enthalpies for the bonds that are broken and formed. Bond enthalpies strictly apply only to gas-phase reactions.

  → Recall:
  $$\Delta H° \approx \sum_{reactants} n\Delta H_B(\text{bonds broken}) - \sum_{products} n\Delta H_B(\text{bonds formed})$$

- **Mechanism of addition reactions**

  → Double bonds contain a high density of high-energy electrons associated with the π-bond. This region of high electron density is attractive to positively charged reactants.

  → An *electrophile* is a reactant that is attracted to a region of high electron density. It may be a positively charged species or one that has or can acquire a *partial* positive charge during the reaction.

  → An example treated in the text is the bromination of ethene to form dibromoethane:
  $$H_2C=CH_2 + Br_2 \rightarrow CH_2BrCH_2Br$$

  → Bromine molecules are polarizable; a partial positive charge builds up on the bromine atom closest to the double bond. Bromine acts as an *electrophile*. The partial charge becomes a full charge and a bromine cation attaches to the double bond, leaving a Br⁻ ion behind. The cyclic intermediate is called a *bromonium ion*.

  → A Br⁻ ion is attracted by the positive charge of the bromonium ion. It forms a bond to one C atom, and the bromine atom in the cyclic ion forms another bond, giving 1,2-dibromoethane.

- **Hydrogenation reaction**
  - → Hydrogen can be added across a double bond with the help of a solid-state catalyst:

$$H_2(g) + \cdots C = C \cdots \xrightarrow{\text{catalyst}} \cdots CH - CH \cdots$$

  - → This reaction is used in the food industry, for example, to convert liquid oils to solids. Recall that alkenes have lower melting points than alkanes, other things equal.

Topic 11C: AROMATIC COMPOUNDS

11C.1 Nomenclature

- **Definition**
  - → Aromatic hydrocarbons are called *arenes*.

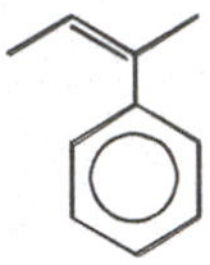

  - → *Benzene* (C_6H_6) is the parent compound. The benzene ring is also called the *phenyl* group, as in 2-phenyl-*trans*-2-butene ($CH_3C(C_6H_5)=CHCH_3$), shown on the right.

  - → Aromatic compounds include those with fused benzene rings, such as naphthalene ($C_{10}H_8$), anthracene ($C_{14}H_{10}$), and phenanthrene ($C_{14}H_{10}$).

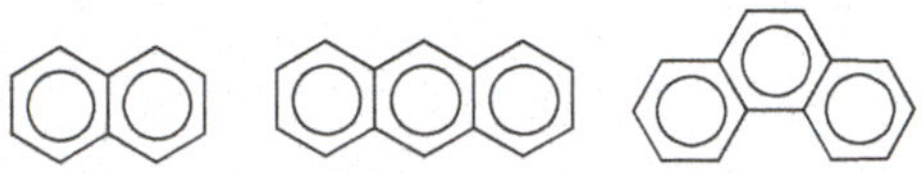

  Naphthalene Anthracene Phenanthrene

- **Benzene ring numbering**
  - → Benzene ring substituents are designated by numbering the C atoms from 1 to 6 around the ring. Compounds are named by counting around the ring in the direction that gives the smallest numbers to the substituents. The prefixes *ortho-*, *meta-*, and *para-* are commonly used to denote substituents at C atoms 2, 3, and 4, respectively, relative to another substituent at C atom 1.

  Example: The three possible dichlorobenzene isomers are shown on the right.

  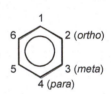

  1,2-Dichlorobenzene
  ortho-Dichlorobenzene

  1,3-Dichlorobenzene
  meta-Dichlorobenzene

  1,4-Dichlorobenzene
  para-Dichlorobenzene

  - → Numbering carbon atoms in fused ring systems is somewhat more complicated and is not covered in the text.

11C.2 Electrophilic Substitution

- **Substitution reactions**

  → *Arenes* have delocalized π-electrons; but unlike alkenes, arenes undergo predominantly *substitution* reactions, with the π-bonds of the ring unaffected.

  Example: The reaction of benzene (C_6H_6) with chlorine produces chlorobenzene when one chlorine atom substitutes for a hydrogen atom: $C_6H_6 + Cl_2 \xrightarrow{\ \text{Fe}\ } C_6H_5Cl + HCl$

- **Electrophilic substitution**

  → If the mechanism of substitution of a benzene ring involves electrophilic attack, the reaction is called *electrophilic substitution*.

- **Halogenation**

  → In the halogenation of benzene, an iron catalyst is used. The iron is converted to iron(III) halide (FeX_3). Iron(III) halide reacts further to polarize the X_2 molecule.

  → The commonly accepted mechanism for aromatic halogenation, X_2, is then

  $$FeX_3 + X\text{--}X \rightleftharpoons X_3Fe^{\delta-}\cdots X\text{--}X^{\delta+} \qquad\qquad \textit{fast equilibrium}$$

  $$X_3Fe^{\delta-}\cdots X\text{--}X^{\delta+} + C_6H_6 \rightarrow FeX_4^- + C_6H_6X^+ \qquad \textit{slow}$$

  $$C_6H_6X^+ + FeX_4^- \rightarrow C_6H_5X + HX + FeX_3 \qquad \textit{fast}$$

  → In the last step, the hydrogen atom is easily removed from the ring, for in that way the π-electron delocalization and stabilization, lost in the slow step, are regained.

- **Nitration**

  → A mixture of HNO_3 and concentrated H_2SO_4 slowly converts benzene into nitrobenzene.

  → The nitronium ion, NO_2^+, is the electrophile and the nitrating agent.

  → The commonly accepted mechanism for aromatic nitration is:

  $$HNO_3 + 2\,H_2SO_4 \rightleftharpoons NO_2^+ + H_3O^+ + 2\,HSO_4^-$$

  $$NO_2^+ + C_6H_6 \rightarrow C_6H_6NO_2^+ \qquad\qquad\qquad \textit{slow}$$

  $$C_6H_6NO_2^+ + HSO_4^- \rightarrow C_6H_5NO_2 + H_2SO_4 \quad \textit{fast}$$

  → In the last step, the hydrogen atom is removed from the ring, restoring aromaticity to it.

- ***Ortho-* and *para*-directing activators**

  → Electrons are donated into the delocalized molecular orbitals of the ring (*resonance effect*).

  → The reaction rate is *much* faster than in unsubstituted benzene, and the substituent is called an *activator*.

  → Products of substitution reactions favor the *ortho* and *para* positions of the ring because these locations have more electron density than the *meta* positions.

  → Examples of *ortho-* and *para*-directing activators include –OH, –NH$_2$, and substituted amines.

  → The atom bonded to the benzene ring has a *nonbonding* pair of electrons.

Note: Alkyl groups have no nonbonding electron pair, yet they are also *ortho*- and *para*-directing activators. The mechanism in this case is a different one (see any organic chemistry text).

- *Ortho*- and *para*-**directing deactivators**
  - → Electrons are donated into the delocalized molecular orbitals of the ring (*resonance effect*), but the substituent is very electronegative.
  - → The reaction rate is *slightly* slower than in unsubstituted benzene, and the substituent is called a *deactivator*. Deactivation occurs when the substituent is *highly* electronegative and withdraws some electron density from the ring.
  - → Products of a substitution reaction still favor the *ortho* and *para* positions of the ring because these positions have *relatively* more electron density than the *meta* positions.
  - → The *only* examples are –F, –Cl, –Br, and –I.
  - → The atom bonded to the benzene ring has a *nonbonding* pair of electrons.

- **Meta-directing deactivators**
  - → Highly electronegative substances that can withdraw electrons partially and/or a substituent that *removes* electrons by resonance
  - → The reaction rate is *much* slower than in unsubstituted benzene, and the substituent is called a *deactivator*.
  - → Products of a substitution reaction favor the *meta* position because the *ortho* and *para* positions have greatly decreased electron density.
  - → Examples include –COOH, $-NO_2$, $-CF_3$, and $-C\equiv N$.
  - → *All* electron pairs on the atom bonded to the benzene ring are *bonding* pairs.

Topic 11D: COMMON FUNCTIONAL GROUPS

11D.1 Haloalkanes

- → Alkanes in which a *halogen atom*, X, replaces one or more H atoms (X = F, Cl, Br, or I)
- → Also called alkyl halides
- → Insoluble in water
- → Some are highly toxic and environmentally unfriendly.

  Example: The chlorofluorocarbon (CFC) 1,2-dichloro-1-fluoroethane, $HFClC–CClH_2$, is partly responsible for depletion of the ozone layer.

- → The C–X bonds in haloalkanes are *polar*; C carries a partial positive charge and the halogen, X, a partial negative one.
- → C–X bond polarity governs much of the chemistry of haloalkanes.
- → Haloalkanes are susceptible to *nucleophilic substitution*, in which a reactant that seeks out centers of positive charge in a molecule (a *nucleophile*) replaces a halogen atom.

→ *Nucleophiles* include anions such as OH^- and Lewis bases with lone pairs such as NH_3.
Example: Hydroxide acts as a nucleophile in the following *hydrolysis* reaction:

$$CH_3Br + OH^- \rightarrow CH_3OH + Br^-$$

11D.2 Alcohols

- **Definitions**

  → *Hydroxyl group:* An –OH group covalently bonded to a C atom

  → *Alcohol:* An organic compound that contains a *hydroxyl* group not directly bonded to an aromatic ring or to a carbonyl group (shown on the right)

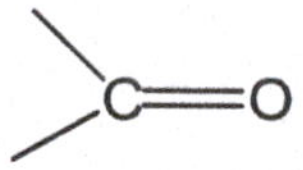

  → Alcohols are named by adding the suffix *–ol* to the stem of the parent hydrocarbon.
  Examples: Methanol for CH_3OH; ethanol for CH_3CH_2OH

  → The number of the C atom attached to it can aid in locating the –OH group.
  Examples: 1-propanol for $CH_3CH_2CH_2OH$; 2-propanol for $CH_3CH(OH)CH_3$

  → A *diol* is an organic compound with two hydroxyl groups.
  Example: 1,2-ethanediol for $HOCH_2CH_2OH$ (common name is ethylene glycol)

- **Classes of alcohols**

  → *Primary alcohol:* RCH_2–OH, where R can be any group
  Examples: Methanol and 1-propanol

  → *Secondary alcohol:* R_2CH–OH, where the R groups can be the same or different
  Examples: 2-propanol, $(CH_3)_2CH$–OH (same R groups); 2-butanol, $CH_3CH_2CH(-OH)CH_3$ (different R groups)

  → *Tertiary alcohol:* R_3C–OH, where the R groups can be the same or different
  Examples: 2-methyl-2-propanol, $(CH_3)_3C$–OH (*same* R groups, CH_3–); also called tertiary butyl alcohol or *t*-butyl alcohol for short.
  3-methyl-3-pentanol, $CH_3CH_2C(CH_3)(-OH)CH_2CH_3$ (*different* R groups, CH_3CH_2– and CH_3–): the structure is shown on the right: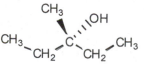

- **Properties of alcohols**

  → Alcohols are polar molecules that can lose the –OH proton in certain solvents, but typically not in water.

  → Alcohols have relatively high boiling points and low volatility because of hydrogen bond formation via the –OH group. In this way, they are similar to water.

  → Compare the normal boiling point of ethanol (78.2 °C), with a molar mass of 46.07 g·mol^{-1}, to that of pentane (36.0 °C), with a molar mass of 72.14 g·mol^{-1}.

  → Boiling points are consistent with hydrogen bonding in ethanol and only London interactions in pentane.

  → Methanol is prepared from synthesis gas, a mixture of carbon monoxide and hydrogen:

$$CO(g) + 2\,H_2(g) \xrightarrow{\text{catalyst, 250 °C, 50-100 atm}} CH_3OH(g)$$

  → Catalyst is a mixture of Cu, ZnO, and Cr_2O_3.

→ Ethanol is prepared by carbohydrate fermentation and by hydration of ethene (addition reaction):

$$CH_2=CH_2(g) + H_2O(g) \xrightarrow{\text{catalyst, 300 °C}} CH_3CH_2OH(g)$$

→ Catalyst is phosphoric acid.

11D.3 Ethers

→ Organic compounds of the form R–O–R, where R is any alkyl group and the two R groups may be the same or different

→ More volatile than alcohols with the same molar mass because ethers do not form hydrogen bonds with each other

→ Act, however, as hydrogen-bond acceptors by using the lone pairs of electrons on the O atom

→ Useful solvents for other organic compounds because they have low polarity and low reactivity

→ Quite flammable and must be handled with care

Examples: Diethyl ether (CH_3CH_2–O–CH_2CH_3);
1-butyl methyl ether ($CH_3CH_2CH_2CH_2$–O–CH_3)

- **Crown ethers**

→ Cyclic polyethers of formula $+CH_2CH_2-O+_n$

→ Name reflects the crownlike shape of the molecules.

→ Bind strongly to alkali metal ions such as Na^+ and K^+, allowing inorganic salts to be dissolved in organic solvents

Example: The common oxidizing agent potassium permanganate ($KMnO_4$) is insoluble in nonpolar solvents such as benzene (C_6H_6). In the presence of [18]-crown-6, it dissolves in benzene according to the reaction on the right. The name indicates an 18-member ring with 6 ether oxygen atoms.

[18]-Crown-6

11D.4 Phenols

→ Organic compounds with a *hydroxyl group* attached *directly* to an aromatic ring

→ The unsubstituted compound, *phenol*, is a white, crystalline molecular solid.

Phenol, C_6H_5OH: or

Melting point: 40.9 °C

→ Substituted phenols occur naturally, and some are responsible for the fragrances of plants.

Examples: *Thymol* is the active ingredient in oil of thyme, and *eugenol* provides the scent and flavor in oil of cloves.

Thymol

Eugenol

- **Acid–base properties of phenols**
  - → Phenols are generally *weak* acids, in contrast to alcohols, which typically are *not* acidic.
  - → The acidity of phenols can be understood on the basis of resonance stabilization (delocalization) of the negative charge of the conjugate base of phenol ($C_6H_5O^-$).

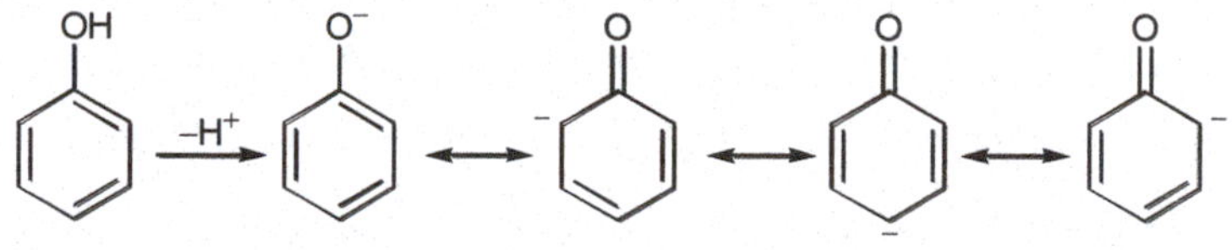

  - → Because of resonance stabilization, the phenoxide ($C_6H_5O^-$) anion is a weaker conjugate base than the corresponding conjugate bases of typical alcohols.

    Example: The ethoxide ($CH_3CH_2O^-$) anion is a stronger base than the phenoxide anion.

11D.5 Aldehydes and Ketones

- **Aldehydes** 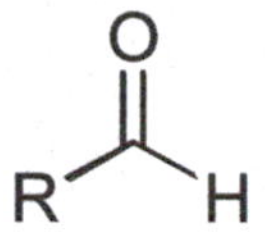
  - → Organic compounds of the form shown on the right, where R is a H atom, an aliphatic group, or an aromatic group
  - → The group characteristic of aldehydes is written as –CHO, as in formaldehyde (HCHO), the first member of the family in which R = H.
  - → Named systematically by replacing the ending *–e* by *–al*, but their many common names include formaldehyde for methanal and acetaldehyde for ethanal
  - → The C atom on the carbonyl group is *included* in the count of C atoms when determining the alkane from which the aldehyde is derived.
  - → A few common aldehydes:

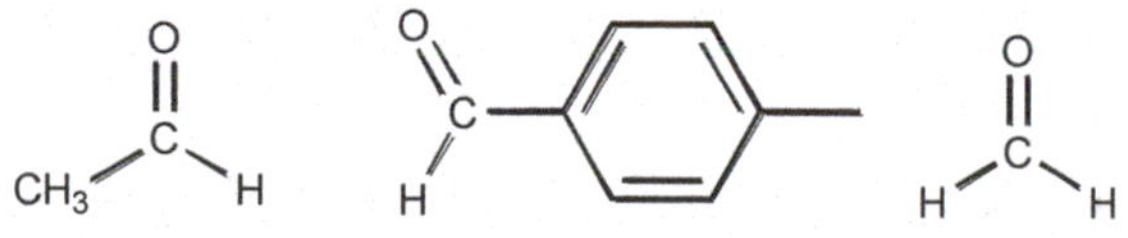

    Ethanal   Benzaldehyde   Methanal

  - → Aldehydes generally contribute to the flavor of fruits and nuts and to the odors of plants.
    Example: Benzaldehyde provides part of the aroma of almonds and cherries.
  - → Aldehydes can be prepared by the *mild* oxidation of *primary* alcohols.
    Example: Formaldehyde is prepared industrially by oxidizing methanol with a Ag catalyst. The overall reaction is:

$$2\,CH_3OH(g) + O_2(g) \xrightarrow{600\,°C,\ Ag} 2\,HCHO(g) + 2\,H_2O(g)$$

- **Ketones**

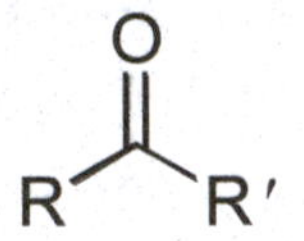

    → Organic compounds of the form shown on the right; the R groups (alkyl or aryl) may be the same or different.

    → The *carbonyl group*, –C=O, characteristic of ketones, is written –CO, as in propanone, CH_3COCH_3 (acetone), the first member of the family.

    → Named systematically by replacing the ending *-e* by *-one*, but many common names include acetone or dimethyl ketone for propanone and methyl ethyl ketone for butanone.

    → The C atom on the carbonyl group is *included* in the count of C atoms when determining the alkane from which the ketone is derived.

    → A number is used to locate the C atom in the carbonyl group to avoid ambiguity.

    → A few common ketones:

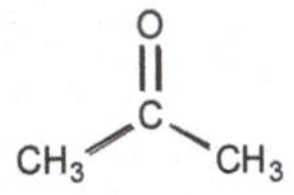

    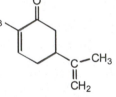

    Propanone 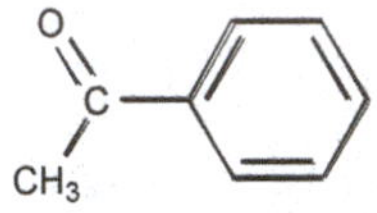   Benzophenone 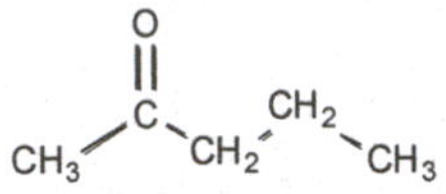   2-Pentanone

    → *Ketones* contribute to flavors and fragrances.

    Example: *Carvone*, shown on the right, is the essential oil in spearmint.

    → Ketones can be prepared by the oxidation of *secondary* alcohols.

    → There is less risk of further oxidation than with aldehydes, so stronger oxidizing agents are used. Dichromate oxidation of secondary alcohols produces ketones with little excess oxidation and in good yields.

    Example: The oxidation of 2-propanol to propanone is shown schematically on the right:

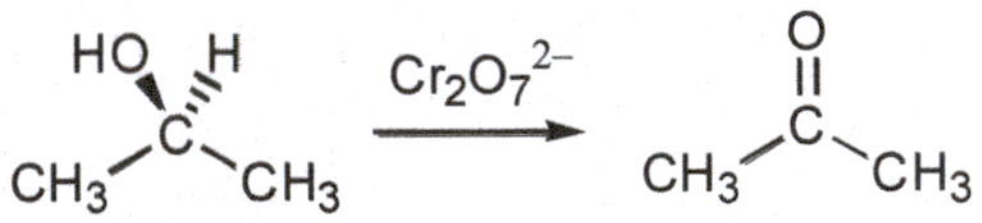

11D.6 Carboxylic Acids

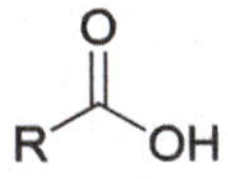

    → Organic compounds of the form shown on the right; R is an H atom, an alkyl group, or an aryl group.

    → The *carboxyl group* characteristic of carboxylic acids is written –COOH, as in formic acid (HCOOH), the first member of the family.

    → The *carboxyl group* contains a *hydroxyl group*, –OH, attached to a *carbonyl group*, –C=O.

    → Named systematically by replacing the ending *-e* by *-oic acid*, but many common names are used, such as formic acid for methanoic acid and acetic acid for ethanoic acid

    → The carbonyl C atom is included in the C atom count to determine the parent hydrocarbon molecule.

→ Some common carboxylic acids:

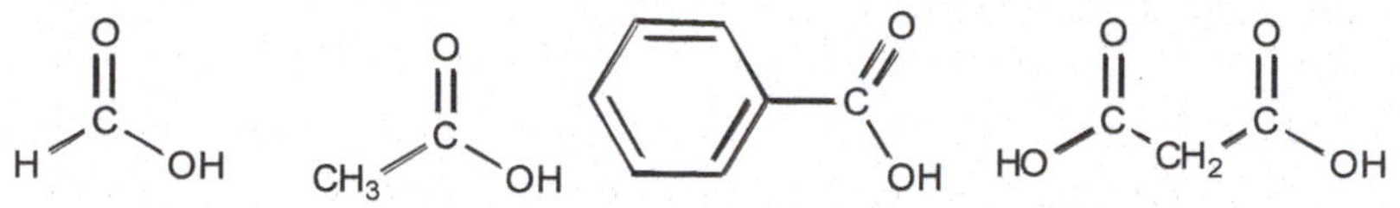

Methanoic acid   Ethanoic acid   Benzoic acid   Malonic acid

Note: Malonic acid contains two –COOH groups; it is called a *diacid*.

→ Carboxylic acids contain hydroxyl groups, –OH, and the carbonyl group –C=O, which can take part in hydrogen bonding.

→ Carboxylic acids can be prepared by oxidation of aldehydes or of *primary* alcohols in an acidified solution containing a strong oxidizing agent such as $KMnO_4$, or $Na_2Cr_2O_7$:

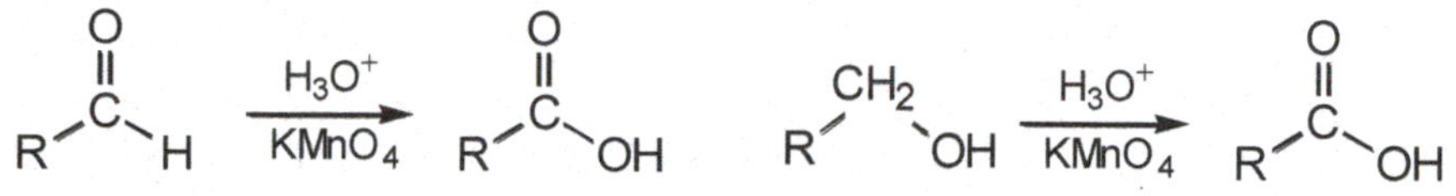

Aldehyde   Primary alcohol

→ In some cases, alkyl groups can be oxidized *directly* to carboxyl groups.

→ An industrial example is the oxidation of the methyl groups on *p*-xylene by a cobalt(III) catalyst to form *terephthalic acid*, which is used in the production of artificial fibers.

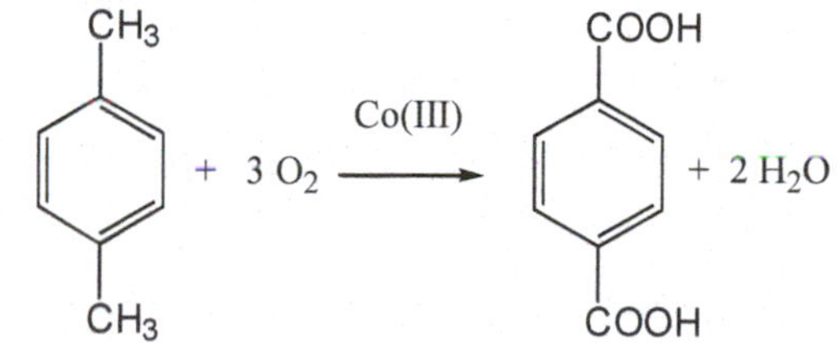

p-Xylene   Terephthalic acid

11D.7 Esters

→ An *ester* is the product of a *condensation* reaction between a carboxylic acid and an alcohol. A water molecule is also produced.

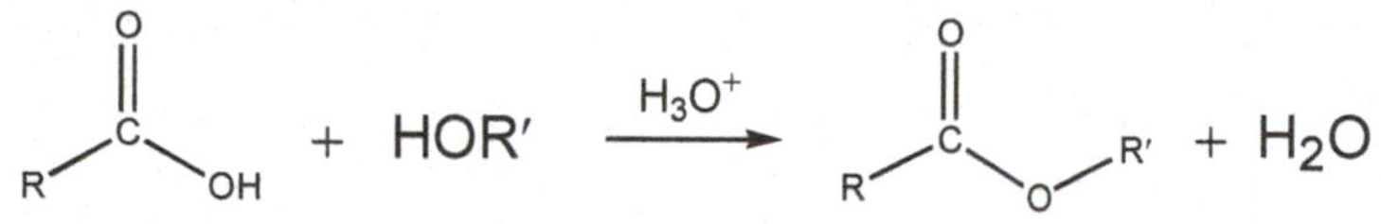

Carboxylic acid   Alcohol   Ester

→ *Esterification* reactions can be acid-catalyzed as shown here.

→ *Condensation* reaction: one in which two molecules combine to form a large one and a small molecule is eliminated

→ Many *esters* have fragrant aromas and contribute to flavors of fruits.

Examples: The esters *n*-amyl acetate and *n*-octyl acetate are responsible for the aromas of bananas and oranges, respectively.

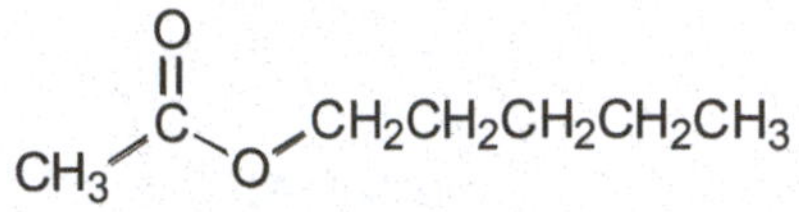

n-Amyl acetate 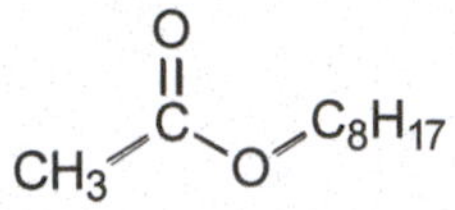   n-Octyl acetate

→ Naturally occurring esters also include fats and oils such as tristearin, a component of beef fat.

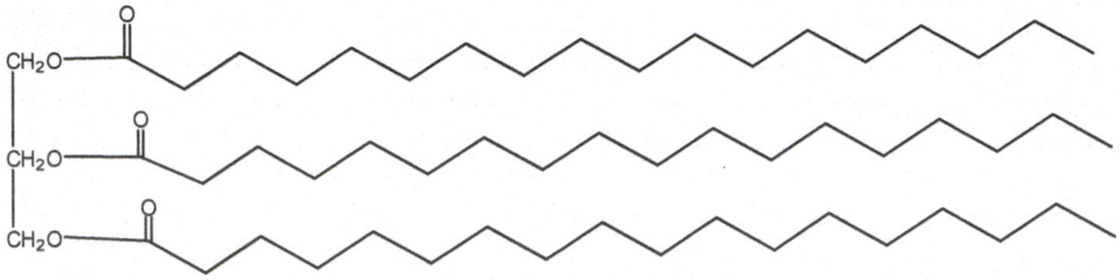

Tristearin, $C_{57}H_{110}O_6$

11D.8 Amines, Amino Acids, and Amides

- **Amines**

  → Derivatives of NH_3, formed by replacing one or more H atoms with organic groups, R

  → The *amino group* ($-NH_2$) is the functional group of *amines*.

  → Named by specifying the groups attached to the nitrogen atom (N) alphabetically, followed by the suffix *amine*

  → Designated as *primary*, *secondary*, or *tertiary*, depending on the number of R groups attached to the N atom

  → Ammonia and several representative amines are shown on the right:

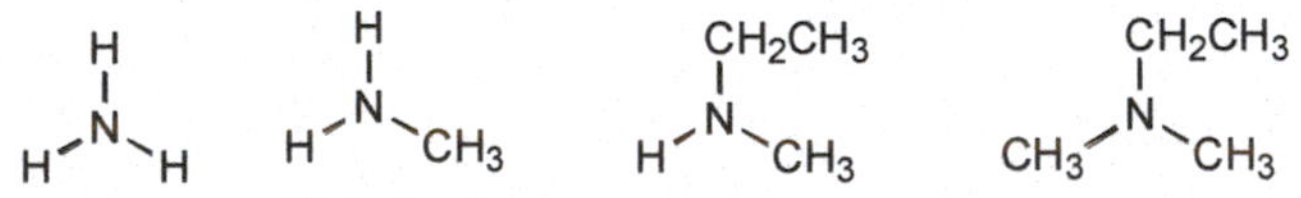

  Ammonia   Methylamine   Ethylmethylamine   Dimethylethylamine
               primary   secondary   tertiary

- **Properties**

  → Characterized by four sp^3 hybrid orbitals on the N atom; three participate in three single bonds and the fourth has a lone pair of electrons.

  → *Amines* are widespread in nature.

  → They often have disagreeable odors.
  Example: *Putrescine*, $NH_2(CH_2)_4NH_2$

  → Similar to ammonia itself, amines are weak bases.

- **Quaternary ammonium ions**

  → Tetrahedral ions of formula R_4N^+

  → Four R groups, which may be the same or different, are bonded to the central nitrogen atom.

  → Negligible acid or base properties and little effect on pH

  → Isolated as quaternary ammonium salts

Example: Dimethylethylpropylammonium chloride, with four R groups, is a quaternary ammonium salt.

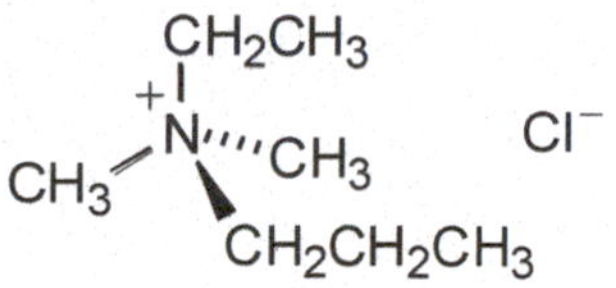

$$CH_3-\overset{\overset{\displaystyle CH_2CH_3}{\overset{+}{|}}}{\underset{\underset{\displaystyle CH_2CH_2CH_3}{}}{N^{\prime\prime\prime}}}\!CH_3 \qquad Cl^-$$

- **Amino acids**
  - → Carboxylic acids with an *amino* group and the *carboxyl* group separated by *one* C atom
  - → Technically, these are α-amino acids. Amino acids with the *amino* and the *carboxyl* groups separated by *two* C atoms are β-amino acids.
  - → A more precise name for an amino acid is an *aminocarboxylic acid*.
  - → Have the general formula $R(NH_2)CH(COOH)$

    Examples: *Glycine, methionine,* and *phenylalanine*

  Note: In amino acids, the central C atom is *stereogenic* and the molecule is *chiral*, except for *glycine*.

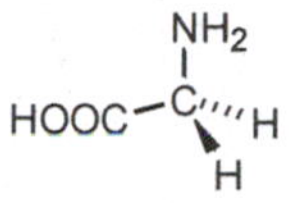

  Glycine 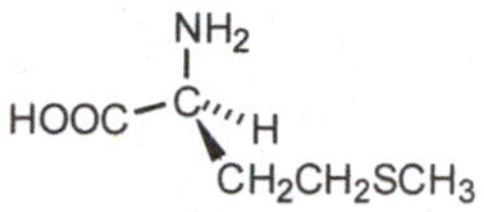   Methionine 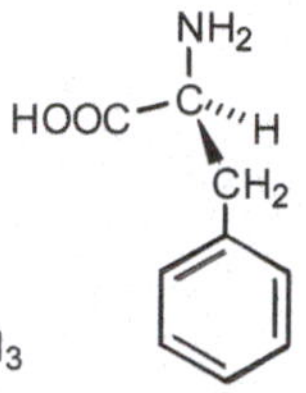   Phenylalanine

- **Properties**
  - → Building blocks of proteins (see **Topic 11E.5** in the text)
  - → Essential for human health
  - → Can form *hydrogen bonds* with *both* the *amino* and *carboxyl* groups
  - → Can form *double ions* (*zwitterions*) by transferring a proton from the *carboxyl* to the *amino* group as shown on the right:

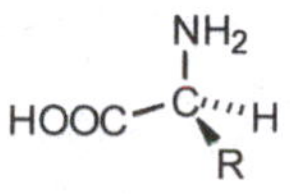

    Amino acid 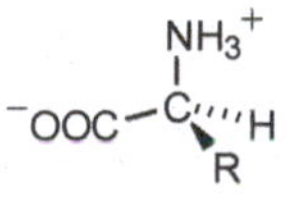   Double ion form

  - → In aqueous solution, amino acids exist in four forms (shown on the right) whose concentrations are pH dependent.

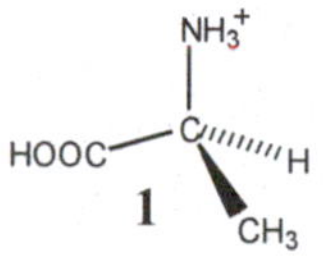

    1 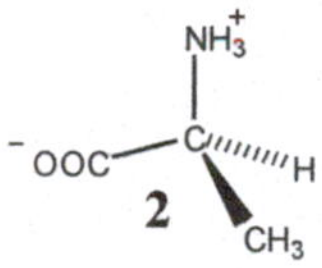   2 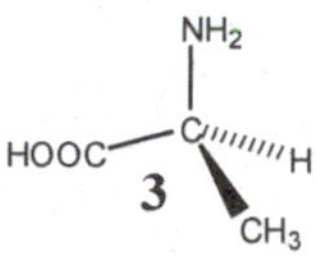   3 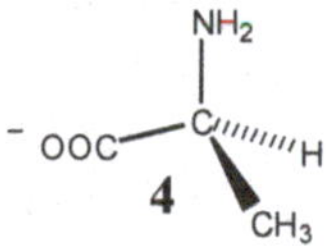   4

  - → The concentrations of these species can be determined as with polyprotic acids (see **Topic 6E** in the text).
  - → Form 1 predominates at low pH, while form 4 predominates at high pH. At neutral pH, form 2, the zwitterion, predominates.
  - → The neutral form 3 exists only in trace amounts in aqueous solutions (see **Figure 11D.5** in the text).

- **Amides**
  - → Molecules with this general formula:

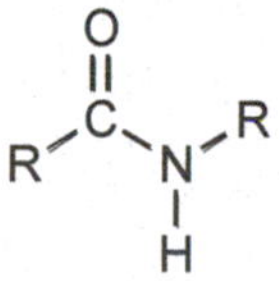

  - → Formed by the condensation reaction of carboxylic acids with amines, eliminating water

→ Near room temperature, the reaction mixture of an amine and a carboxylic acid forms an ammonium salt that reacts further to form the amide at higher temperature.

→ The general reaction:

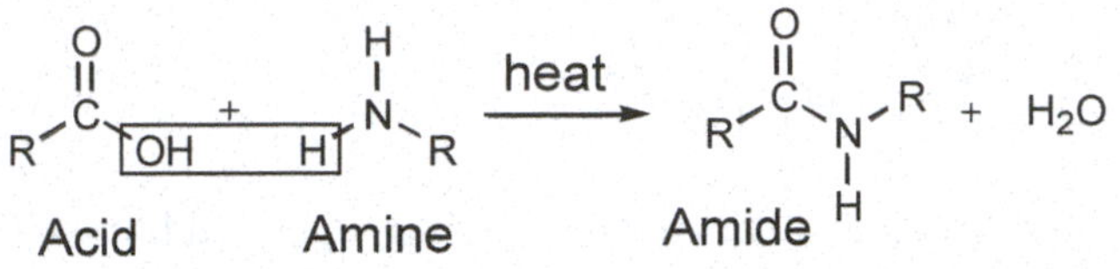

Acid   Amine   Amide

Note: This amide has an N–H group that may participate in *intermolecular* hydrogen bonding. The atoms that form the product water molecule are shown in the box.

Topic 11E: POLYMERS AND BIOLOGICAL MACROMOLECULES

11E.1 Addition Polymerization

- **Addition polymers** (see **Table 11E.1** in the text for common examples)

  → Form when *alkene* monomers react with themselves with no net loss of atoms to form *polymers*

  → Prepared by *radical polymerization,* a radical chain reaction

  → Radical polymerization reactions are started by using an *initiator* such as an organic peroxide, R–O–O–R, which when heated decomposes to form two free radicals.

  → A radical polymerization mechanism is shown for ROOR and CH_2=CHX:

  Initiation: R—O—O—R $\xrightarrow{\text{heat}}$ R—O• + •O—R

  Propagation:

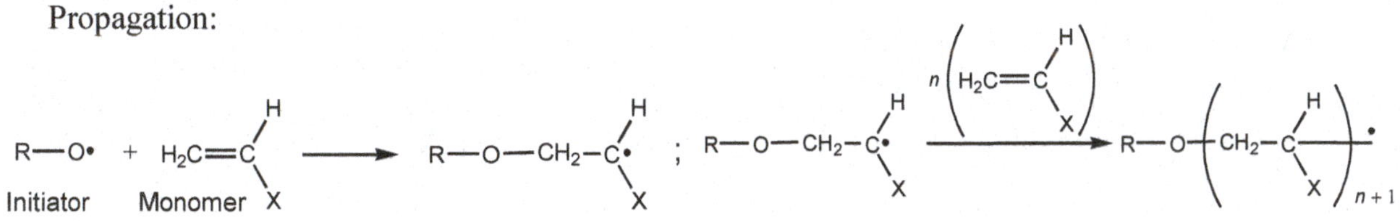

  Termination:

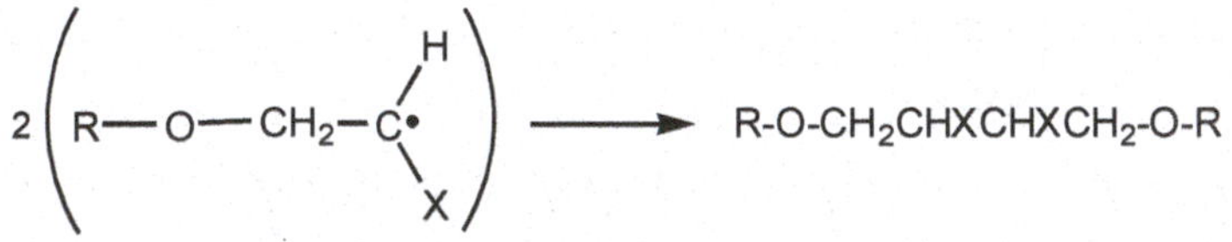

  → Radical polymerization reactions include *initiation* and *propagation* steps as shown here.

  → The reaction terminates when all the monomer is consumed or when two radical chains of any length react to form a single diamagnetic (nonradical) species.

  → One possible *termination* step is shown here.

- **Stereoregular polymer**
  - → Each unit, or pair of repeating units in the polymer, has the same relative orientation.
  - → These polymers pack together well and are relatively strong, dense, and impact-resistant.
- **Isotactic polymer**
  - → A *stereoregular* polymer with all substituents on the same side of the extended carbon chain

    Example: A portion of an isotactic polypropylene chain:

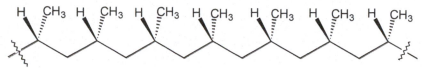

Isotactic configuration - same side

- **Syndiotactic polymer**
  - → A *stereoregular* polymer with substituents alternating regularly on either side of the extended carbon chain

    Example: A portion of a syndiotactic polypropylene chain with alternating methyl groups:

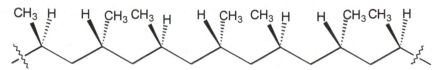

Syndiotactic configuration - alternating sides

- **Atactic polymer**
  - → A polymer with substituents randomly oriented with respect to the extended carbon chain
  - → *Atactic* polymers are *not* stereoregular. They tend to be amorphous and not well suited for many applications.

    Example: An *atactic* polypropylene chain with randomly oriented methyl side groups:

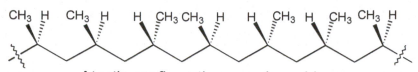

Atactic configuration - random sides

- **Ziegler–Natta catalysts**
  - → Consist of aluminum- and titanium-containing compounds, such as titanium tetrachloride, $TiCl_4$, and triethyl aluminum, $(CH_3CH_2)_3Al$
  - → Used in the production of *stereoregular* polymers, including synthetic rubber
  - → Chemists were unable to synthesize rubber with useful mechanical properties until the discovery of Ziegler–Natta catalysts in 1953.

11E.2 Condensation Polymerization

- **Condensation polymers**
  - → Are formed
    - ⇒ By a series of *condensation* reactions
    - ⇒ From reactants, each of which has *two* functional groups
    - ⇒ From stoichiometric amounts of the reactants
  - → Encompass the classes of polymers known as *polyesters* and *polyamides*
  - → Are characterized by reactions proceeding without the need for an initiator
  - → Typically have shorter chain lengths than addition polymers because each monomer can initiate the reaction

- **Polyesters**
  - → Form from the condensation of a *diacid* and a *diol*
  - → For polyester polymerization, it is necessary to have two functional groups on each monomer and to mix stoichiometric amounts of the reactants.

    Example: Kodel polyester is formed from the esterification of terephthalic acid and 1,4-bis(hydroxymethyl)-cyclohexane, as shown by these reactions:

- **Polyamides**
  - → Are formed from the condensation of a *diacid* with a *diamine*
  - → For polyamide polymerization, it is necessary to have two functional groups on each monomer and to mix stoichiometric amounts of the reactants.

Example: Nylon-66 forms by condensation of 1,6-hexanedioic acid and 1,6-hexanediamine. The reactants first form a salt, which then condenses to the polyamide.

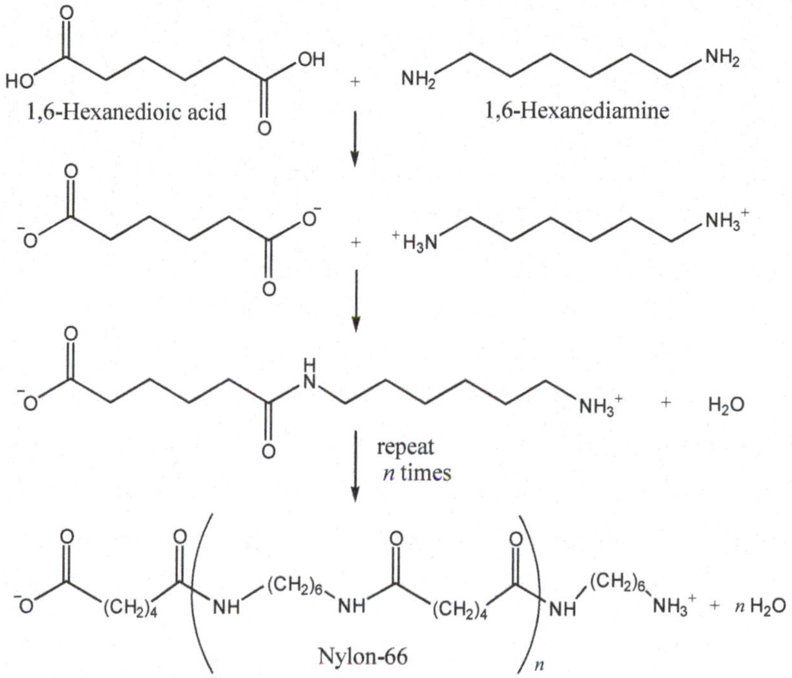

1,6-Hexanedioic acid + 1,6-Hexanediamine

repeat *n* times

Nylon-66

→ N-H···O=C hydrogen bonding between chains strengthens nylons and helps them absorb moisture.

11E.3 Copolymers and Composite Materials

→ Polymers with more than one type of repeating unit

→ Produced from more than one type of monomer

→ Four forms: *alternating, block, random,* and *graft* copolymers (see **Figure 11E.9** in the text)

- **Alternating copolymers**

  → Follow the pattern –A–B–A–B–A–B–A–B–, where A and B are monomer units

  → Nylon-66 (discussed previously) is an *alternating* copolymer. The index (66) indicates the number of C atoms (6) in each type of monomer.

- **Block copolymers**

  → Follow the pattern –A–A–A–A–B–B–B–B–A–A–A–

  → Long segments of one monomer, A, are followed by long segments of monomer B.

  Example: *High-impact polystyrene* is a block copolymer of *styrene* and *butadiene*:

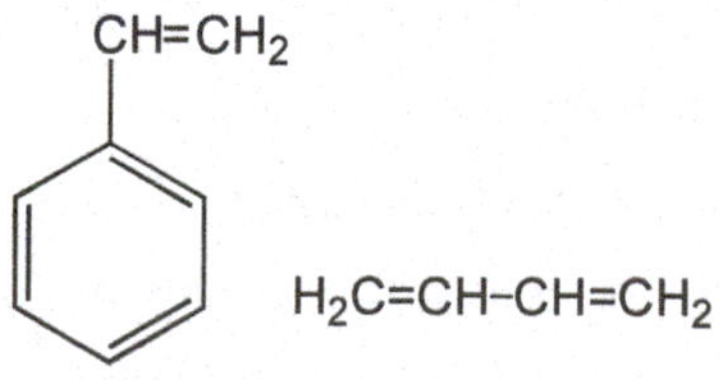

Styrene   1,3-Butadiene

- **Random copolymers**
  - → Follow no particular pattern

    –A–A–B–A–B–A–B–B–A–B–A–B–B–B–A–B–A–A–B–B–B–

    Example: Radical polymerization of the monomers styrene and 3-methylstyrene is expected to form a random copolymer:

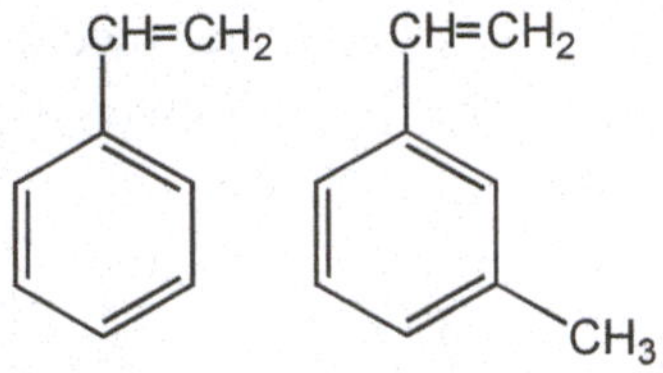

    CH=CH₂   CH=CH₂

    Styrene   3-Methylstyrene

- **Graft copolymers**
  - → Consist of long chains of one monomer, A, with pendant chains of the second monomer, B:

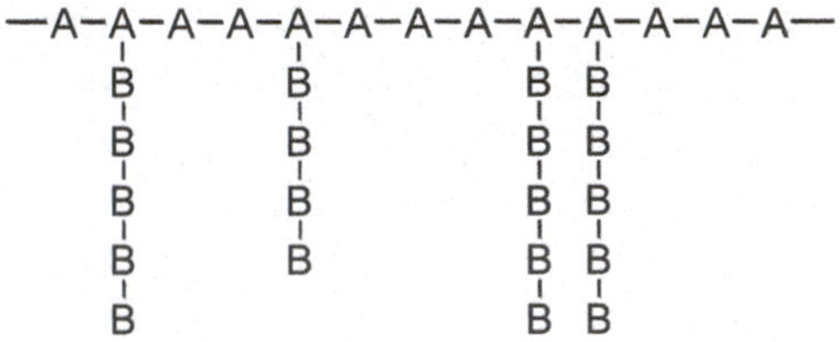

    Example: Soft contact lenses are composed of a graft copolymer that has a backbone of nonpolar monomers but side groups of different water-absorbing monomers.

    —A–A–A–A–A–A–A–A–A–A–A–A–A–A—

    Graft copolymer

- **Properties of copolymers**
  - → The different types of copolymers extend the range of physical properties obtainable for materials.

    Example: Soft polyurethane foams formed from *diisocyanates* and *glycols* are used for insulation and for furniture stuffing. The formation of a *polyurethane*:

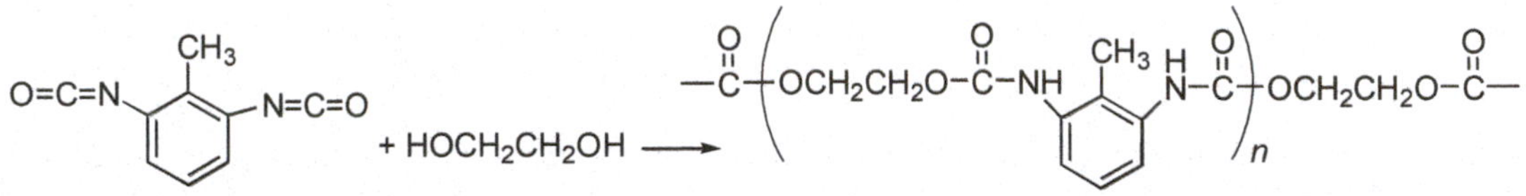

    O=C=N   N=C=O

    CH₃

    Toluene-2,6-diisocyanate   Ethylene glycol

    + HOCH₂CH₂OH ⟶

    A polyurethane

- **Composite materials**
  - → Consist of two or more materials solidified together
  - → Combine the advantages of the component materials and exhibit superior properties
  - → Differ from copolymers because their components retain their identities in what can be large separate regions.

    Example: Fiberglass, a material of great strength and flexibility, consists of inorganic materials in a polymer matrix.

11E.4 Physical Properties of Polymers

- → Polymers can be designed to have specific properties for a specific application.

- **Synthetic polymers**
  - → Have no definite molar mass, only an *average* value
  - → Tend to soften gradually upon heating and have no definite melting point
  - → Contain chains of various lengths mixed together
  - → Longer *average* chain length leads to higher viscosity and a higher softening point.

- **Properties of polymers**
  - → Depend on the *average* chain length
  - → Depend on the *polarity* of the functional groups

    Example: Nylons with polar side groups capable of hydrogen bonding are strong and absorb water; polypropylene, with H and CH_3 side groups, is virtually impervious to water.

  - → Polar groups are associated with stronger intermolecular forces and tend to increase both softening points and mechanical strength. Longer chain length also increases mechanical strength.
  - → Depend on the manner in which chains pack
  - → Long unbranched chains form crystalline regions that lead to strong, dense materials.
  - → Branched-chain polymers exhibit more tangled arrangements; they are less likely to form crystalline regions and tend to be weaker, less dense materials.

- **Elasticity**
  - → Ability of a polymer to return to its original shape after being stretched
  - → *Elastomers* are materials that return easily to their original shape after stretching.
  - → *Elasticity* of natural rubber is improved by vulcanization (heating with S), which forms disulfide (−S−S−) links between chains, increasing the resilience of the polymer (see **Figure 11E.14** in the text).
  - → Extensive cross-linking provides a rigid network of interlinked polymer chains and leads to very hard materials.
  - → Most polymers are electrical insulators, but conducting polymers are known (see **Box 11E.1** in the text).

- **Thermoplastic polymer**
  - → Can be softened after it is molded
  - → Can be recycled by melting and reprocessing
  - → Is often made of addition polymers

    Example: Polyethylene (for recycling codes, see **Table 11E.2** in the text)

- **Thermosetting polymer**
  - → Takes a permanent shape when molded; does not soften when heated.

    Examples: Vulcanized rubber in car tires and urea formaldehyde foam

- **Silicones**
  - → Synthetic polymers based on silicone
  - → Contain −O−Si−O−Si−O− chains with pendant organic groups, such as CH_3 (**Figure 11E.15** in the text).
  - → Used as waterproofing materials, implants, drug delivery, and cosmetics.

11E.5 Proteins

→ Condensation copolymers of *up to* 20 naturally occurring amino acids (see **Table 11E.3** in the text)

→ Nine of the 20 amino acids, known as *essential amino acids,* cannot be produced by the human body and must be ingested.

→ Perform highly specific functions in the human body

→ Enzymes (globular proteins) act as specific and efficient catalysts.

  Example: The enzyme *alcohol dehydrogenase*, a globular protein, oxidizes ethanol to ethanal.

- **Peptides**

  → Molecules formed from two or more amino acids

  → Named starting with the amino acid on the left (N-terminus)

  → The –CO–NH– link is called a *peptide bond*, and each amino acid in a peptide is called a *residue*.

  → Typical proteins contain *polypeptide chains* of more than a hundred residues joined through peptide bonds and arranged in a particular order.

  → *Oligopeptides* are molecules with only a few amino acid residues.

  Example: Reaction of the naturally occurring amino acids aspartic acid (Asp) and phenylalanine (Phe) produces the artificial sweetener aspartame (Asp-Phe), a *dipeptide* that contains two *residues*. Notice the peptide link –CO–NH– shown in the box.

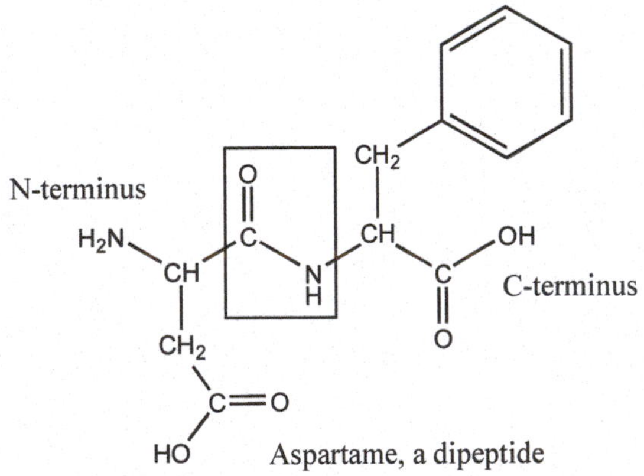

Aspartame, a dipeptide

- **Structure of proteins**

  → *Primary structure*

  ⇒ Sequence of residues in the peptide chain

  Example: The *primary* structure of *aspartame* is Asp-Phe. The *dipeptide* with the N- and C-termini interchanged is Phe-Asp, a *different* oligopeptide.

  → *Secondary structure*

  ⇒ Describes the shape of the polypeptide chain

  ⇒ Controlled by *intramolecular* interactions between different parts of the peptide chain

⇒ Common *secondary* structures include the α *helix* and the β *sheet* (see **Figures 11E.17** and **11E.18**, respectively, in the text).

⇒ The α-*helix* is a helical portion of a polypeptide held together by hydrogen bonding.

⇒ The β-*pleated sheet* is a portion of a polypeptide in which the segments of the chain lie side by side to form nearly flat sheets linked by hydrogen bonds.

→ *Tertiary structure*

⇒ Describes the overall three-dimensional shape of the polypeptide, including the way in which α-helix and β-sheet regions fold together to shape the macromolecule.

⇒ Folding is a consequence of the hydrophobic and hydrophilic interactions between residues lying in different parts of the primary structure. One important link responsible for *tertiary* structure is the *disulfide link* (–S–S–) between amino acids containing sulfur.

⇒ Proteins are classified as *globular* (ball-shaped) or *fibrous* (hair-shaped, with long chains of polypeptides that occur in bundles).

⇒ Globular proteins are soluble in water; fibrous proteins are not. Essentially, all enzymes are globular proteins.

Example: *Hemoglobin* (responsible for oxygen transport in blood, see **Figure 11E.18** in the text) and *cytochrome c* (a component of electron - transport chains in mitochondria and bacteria) are globular proteins; *fibroin* (the protein of silk) is a fibrous protein.

⇒ Spider silk is one of the strongest known fibers. Artificial spider silk is used to make lightweight, thin, and flexible body armor.

→ *Quaternary structure*

⇒ Describes the arrangement of *subunits* in proteins with more than one polypeptide chain

⇒ Not all proteins contain more than one subunit, so not all have a *quaternary* structure.

Example: *Hemoglobin* (see **Figure 11E.18** in the text) contains four polypeptide units and has a *quaternary* structure.

- **Denaturation**

  → Loss of structure of proteins is called denaturation.

  → Occurs when a protein loses quaternary, tertiary, or secondary structure (disruption of *noncovalent* interactions). Denaturation may also include primary structure degradation and the cleavage of peptide bonds.

  → *Denaturation* is caused by heating and other means, is often irreversible, and is accompanied by loss of function of the protein.

11E.6 Carbohydrates

→ Most abundant class of naturally occurring organic compounds

→ Constitute more than 50% of the Earth's biomass

→ Members include starches, cellulose, and sugars.

→ Often have the empirical formula CH_2O (hence the name *carbohydrates*)

- **D-Glucose**
  - → Most abundant carbohydrate
  - → Forms starch and cellulose by polymerization reactions
  - → Has the molecular formula $C_6H_{12}O_6$
  - → Is classified as both an alcohol *and* an aldehyde (*pentahydroxyl aldehyde*)
  - → Exists in an acyclic (open-chain) and in two cyclic (closed-chain) forms called *anomers*
  - → In water, the cyclic structures, also called *glucopyranoses*, are favored.

- **Structure**
  - → The straight-chain structure of D-glucose and the interconversion to the two cyclic *anomers*:

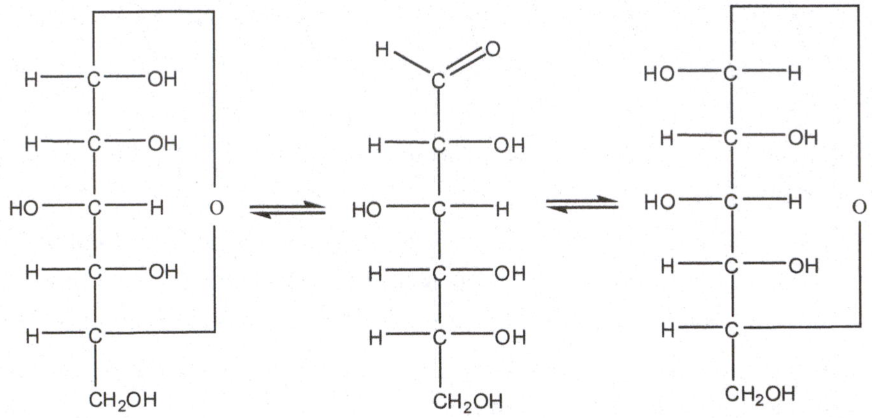

α-D-Glucose	D-Glucose	β-D-Glucose
36%	0.02%	64%

  - → The percentages indicate proportions of the several species present in water solution.
  - → The anomers, α-D-glucose and β-D-glucose, differ in the stereochemistry at the carbon atom derived from the aldehyde C atom.
  - → More realistic representations of the cyclic forms of D-glucose are shown on the right:

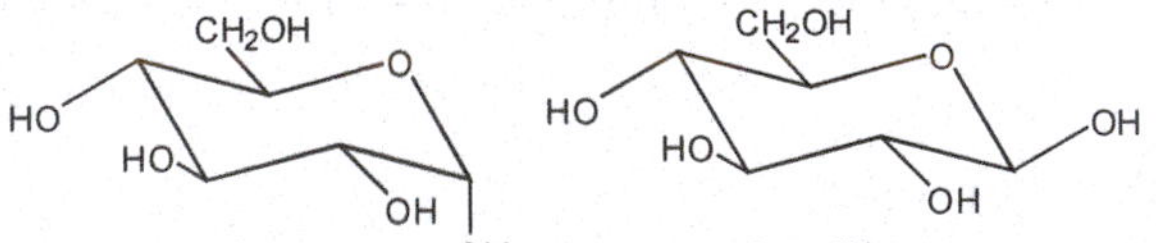

α-D-Glucose   β-D-Glucose

Notes: 1) L-Glucose, not found in nature, is the mirror image of D-glucose, and together they form a pair of enantiomers, which differ in their direction of rotation of polarized light.

2) The symbols D and L refer to a configuration relationship to the compound D-glyceraldehyde and not necessarily to the sign of rotation of light.

- **Polysaccharides**
  - → Polymers of glucose
  - → Include starch (digestible by humans) and cellulose (not digestible by humans)

- **Cellulose**

  → Same units as starch but not linked the same

  → Forms flat ribbonlike strands (see **Figure 11E.27** in the text)

  → Most abundant organic chemical in the world

  → Produced by photosynthesis

    Example: Cellulose, the structural material of plants and wood, is a condensation polymer of β-D-glucose. A three-subunit chain of cellulose is shown below. The polymer is formed formally by elimination of H_2O from glucose monomers.

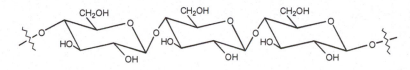

11E.7 Nucleic Acids

- **DNA**

  → Abbreviation for *deoxyribonucleic acid*

  → DNA molecules

    ⇒ Carry genetic information from generation to generation

    ⇒ Control the production of proteins

    ⇒ Serve as the template for the synthesis of RNA

- **Structure of DNA**

  → DNA molecules are condensation copolymers of enormous size.

  → DNA is composed of a sugar phosphate backbone and pendant bases:

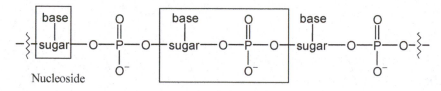

  → A base–sugar unit is called a *nucleoside*.

  → A base–sugar–phosphate grouping is called a *nucleotide*.

- **Sugar and bases**

  → In DNA, the sugar is *deoxyribose*.

  → There are four possible bases: *cytosine* (C), *guanine* (G), *adenine* (A), and *thymine* (T).

→ The sugar and four bases:

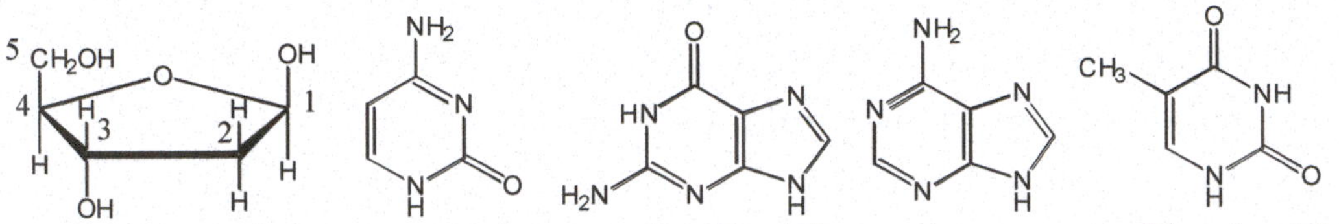

Deoxyribose   Cytosine (C)   Guanine (G)   Adenine (A)   Thymine (T)

Nucleosides

→ Formed in DNA by condensation involving the OH group at carbon 1 in deoxyribose and an appropriate amine hydrogen atom of a base

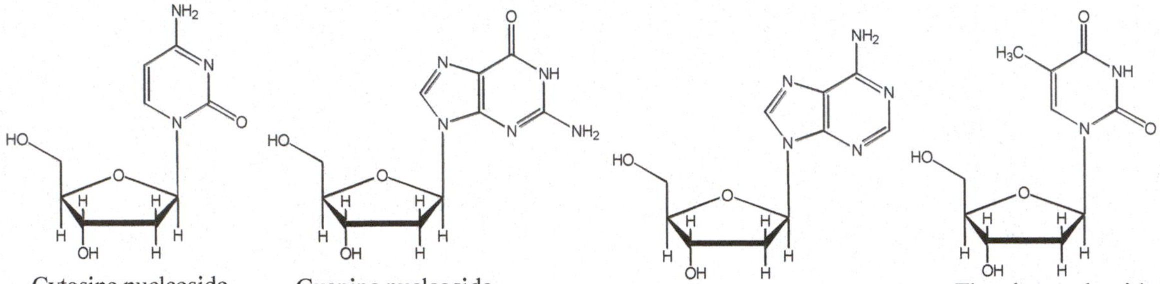

Cytosine nucleoside   Guanine nucleoside   Adenine nucleoside   Thymine nucleoside

- **Nucleotides**

  → Condense at carbon 3 and carbon 5, eliminating water, to form the DNA molecule (nucleic acid), a *polynucleotide*

  → A trinucleotide segment of a polynucleotide is shown on the right:

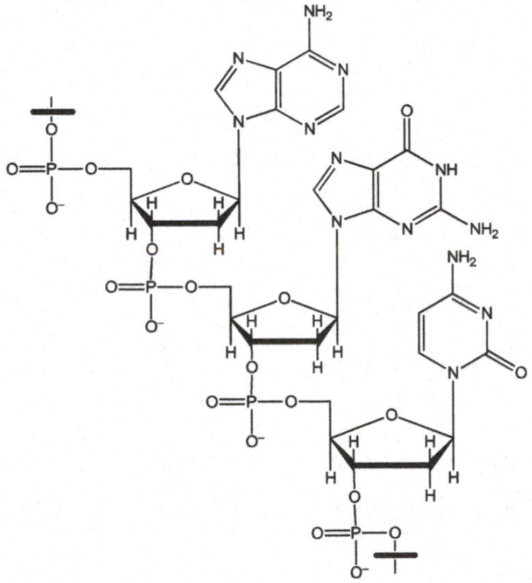

- **DNA double helix**

  → Condensation of nucleotides to a dinucleotide is shown in **Figure 11E.24** in the text.

  → Polynucleotide (nucleic acid) strands link to each other in pairs to form the well-known double helix structure (see **Figure 11E.25** in the text for a depiction of base pairing).

  → Association between strands of two polynucleotides occurs when hydrogen bonds between bases on *different* strands are formed.

  → In DNA, only two types of base pairs (G with C and A with T) occur.

→ Hydrogen bonds indicated by dashed lines:

Cytosine   H   Guanine   Thymine   Adenine

Note: CG and AT base pairs have approximately the same size and shape, which reduces distortions in the double-helix structure.

- **RNA** (not covered in the text)
  - → Abbreviation for *ribonucleic acid*
  - → A polynucleotide, similar to DNA
  - → Shorter than DNA, generally single-stranded

- **Three types**
  - → *Messenger* RNA (mRNA) carries genetic information from the cell nucleus to the cytoplasm, where translation to protein takes place.
  - → *Ribosomal* RNA (rRNA) appears to act as a catalyst in the biosynthesis of proteins.
  - → *Transfer* RNA (tRNA) is used to transport amino acids during protein synthesis.

- **Structure**
  - → RNA contains four base pairs, as in DNA, but *uracil* (U) replaces *thymine* (T).
  - → *Uracil* lacks the methyl group of *thymine* at the carbon-5 position.

Thymine (T)   Uracil (U)

Note: The position of the methyl group on thymine does not interfere with its hydrogen bonding with adenine (A) (see figure of an AT base pair shown previously).

→ The sugar in RNA is ribose. (Deoxyribose is shown for comparison.)

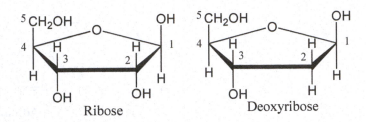

Ribose   Deoxyribose

→ For each sugar, the carbon atoms are numbered 1 to 5 for reference purposes.

→ Although RNA molecules are generally single-stranded, they contain regions of a double helix produced by the formation of *hairpin loops*.

→ In these regions, the usual base pairing is A with U and G with C.

→ However, in RNA, imperfections in base pairing are common.

INTERLUDE

Impact on Technology: Fuels

Gasoline

→ Derived from petroleum by fractional distillation (see **Topic 5C.3**) and further processing

→ Contains primarily C_5 to C_{11} hydrocarbons

→ Petroleum is refined to increase the quantity and quality of gasoline.

• **Quantity**

→ The quantity of gasoline is increased by *cracking* (breaking down long hydrocarbon chains) and by *alkylation* (combining small molecules to make larger ones).

Examples: 1) Cracking fuel oil yields an octane–octane isomeric mixture.

2) Alkylation converts a butane–butene mixture to octane.

$$C_4H_{10} + C_4H_8 \xrightarrow{\text{catalyst}} C_8H_{18}$$

• **Quality**

→ Octane rating: used to measure the quality of gasoline

→ Other things equal, branched hydrocarbons have higher octane ratings than unbranched ones.

→ *Isomerization* and *aromatization* are used to improve octane rating of gasoline.

→ *Isomerization* is used to convert straight–chain hydrocarbons to branched ones:

$$CH_3(CH_2)_6CH_3 \xrightarrow{AlCl_3} CH_3C(CH_3)_2CH_2CH(CH_3)CH_3$$

→ *Aromatization* converts alkanes to arenes, as in the conversion of heptane to methylbenzene:

$$CH_3(CH_2)_5CH_3 \xrightarrow{\text{AlCl}_3,\ \text{Cr}_2\text{O}_3} CH_3C_6H_5 + 4\,H_2$$

→ Ethanol, a renewable fuel, also increases the octane rating of gasoline.

Coal

- **Uses**

  → Coal contains a much lower ratio of hydrogen to carbon than petroleum and is harder to purify, to work, and to transport.

  → End product of anaerobic decay of vegetable matter

  → Less environmentally friendly than gasoline

  → When it burns, coal also releases a large amount of pollution in the form of particulate matter (primarily ash) and sulfur and nitrogen oxides.

  → Contains many aromatic rings and is primarily aromatic in nature

  → Coal fragments, when heated in the absence of oxygen, yield *coal tar*, which contains many aromatic hydrocarbons and their derivatives.

  → Consequently, coal is used as a raw material for other chemicals.

  → Many pharmaceuticals, fertilizers, and dyes are derived from *coal tar*

    Examples: Naphthalene for making indigo dyes (for blue jeans) and benzene, used to make nylon, detergents, and pesticides

- **Coal gas**

  → When coal is destructively distilled—heated in the absence of oxygen so that it decomposes and vaporizes—its sheetlike molecules break up, and the fragments include the aromatic hydrocarbons and their derivatives.

  → *Coal gas,* which is given off first, contains carbon monoxide, hydrogen, methane, and small amounts of other gases. The complex liquid mixture that remains is called *coal tar.*

  → *Coal gas* is the starting point for a number of alternative fuels. The mixture of carbon monoxide and hydrogen can be used as is, but that mixture is highly toxic.

  → Therefore, coal gas is used to make gaseous fuels such as methane and hydrogen and liquid fuels such as methanol.

  → These fuels are highly desired because they burn cleanly, producing little air pollution.